OT 25:
Operator Theory: Advances and Applications
Vol. 25

Birkhäuser Verlag
Basel · Boston

MATH

Georgii S. Litvinchuk
Ilia M. Spitkovskii

Factorization of Measurable Matrix Functions

Edited by Georg Heinig

Translated from the Russian by Bernd Luderer

With a foreword by Bernd Silbermann

1987

Birkhäuser Verlag
Basel · Boston

Authors Address:
Prof. Dr. Georgii S. Litvinchuk
Dr. Ilia M. Spitkovskii
Institute of Hydromechanics
Academy of Science
USSR–170100 Odessa, Ukraine

Library of Congress Cataloging in Publication Data

Litvinchuk, G. S. (Georgii Semenovich), 1931–
Factorization of measurable matrix functions.
(Operator theory, advances and applications ; v. 25)
Bibliography: p.
Includes index.
1. Matrices. 2. Factorization (Mathematics)
I. Spitkovskii, Ilia M. 1953– . II. Heinig, Georg.
III. Title. IV. Series.
QA188.L58 1987 512.9'434 87–6632
ISBN 0-8176-1883-X (U.S.)

CIP-Kurztitelaufnahme der Deutschen Bibliothek

Litvinchuk, Georgii S.:
Factorization of measurable matrix functions /
Georgii S. Litvinchuk ; Ilia M. Spitkovskii. Ed. by
Georg Heinig. Transl. from the Russian by Bernd
Luderer. With a foreword by Bernd Silbermann. –
Basel ; Boston ; Stuttgart : Birkhäuser, 1987.
(Operator theory ; Vol. 25) (Basel . . .)
ISBN 3-7643-1883-X
ISBN 0-8176-1883-X (Boston)
NE: Spitkovskii, Ilia M.

Licensed edition for the distribution in all non-socialistic countries by Birkhäuser Verlag, Basel 1987
Printed in GDR
ISBN 3-7643-1883-X
ISBN 0-8176-1883-X

Foreword

In treating mathematical problems, it is often necessary to factorize matrix-valued (or even operator-valued) functions in a certain manner. Factorizations closely connected with the notion of Wiener-Hopf factorization play a prominent role. The development of factorization theory was, among other things, stimulated by needs arising in the theory of singular integral operators and its manifold modifications.

Richness and profundity of the results obtained as well as their significance for the theory of singular integral operators and other branches of mathematics required explanation of this theory in monographs. After the book "Factorization of Matrix Functions and Singular Integral Operators" (Birkhäuser Verlag, Basel 1981), excellently written by K. CLANCEY and I. GOHBERG, the present monograph of G. S. LITVINCHUK and I. M. SPITKOVSKIĬ "Factorization of Measurable Matrix Functions" is now the second work available on problems of Wiener-Hopf factorization.

In addition to problems already discussed in the book of CLANCEY and GOHBERG, it contains a series of more recent results partly due to the authors and of considerable interest. It should be emphasized that the monograph reflects not only the state of the art in Wiener-Hopf factorization of matrix functions, but it also focusses the reader's attention on many problems which remain to be solved. The rich bibliography containing both theoretical papers and work on practical applications is a further advantage of the book, which will greatly benefit all those interested in factorization theory.

B. Silbermann

Preface

A series of reviews and original papers on factorization theory, which were read at the Odessa city seminar on boundary value problems and singular integral equations in 1973-77, served as a source of the present book. Furthermore, lectures given by the authors at the Department of Mechanics and Mathematics of the Odessa University have been incorporated.

The authors are aware that their contribution to factorization theory, although it is very comprehensive (ranging from existence problems of a factorization, questions of estimates and stability of partial indices, to applications to the theory of the generalized Riemann boundary value problem), looks much more modest than the successes of their predecessors. Nevertheless, the necessity of creating a sufficiently complete review of factorization problems of measurable matrix functions explained from a unified point of view seemed so evident to the authors that they decided to write this booklet.

It may be considered complementary to the well-known monographs of N. I. MUSKHELISHVILI and N. P. VEKUA, which are devoted to the classical factorization theory of matrix functions with Hölder elements. We wish to emphasize that the subject matter will be discussed along lines meant not only for specialists familiar with the material, but also for readers who first become acquainted with the topic as well as for persons whose interest focusses, above all, on applications. The reader's judgment will show in how far this attempt has been successful.

The authors

CONTENTS

INTRODUCTION

The notion of factorization is widely used in different branches of mathematics. The term factorization itself simply means the representation of some mathematical object (number, function, matrix, operator, etc.) as a product of objects of the same type, but having certain additional properties. According to the type of the objects and the character of the problem studied (multiplication of numbers, pointwise multiplication of functions, their composition, etc.) this product can be understood in a different manner.
Many mathematical propositions are, in fact, theorems on the existence and properties of a certain factorization. In principle, even the fundamental theorem of arithmetic (on the factorization of natural numbers in prime factors) and the fundamental theorem of algebra (on the polynomial factorization over the complex field) can be considered as examples of factorization theorems. The importance of these theorems needs no comments.

As more special examples we mention the representation of matrices and operators in a Hilbert space in the form of a product of isometric by positively definite ones (so-called polar representation), the representation of functions of Hardy classes as a product of an inner and an outer one and its generalization to the case of matrix and operator functions, but also the representation of entire functions as product of other entire functions.

Many methods for solving equations and boundary value problems of a different kind are based on the idea of factorization. The scheme for using factorization in solving equations may be described in the following way: Let some operator A be represented in the form

$$A = A_+^{-1} A_- . \tag{1}$$

Then the equation $y = Ax$ is equivalent to the equation $A_+ y = A_- x$. For a suitable choice of representation (1), the latter relation turns out to be more convenient for investigation, and sometimes it can be solved explicitly. Simultaneously, the original equation will also be solved.

For instance, the representation of a square matrix A with principal minors different from zero as the product of non-singular triangular matrices allows us to reduce the solution of a system of linear equations with matrix A to a triangular system. There exists a continuous analogue of the mentioned result for Fredholm integral operators and a more general result on the factorization of an operator acting in a Hilbert space along a chain of projections as well (GOHBERG, KREǏN [5], BARKAR', GOHBERG [1,2]).

The factorization of matrix functions defined on some contour, which is studied in the present book also belongs to factorizations of this type. It appeared in connection with the study of systems of singular integral equations and boundary value problems for analytic functions.

Throughout the book, by a factorization of the matrix function G defined on a closed contour Γ, which divides the closed complex plane into parts $D^+(\ni 0)$ and $D^-(\ni \infty)$, we understand a representation of G in the form

$$G(t) = G_+(t)\Lambda(t)\,G_-(t) \qquad (t \in \Gamma),$$

where $G_\pm$ are boundary values of a matrix function analytic and non-singular in $D^\pm$, and

$$\Lambda(t) = \mathrm{diag}(t^{\varkappa_1},\dots,t^{\varkappa_n}).$$

Under sufficiently general assumptions, the integers $\varkappa_1,\dots,\varkappa_n$ are uniquely defined (up to order) by the matrix function G and are called its partial indices[1].

The factorization of matrix functions is the main part in the theory of vector-valued Riemann boundary value problems. It permits us to reduce the general case of this problem to the so-called jump problem[2]. In this aspect, the problem of factorization has been studied under different restrictions on the smoothness of the contour Γ and the matrix function G, starting from the end of the previous century by several generations of mathematicians. The result of this work for the case of piecewise Hölder matrix functions and piecewise smooth contours are summed up in the well-known monographs of MUSKHELISHVILI [1], VEKUA [4], in the survey of GAKHOV [3] and the paper of KHVEDELIDZE [1].

Since the late fifties, the further development of the theory of the Riemann boundary value problem, systems of singular integral equations with Cauchy kernel and systems of Wiener-Hopf equations has stimulated

---

[1] A more complete and exact definition of factorization will be given at the beginning of the second chapter.

[2] Incidentally, we note that to vector boundary value problems with a shift (in particular, to the Gazemann problem) there correspond definitions of factorization different from that one given above. In the present book we do not study factorizations of such a type, since in the vector case the theory has not been sufficiently elaborated hitherto. The corresponding factorizations for the scalar case can be found in the monograph of LITVINCHUK [3].

the creation of new directions in the factorization problem. The factorization of matrix functions with elements from the Wiener algebra (GOHBERG, KREĬN [3]), the factorization of continuous (SIMONENKO [2]) and piecewise continuous (GOHBERG, KRUPNIK [1], MANDZHAVIDZE [6]) matrix functions has been studied, and a number of results concerning the factorization of measurable bounded matrix functions was obtained (DANILYUK [4,5], SIMONENKO [4,6], DOUGLAS [1]). At the same time, it became clear that even in case of a continuous non-singular matrix function G a factorization with continuous factors $G_{\pm}$ does not always exist. On the one hand, this gave rise to the search for new classes of matrix functions admitting a factorization with continuous factors (see BUDYANU, GOHBERG [3,4]) and, on the other hand, it led to the creation of the notion of factorization in classes L_p, $1 < p < \infty$. In case of continuous non-singular matrix functions, this factorization does not depend upon the choice of p. However, such a dependence holds already for piecewise continuous matrix functions. In this connection, there arises the task of studying the set of parameter values p, for which a given matrix function G can be factored.

Moreover, in the Riemann boundary value problem theory in L_p spaces, a situation may emerge where the problem has a finite index, but is not normally solvable. Owing to this, we have to impose additional demands upon the factorization in order to guarantee the normal solvability of the boundary value problem. In the following, a factorization satisfying these requirements will be called a Φ-factorization.

There were also studies on the enlargement of the class of contours Γ, for which factorization theorems continue to be valid. The crucial idea of these studies reduces to the development of sufficient conditions for the boundedness of the operator of singular integration on Γ (see e.g. KHVEDELIDZE [4]).

In connection with the problem of calculating defect numbers and the approximate solution of the vector-valued Riemann boundary value problem and singular integral equations, the questions of stability of the factorization factors $G_{\pm}$ and Λ as well as of finding estimates for partial indices are not yet well understood, even under the classic assumptions made in VEKUA [4], GAKHOV [3], MUSKHELISHVILI [1].
In the last 15 years a number of new results has been obtained in this direction (sufficient stability conditions, which can effectively be verified, sign preservation of all partial indices, which is also necessary in some cases (cf. VERBITSKIĬ, KRUPNIK [1,2], SPITKOVSKIĬ [5, 10], POUSSON [2], RABINDRANATHAN [1] and others)).
In obtaining these results, the notion and properties of the numerical

range (or Hausdorff set) of a matrix (MARCUS, MINC [1]) is used, i.e. questions related to the formula for the norm of a Hankel operator (see ADAMYAN, AROV, KREĬN [1]) and the operator of singular integration (VERBITSKIĬ, KRUPNIK [1], KRUPNIK [4]).

We would also like to note that the factorization problem for the spectral density of a stationary random process (ROZANOV [1]) which arose in probability theory, some problems of projection methods for the solution of systems of singular integral and convolution type equations (GOHBERG, FELDMAN [3]) and the theory of canonical systems of linear differential equations (KREĬN, MELIK-ADAMYAN [1,2]) show that the importance of the factorization problem for matrix functions is not limited by its role in the theory of the vector-valued Riemann boundary value problem.
The factorization problem for matrix functions arises also in a natural way in the study of convolution type equations on a finite interval and on systems of intervals (GANIN [1], KARLOVICH, SPITKOVSKIĬ [1], KOMYAK [1-4], NOVOKSHENOV [1], PAL'TSEV [2-6]. SPITKOVSKIĬ [15]). We emphasize that here(in contrast to the classical case of the half-line) even scalar equations yield a matrix problem of factorization.

All this increases interest in the peculiarities of factorization of special type matrix functions (unitary, Hermitian, functionally commutable , etc.), information on which is scattered in various papers (GORDIENKO [1], GOHBERG, KREĬN [3], KREĬN, MELIK-ADAMYAN [1,2], NIKOLAĬCHUK, SPITKOVSKIĬ [1,2], CHEBOTAREV [1,2], SHMUL'YAN [1,2] and others).

At present there do not exist monographs or reviews on the factorization of matrix functions, in which the problems mentioned above would be discussed in a sufficiently complete manner[1]). There are only three papers of a general nature by BOJARSKI [3], GOHBERG, KREĬN [3], KHVEDELIDZE [4] and the monographs of GOHBERG, FELDMAN [3], DANILYUK [6], PRÖSSDORF [1] and PRÖSSDORF, MICHLIN [1], in which some questions on the factorization of matrix functions are explained. The authors hope that the present book will contribute towards filling the gaps existing in the literature concerning this topic.
We now intend to briefly describe the contents of the book. The first chapter is of auxiliary character. It contains propositions from operator theory in Banach spaces and information on different classes of analytic functions (mainly on Smirnov classes) necessary for understanding the material contained in the main part of the book. In Chapter 2 the notions of factorization and the domain of factorability of matrix

[1]) While this manuscript was in preparation for publication, the monograph CLANCEY, GOHBERG [3] came out. Its contents is essentially covered by the present book.

functions are introduced, and their simplest properties are established. Besides, the factorization problem for matrix functions meromorphic in D^+ or D^- (in particular, rational matrix functions) is treated there, and the problem of the change of partial indices when being multiplied by matrix functions of the same kind is studied.
The connection between the factorization problem for matrix functions and the vector-valued Riemann boundary value problem is regarded in Chapter 3. Φ-factorization properties resulting from this connection and corresponding propositions of the theory of Fredholm operators are established.

Chapter 4 is devoted to triangular and functionally commutable matrix functions. In this chapter the criterion of Φ-factorability and estimates for partial indices (but for second-order matrix functions even exact formulae) are derived.

In Chapter 5 necessary and sufficient conditions of Φ-factorability of continuous and piecewise continuous matrix functions, but also for matrix functions of the class $L^{\pm}_{\infty} + C$ are established. The factorization problem in decomposing algebras of functions including the factorization problem of Hölder matrix functions on a piecewise smooth contour and matrix functions from the Wiener algebra on the circle will be studied.

The criterion for stability of partial indices, sufficient conditions for their non-positivity, non-negativity, coincidence or equality to zero are explained in Chapter 6. In addition, we shall investigate the behaviour of the factorization factors $G_{\pm}$ under small perturbations of the matrix function G.

Chapter 7 is dedicated to the factorization problem on the unit circle. Furthermore, the factorization of Hermitian and unitary matrix functions will be studied. The necessity of several sufficient conditions for non-negativity (non-positivity, equality to zero, etc.) of partial indices established in Chapter 6 will be proved.

Chapter 8 is concerned with sufficient conditions for Φ-factorability based on local principle and results of Chapters 5 - 7.

Finally, in Chapter 9 the theory developed in Chapters 3 - 8 is used for the study of the generalized Riemann boundary value problem, which has manifold applications in geometry, mechanics, and physics (see BUDYANU [1], GAKHOV [2]).

The last section of each chapter contains information and comments on results, which were not reflected in the main part of the chapter, but are related to it in certain ways. Some unsolved problems are formulated. The first chapter makes an exception. For the reader's convenience, comments on references accompany every proposition (or a group of closely connected propositions).

Lastly, we want to remark that in the book we do not focus our attention on such problems as the factorization problem relative to a contour lying on a Riemann surface (cf. L.F. ZVEROVICH [2], E.I. ZVEROVICH [1], ZVEROVICH, PANCHENKO [1], KVITKO [1], KRUGLOV [1], PANCHENKO [1], RODIN [1,2], RODIN, TURAKULOV [1], SELEZNEV [2], KOPPELMAN [1], RÖHRL [1]) or with respect to complex (GOHBERG, FELDMAN [1], MANDZHAVIDZE [7,8]) and Jordan non-rectifiable (among them quasiconformal) contours (see KATS [3], SELEZNEV [1-3]). Furthermore, we do not deal with the natural generalization of the factorization problem for matrix functions - the question on factorability of operator functions. Concerning this topic the reader is referred to Chapter 6 in the monograph of CLANCEY and GOHBERG [3] and the literature cited there, but also to the papers of BUDYANU [1], LEITERER [1], SERGEEV [1], GOHBERG, KAASHOEK, van SCHAGEN [1], van der MEE [1,2], MUCKENHOUPT [1], NAKAZI TAKAHIKO [1], SUCIU [1].[1)] We do not dwell on various applications of the factorization problem. We only refer to GUSAK [1,2], DUKHOVNYĬ [1], MALYSHEV [1], ROZANOV [1], where methods and results of the factorization theory are applied to probability-theoretic problems. Applications to mechanics one can find in the monograph of VOROVICH, ALEKSANDROV, BABESHKO [1] as well as the papers BABESHKO [1,2], BELOKOPYTOVA, IVANENKO, FIL'SHTINSKIĬ [1], VOROB'EV [1], KULIEV, SADYKHOV [1], MOSSAKOVSKII, MISHCHISHIN [1], TOLOKONNIKOV, PEN'KOV [1], KHRAPKOV [1], TSITSKISHVILI [3,4], CHEREPANOV [1]. Concerning applications to difraction problems, see the monograph VIROZUB, MATSAEV [1] and the article POPOV [1], CHAKRABARTI [2], HEINS [1,2], IDEMEN [2]. Finally, for applications of the factorization theory to the method of inverse problems and other questions of theoretical physics, the reader is referred to the monograph NOVIKOV (ed.) [1] and the papers of EMETS [1], ZAKHAROV, SHABAT [1], ITS [1], KRICHEVER [1,2], MASKUDOV, VELIEV [1], MEUNARGIYA [1], NOVOSHENOV [2,3], TROITSKIĬ [1], DUVACZ, ZELAZNY [1], IDEMEN [1], MEISTER [1], RAJAMAKI [1], REYMAN, SEMENOV-TIAN-SHANSKY [2], SIEWERT, KELLEY, GARCIA [1].

---

1) see also GOHBERG, LEITERER [1-2].

CHAPTER 1. BACKGROUND INFORMATION

This chapter is of an auxiliary character. Its first section contains facts necessary for the further explanations concerning lineals of finite codimension as well as semi-Fredholm and Fredholm operators in Banach spaces. Most of these statements are well-known and have been published in several recent monographs.[1] However, Theorem 1.12 on the Fredholmness of block operators and the local principle in the theory of Fredholm operators make an exception. The latter was proved in the form we need only in SIMONENKO's paper [5]. Therefore, Theorems 1.12 and 1.14 will be given with detailed proofs. The remaining theorems in Section 1.1 we state without proof, referring the reader in every case to available sources. We do not intend to explain the history of semi-Fredholm and Fredholm operator theory, since this has already been excellently described (see, e.g. KATO [1]).

In the second section of the present chapter, above all, known theorems on analytic functions of Smirnov classes are stated, whereas in Section 1.3 properties of singular integral operators related to the results of Section 1.2 are described. Here as well as in Section 1.1, we give proofs only for those statements, which we did not find in an appropriate form (Theorems 1.26, 1.27, 1.29, 1.31, 1.35) or which are published only in several articles (Theorems 1.20, 1.32, 1.33).

Since the genesis of the formation of many results of the theory of Smirnov classes has received little attention in modern monographic literature, we give some historic information and comments on the material of Sections 1.2 and 1.3.

1.1. SOME FACTS ON THE GEOMETRY OF BANACH SPACES AND OPERATOR THEORY IN THESE SPACES

A linear manifold in a Banach space not necessarily closed we shall call a lineal. For a closed lineal we use the term subspace. The dimension of the factor-space $\mathcal{B}|\mathcal{L}$ is said to be the codimension of the lineal $\mathcal{L}$ being located in a Banach space $\mathcal{B}$:

$$\operatorname{codim} \mathcal{L} = \dim(\mathcal{B}|\mathcal{L}).$$

1) See e.g. GOHBERG, KRUPNIK [4], S.KREIN [1], MICHLIN, PRÖSSDORF [1]

THEOREM 1.1. Suppose $\mathcal{L}_1$ to be a subspace of the Banach space $\mathcal{B}$, and let $\mathcal{L}_2 \supseteq \mathcal{L}_1$ be a lineal. Then $\dim \mathcal{L}_2|\mathcal{L}_1 < \infty$ implies that $\mathcal{L}_2$ is also a subspace. On the other hand, if $\mathcal{L}_2$ is a subspace and $\mathcal{L}_1 \subseteq \mathcal{L}_2$ is the range of a closed operator, then $\dim \mathcal{L}_2|\mathcal{L}_1 < \infty$ implies that $\mathcal{L}_1$ is a subspace, too.

The first part of the theorem is an elementary fact (we refer to GOHBERG, KRUPNIK [4], p. 50 for its proof). The second part is a consequence of the closed graph theorem (see, e.g., PALAIS [1], Ch. 7, Th. 1).

DEFINITION 1.1. The set of all elements f of the dual space $\mathcal{B}^*$ with $f(x) = 0$ for all $x \in \mathcal{L}$ is called the orthogonal complement $\mathcal{L}^\perp$ to the lineal $\mathcal{L}$ ($\subseteq \mathcal{B}$).

Obviously, $\mathcal{L}^\perp$ is a subspace of the space $\mathcal{B}^*$. The orthogonal complement to the lineal $\mathcal{L}$ coincides with the orthogonal complement to its closure $\overline{\mathcal{L}}$. There is a natural isomorphism between $\mathcal{L}^\perp$ and the space dual to $\mathcal{B}|\overline{\mathcal{L}}$. As a corollary of this isomorphism, we mention

THEOREM 1.2. For any lineal $\mathcal{L}$ ($\subseteq \mathcal{B}$), the dimension of the orthogonal complement coincides with the codimension of its closure (in the sense that either both values are infinite or both values are finite, and then they are equal to each other).

DEFINITION 1.2. A subspace $\mathcal{L}_0$ of the Banach space $\mathcal{B}$ is called complementable, if there exists a subspace $\mathcal{L}_1(\subseteq \mathcal{B})$ such that each vector $x \in \mathcal{B}$ admits a unique representation in the form $x = x_0 + x_1$ ($x_j \in \mathcal{L}_j$, $j = 0,1$). In this case, the subspace $\mathcal{L}_1$ is said to be the direct complement to $\mathcal{L}_0$ (with respect to $\mathcal{B}$), and the space $\mathcal{B}$ is said to be decomposed in the direct sum of its subspaces $\mathcal{L}_0$ and $\mathcal{L}_1$ (written $\mathcal{B} = \mathcal{L}_0 \dot{+} \mathcal{L}_1$).

The notion of the direct sum is transferred to the case of an arbitrary finite number of summands in an obvious manner.

Every subspace in a Hilbert space can be complemented; one of its direct complements is the orthogonal complement. In every Banach space not isomorphic to a Hilbert one, there are non-complementable subspaces

(LEE, SARASON [1]). Nevertheless, there holds

THEOREM 1.3 (S. KREĬN [1], p. 31). Let $\mathfrak{B}$ be a Banach space and $\mathfrak{B}_0$ a subspace of finite codimension. Then $\mathfrak{B}_0$ is complementable, and the dimension of all its direct complements is equal to $\operatorname{codim} \mathfrak{B}_0$. If, moreover, $\mathcal{L}$ is a lineal dense in $\mathfrak{B}$, then the lineal $\mathcal{L} \cap \mathfrak{B}_0$ is dense in $\mathfrak{B}_0$, and there exists a direct complement to $\mathfrak{B}_0$ with respect to $\mathfrak{B}$ lying in $\mathcal{L}$.

We shall say that a linear (generally speaking, unbounded) operator **A** acts in the Banach space $\mathfrak{B}$, if its domain dom **A** and image im **A** lie in $\mathfrak{B}$ [1].

The kernel of the operator **A**, i.e., the solution set of the homogeneous equation $\mathbf{A}x = 0$, will be denoted by ker **A**.

Owing to the linearity of the operator **A**, the sets dom **A**, im **A** and ker **A** are lineals in $\mathfrak{B}$. By $[\mathfrak{B}]$ we denote the class of all linear bounded operators acting in $\mathfrak{B}$, the domain of which coincides with $\mathfrak{B}$. This class of operators supplied with the norm $\|\mathbf{A}\| = \sup\{\|\mathbf{A}x\| : x \in \mathfrak{B}, \|x\| \leq 1\}$ is itself a Banach space. Furthermore, the class $[\mathfrak{B}]$ is a Banach algebra, i.e. such a Banach space, in which the operation of multiplication is defined (in the case under study, its role plays the composition of operators) consistent with the linear structure and possessing the property $\|\mathbf{A}\mathbf{B}\| \leq \|\mathbf{A}\| \cdot \|\mathbf{B}\|$ [2].

DEFINITION 1.3. The operator B is referred to as a part of the operator **A**, if $\operatorname{dom} B \subseteq \operatorname{dom} \mathbf{A}$ and $\mathbf{A}x = Bx$ for all $x \in \operatorname{dom} B$.

Under the conditions of Definition 1.3, the operator B is sometimes also said to be the restriction of A on dom B (written $B = \mathbf{A}|_{\operatorname{dom} B}$).

DEFINITION 1.4. The values $\alpha(\mathbf{A}) = \dim \ker \mathbf{A}$ and $\beta(\mathbf{A}) = \dim(\mathfrak{B} | \overline{\operatorname{im} \mathbf{A}})$ are called defect numbers of the operator **A**, and their difference

---

1) Most of the definitions and results stated below can be transferred to operators acting from one space into another. However, for our aims it is sufficient to regard operators acting in one space.

2) A short survey of the theory of Banach algebras can be found, for example, in the appendix written by V.M. TIKHOMIROV to the book KOLMOGOROV, FOMIN [1].

ind $A = \alpha(A) - \beta(A)$ is the <u>index</u> of the operator A.

The index is defined, if at least one of the defect numbers is finite. If both are finite, then A is called an operator with finite index.

In case the image of the operator A is closed, the equation $Ax = y$ is usually said to be normally solvable. For the sake of brevity, in this case the operator A itself is also called <u>normally solvable.</u>

<u>DEFINITION 1.5.</u> A closed normally solvable operator A is called <u>Fredholm</u> iff $\alpha(A)$ and $\beta(A)$ are finite. In case that at least one of these numbers is finite, A is referred to as <u>semi-Fredholm.</u> If $\alpha(A) < \infty$, A is called a Φ_+- and, if $\beta(A) < \infty$, A is called a Φ_--operator.

It follows from Theorem 1.1 that $A \in [\mathfrak{B}]$ is a Φ_--operator if and only if its image has a finite codimension.

The following five theorems are a collection of fundamental facts on Φ_+-, Φ_-- and Fredholm operators. Their proofs can be found e.g. in GOHBERG, KREĬN [1], KATO [1], S. KREĬN [1]; for the case of bounded operators we also refer to Ch. 4 in GOHBERG, KRUPNIK [4], Ch. 2 in DANILYUK [6] and Section 8 of Ch. 1 of SHUBIN [3].

<u>THEOREM 1.4.</u> For every Φ_+- (Φ_--, Fredholm) operator A acting in $\mathfrak{B}$, there is an $\varepsilon > 0$ such that if $B \in [\mathfrak{B}]$ and $\|B\| < \varepsilon$, then

1) the operator $A + B$ is also a Φ_+- (Φ_--, Fredholm) operator,
2) ind $(A+B)$ = ind A,
3) $\alpha(A+B) \leq \alpha(A)$, $\beta(A+B) \leq \beta(A)$.

Theorem 1.4 means the stability of semi-Fredholmness or Fredholmness of the index as well as the semi-stability of defect numbers under perturbations of the operator with a sufficiently small norm.

<u>THEOREM 1.5.</u> If the operator A acting in $\mathfrak{B}$ is a Φ_+- (Φ_--, Fredholm) operator and the operator $B \in [\mathfrak{B}]$ is compact, then $A + B$ is also a Φ_+- (Φ_--, Fredholm) operator and ind $(A+B)$ = ind A.

Theorem 1.5 expresses the stability of semi-Fredholmness and Fredholmness and of the index of the operator A under compact perturbations.

Sometimes Theorems 1.4 and 1.5 are called the first and second stability theorem.

THEOREM 1.6. If the operators A and B acting in $\mathfrak{B}$ are Fredholm and B is densely defined, then the operator BA is also Fredholm and ind BA = ind A + ind B.

DEFINITION 1.6 (see KATO [1], p. 212). The operators A and B acting in $\mathfrak{B}$ and $\mathfrak{B}^*$ respectively, are referred to as adjoint to each other, if for all $x \in$ dom A and $f \in$ dom B , there holds $(Bf)(x) = f(Ax)$.

Generally speaking, there are many operators acting in $\mathfrak{B}^*$, which are adjoint to A . But if the operator A is densely defined, then all operators adjoint to it are parts of one and the same uniquely defined operator, which will be denoted by A^* . The operator A^* is always closed.

THEOREM 1.7. Let the operator A acting in $\mathfrak{B}$ be closed and densely defined. Then $(\mathrm{im}\ A)^{\perp} = \ker A^*$ and $(\ker A)^{\perp} \supseteq \mathrm{im}\ A^*$, where in the last inclusion equality holds if and only if the operator A (or, equivalently, A^*) is normally solvable.

COROLLARY 1.2. The closed and densely defined operator A is a Φ_+- (Φ_--, Fredholm) operator iff the operator A^* is a Φ_-- (Φ_+-, Fredholm) operator. In this case we have $\alpha(A) = \beta(A^*)$, $\beta(A) = \alpha(A^*)$.

DEFINITION 1.7. The operator $M \in [\mathfrak{B}]$ is called a left (right) regularizer of the operator A ($\in [\mathfrak{B}]$), if the operator $MA - I$ ($AM - I$) is compact. If M is simultaneously a left and right regularizer, then it is called a (two-sided) regularizer.

It follows from the definition that if an operator has left and right regularizers, then all of them are regularizers and differ from each other by compact summands.

THEOREM 1.8. The operator $A \in [\mathfrak{B}]$ admits a regularization from the left (right) if and only if it is a Φ_+- (Φ_--) operator and the subspace im A (ker A) is complementable. Among all operators of class $[\mathfrak{B}]$, the Fredholm operators and only they admit a regularization.

A simple consequence of Theorem 1.8 is

THEOREM 1.9. If the product BA of operators A and B of the class [𝔅] is Fredholm, then A is a Φ_+- and B a Φ_--operator. If in this case B is a Φ_+- (A is a Φ_--) operator, then A (respectively B) is Fredholm.

DEFINITION 1.8. The factor algebra of the Banach algebra [𝔅] with respect to the ideal $\mathcal{S}$ of compact operators is called Calkin algebra.

According to this definition, the elements of the Calkin algebra are classes of equivalent operators, where two operators A and B are said to be equivalent, if $A-B \in \mathcal{S}$. In all what follows, we denote this equivalence by $A \sim B$. The norm of the image $\hat{A}$ of the operator $A \in [𝔅]$ in the Calkin algebra is called essential norm of A (written $|A|$). In other words: $|A| = \inf\{\|A+T\| : T \in \mathcal{S}\}$.

Owing to the closedness of the ideal of compact operators, the Calkin algebra provides a Banach algebra.

From Theorem 1.8 we deduce that the operator $A \in [𝔅]$ is Fredholm if and only if its image $\hat{A}$ is invertible in the Calkin algebra, and the inverse of $\hat{A}$ is that element of the Calkin algebra, which consists of all regularizers of A (or, what is the same, of any other operator from the class $\hat{A}$).

Next we formulate a simple and well-known result from the theory of Banach algebras, which we need below both in general form, and for the special case of the Calkin algebra.

THEOREM 1.10. Let $\{x_k\}$ be a sequence of invertible elements of the Banach algebra $\mathcal{B}$ converging in $\mathcal{B}$ to x. Then the relation $\sup_k \|x_k^{-1}\| < \infty$ is necessary and sufficient for the invertibility of x.

Applying Theorem 1.10 to the Calkin algebra we obtain

<u>THEOREM 1.11.</u> Let $\{A_k\} \subset [\mathfrak{B}]$ be a sequence of Fredholm operators converging to the operator A. The operator A is Fredholm if and only if there exists a sequence $\{B_k\}$ of regularizers of the operators A_k such that $\sup\|B_k\| < \infty$.

Let the Banach space $\mathfrak{B}$ be represented as the direct sum of its subspaces: $\mathfrak{B} = \mathfrak{B}_1 \dot{+} \mathfrak{B}_2$. Then every operator $A \in [\mathfrak{B}]$ may be written in the form of a block-matrix

$$A = \begin{bmatrix} A_{11} & A_{12} \\ A_{21} & A_{22} \end{bmatrix} \qquad (1.1)$$

where A_{ij} is a linear bounded operator from $\mathfrak{B}_j$ into $\mathfrak{B}_i$ $(i,j = 1,2)$. The equation (1.1) must be understood in the sense that the decomposition $x = x_1 + x_2$ $(x_j \in \mathfrak{B}_j)$ corresponds to the decomposition $Ax = y_1 + y_2$ $(y_j \in \mathfrak{B}_j)$, where $y_1 = A_{11}x_1 + A_{12}x_2$, $y_2 = A_{21}x_1 + A_{22}x_2$.

Supposing that in the representation (1.1) one of the blocks A_{ij} is compact we can formulate the criterion of Fredholmness of the operator A in terms of its remaining blocks. For the sake of definiteness, we assume the block A_{12} to be compact.

<u>THEOREM 1.12.</u> Let in the representation (1.1) of the operator $A \in [\mathfrak{B}]$ the block A_{12} be compact. Then for the Fredholmness of the operator A it is necessary and sufficient that the following conditions hold:

1) A_{11} is a Φ_--operator,
2) A_{22} is a Φ_+-operator,
3) $\gamma_1 = \dim \mathcal{L}_1 < \infty$, where $\mathcal{L}_1 = \ker A_{11} \cap \{x \in \mathfrak{B}_1 : A_{21}x \in \operatorname{im} A_{22}\}$
4) $\gamma_2 = \operatorname{codim} \mathcal{L}_2 < \infty$, where $\mathcal{L}_2 = \operatorname{im} A_{22} + A_{21} \ker A_{11}$.

If the conditions 1) - 4) are satisfied, then the index of the operator A is calculated by the formula

$$\operatorname{ind} A = \alpha(A_{22}) - \beta(A_{11}) + \gamma_1 - \gamma_2. \qquad (1.2)$$

Moreover, if $A_{12} = 0$, then

$$\alpha(A) = \alpha(A_{22}) + \gamma_1 \,, \quad \beta(A) = \beta(A_{11}) + \gamma_2. \qquad (1.3)$$

Proof. Owing to Theorem 1.5 it suffices to study the case $A_{12} = 0$. First let us assume that the conditions 1) – 4) are fulfilled. We show that A is Fredholm and (1.2) holds. Let P_1 denote the projection of $\mathcal{B}$ onto $\mathcal{B}_1$ along $\mathcal{B}_2$ and P_1^0 its restriction on $\ker A$. It is easily verified that

$$\operatorname{im} P_1^0 = \mathcal{L}_1 \quad \text{and} \quad \ker P_1^0 = \ker A_{11}.$$

Since $\ker A|\ker P_1^0$ is isomorphic to $\operatorname{im} P_1^0$, this implies $\alpha(A) < \infty$ and

$$\dim \ker A|\ker P_1^0 = \alpha(A) - \alpha(A_{11}) = \dim \mathcal{L}_1 = \gamma_1.$$

Therefore, the first relation (1.3) is proved.

Next we prove that $\operatorname{im} A$ has finite codimension and (1.3) holds. By assumption, there exist direct complements M_1 to $\operatorname{im} A_{11}$ in $\mathcal{B}_1$ and M_2 to $\mathcal{L}_2$ in $\mathcal{B}_2$. We show that $M = M_1 \dotplus M_2$ is a direct complement of $\operatorname{im} A$ in $\mathcal{B}$. Let $y \in \operatorname{im} A \cap M$. Since $P_1 \operatorname{im} A = \operatorname{im} A_{11}$, we have $y_1 = P_1 y \in \operatorname{im} A_{11}$ and, on the other hand, $y_1 \in M_1$. Hence $y_1 = 0$. From this we get $y \in \operatorname{im} A \cap M_2$. In view of $\mathcal{L}_2 = \operatorname{im} A \cap \mathcal{B}_2$ this means $y \in \mathcal{L}_2 \cap M_2$, i.e. $y = 0$. Thus $\operatorname{im} A \cap M = \{0\}$.

Taking now $y = y_1 + y_2$ $(y_j \in \mathcal{B}_j,\ j = 1,2)$ arbitrarily, we have a decomposition

$$\begin{aligned} y_1 &= A_{11}x_1 \qquad\qquad + f_2 \\ y_2 &= A_{21}u_1 + A_{22}x_2 + f_2, \end{aligned}$$

where $u_1 \in \ker A_{11}$, $f_j \in M_j$, $x_j \in \mathcal{B}_j$ $(j = 1,2)$.
We have, furthermore, a decomposition

$$A_{21}x_1 = A_{21}u_1' + A_{22}x_2' + f_2'$$

with $u_1' \in \ker A_{11}$, $x_2' \in \mathcal{B}_2$, $f_2' \in M_2$. Putting $x = x_1 + u_1 - u_1' + x_2 - x_2'$ we obtain

$$Ax = y_1 - f_1 + y_2 - f_2 + f_2',$$

which is $y = Ax + f$, where $f = f_1 + f_2 - f_2' \in M$.
We proved $\mathcal{B} = \operatorname{im} A \dotplus M$. In particular, this implies that $\operatorname{im} A$ has a finite codimension equal to $\beta(A_{11}) + \gamma_2$. Since by Theorem 1.1 $\operatorname{im} A$ is closed, the second relation (1.3) results. The index formula (1.2) is a consequence of (1.3).

Now let us assume that A is Fredholm and

$$R = \begin{bmatrix} R_{11} & R_{12} \\ R_{21} & R_{22} \end{bmatrix}$$

is a regularizer of A. Then, obviously, R_{11} is a right regulizer of A_{11} and R_{22} a left regularizer of A_{22}. Hence 1) and 2) are fulfilled. Since, as proved above, $\ker A|\ker A_{22}$ is isomorphic to $\mathcal{L}_1$, we have $\gamma_1 < \infty$. Further, the finiteness of $\beta(A)$ implies the finiteness of the codimension of $\mathcal{L}_2 = \operatorname{im} A \cap \mathfrak{B}_2$ in $\mathfrak{B}_2$. Hence 3) and 4) are satisfied, too, and the theorem is proved. ≡

Theorem 1.12 is formulated in SPITKOVSKIĬ [11] (see also SPITKOVSKIĬ [12], where this theorem was presented with proof). Formulae (1.2) allow for obtaining in case $A_{12} = 0$, a criterion of injectivity and surjectivity of the operator A. This result was obtained independently by STOROZH [1]. It is omitted here, since we do not need it. Of course, various special cases of Theorem 1.12 were known previously. One of them important for the following is the case that some of the diagonal blocks of the operator A is Fredholm. This case can be investigated (even without compactness assumption for A_{12}) with the help of Frobenius' formulae (see KIRILLOV, GVISHIANI [1] p. 85 as well as PETROV [1]). However, we need only the case if A_{12} is compact. Thus, we obtain the corresponding result as a consequence of Theorem 1.12.

<u>COROLLARY 1.3.</u> Let in the representation (1.1) of the operator A the block A_{12} be compact, and let one of the diagonal blocks A_{11} and A_{22} be Fredholm. Then for the Fredholmness of the operator A it is necessary and sufficient that the second block is also Fredholm. If this condition is fulfilled, then

$$\operatorname{ind} A = \operatorname{ind} A_{11} + \operatorname{ind} A_{22}.$$

If, in addition, $A_{12} = 0$, then

$$\alpha(A) = \alpha(A_{11}) + \alpha(A_{22}) - m,$$
$$\beta(A) = \beta(A_{11}) + \beta(A_{22}) - m,$$

where $m = \dim(A_{21}\ker A_{11})|(\operatorname{im} A_{22} \cap A_{21} \ker A_{11})$.

In fact, if e.g. the operator A_{11} is Fredholm, then conditions 1) and 3) of Theorem 1.12 are automatically satisfied, and thanks to the finite dimension of the lineal A_{21} ker A_{11}, condition 4) is equivalent to the condition that A_{22} is a Φ_--operator.

The validity of the formulae for the index and the defect numbers follows from the corresponding formulae of Theorem 1.12, taking into account that for Fredholm A_{11} and A_{22} we have

$$\dim(\ker A_{11} \cap \mathcal{L}_1) = \dim \ker A_{11} - m,$$

$$\dim(\mathfrak{B}_2|\mathcal{L}_2) = \dim(\mathfrak{B}_2|\operatorname{im} A_{22}) - m.$$

In view of Corollary 1.3, the condition of Fredholmness of the diagonal blocks of the operator (1.1) with compact blocks outside the diagonal is sufficient for the Fredholmness of the operator A itself. In the general case, this condition is not necessary. A simple example is provided by the operator of two-sided shift of infinite multiplicity acting in a Hilbert space $\mathfrak{H}$. This operator is invertible, but, decomposing $\mathfrak{H}$ in a suitable manner in an orthogonal sum, it can be represented in the form (1.1), where $A_{12} = 0$ and A_{11} is the operator of one-sided shift with an infinite dimensional kernel. In Chapter 4 we shall consider some examples in detail, which illustrate the same situation, but are directly related to the subject of the present book.

<u>COROLLARY 1.4.</u> Let the Banach space $\mathfrak{B}$ be represented as the direct sum of its subspaces $\mathfrak{B}_j$ $(j = 1,\ldots,k)$ in such a way that the operator $A \in [\mathfrak{B}]$ commutes up to a compact summand with any operator P_j projecting $\mathfrak{B}$ on $\mathfrak{B}_j$ parallel to the direct sum of the remaining subspaces: $AP_j - P_jA \in \mathcal{S}$. Then for the Fredholmness of A it is necessary and sufficient that the operators $A_{jj} = P_jA|_{\mathfrak{B}_j}$ considered in $\mathfrak{B}_j$ be Fredholm. In this case $\operatorname{ind} A = \sum_{j=1}^{k} \operatorname{ind} A_{jj}$.

Corollary 1.4 can be proved by induction on k with the help of Theorem 1.12. However, we do not intend to verify this assertion, since the result of Corollary 1.4 can be easily obtained directly and is well-known (see e.g. GOHBERG, KREĬN [3]).

As a rule, in the present book the role of the Banach space $\mathfrak{B}$ will be played by the space $L_p^n(\Gamma)$ (briefly L_p^n) of n-dimensional vector functions given on the rectifiable contour Γ , each component of which is summable in the pth power $(1 \leq p < \infty)$. In addition to the general theorems given above, we need the following propositions connected with the specific nature of the space L_p^n .

THEOREM 1.13 (M. RIESZ-TORIN). If the operator $A \in [L_{p_1}^n]$ maps the subspace $L_{p_0}^n$ $(1 < p_1 < p_0 < \infty)$ into itself, then, for all $p \in (p_1, p_0]$, the operator $A|_{L_p^n}$ belongs to the class $[L_p^n]$.

The proof of this well-known theorem can be found, for instance, in KRASNOSEL'SKIĬ et al. [1].

Now we are going to explain the local principle of the theory of Fredholm operators by changing the abstract local principle for Banach algebras described in Ch. 12 of the book GOHBERG, KRUPNIK [4] in an appropriate manner. The formulation of the local principle is premised by several definitions.

We denote by $C(\Gamma)$ $(=C)$ the space of continuous functions on Γ, and by R the lineal of rational functions with poles off Γ .

DEFINITION 1.9. The operator $A \in [L_p^n]$ is called an operator of local type, if it commutes up to a compact summand with the operator of multiplication by an arbitrary function of class R .

Since the lineal R is dense in C with respect to the uniform norm (see WALSH [1] p. 67), local type operators possess, for any $a \in C$, the property $aA - AaI \in \mathcal{S}$.

We denote the class of continuous functions on Γ equal to one in some neighbourhood (depending on the function) of the point $t \in \Gamma$ by C_t.

DEFINITION 1.10. The operators A, $B \in [L_p^n]$ are said to be equivalent at the point $t_0 (\in \Gamma)$, if, for every $\varepsilon > 0$, one can find a function $a \in C_{t_0}$ such that $|(A-B)aI| < \varepsilon$, $|a(A-B)| < \varepsilon$.

DEFINITION 1.11. The operator B_l (B_r respectively) $\in [L_p^n]$ is referred to as a left (right) local regularizer of the operator $A \in [L_p^n]$ at the point $t_0 \in \Gamma$, if there exists a function $a \in C_{t_0}$ such that $aI - B_l\ AaI \in \mathcal{S}$ ($aI - aAB_r \in \mathcal{S}$).

An operator A having a left and a right local regularizer at some point $t_0 \in \Gamma$ will be called locally Fredholm at this point.

THEOREM 1.14 (I.B. SIMONENKO). Let $A \in [L_p^n]$ be a local type operator. If we can, for every $t \in \Gamma$, specify an operator A_t locally Fredholm at the point t and equivalent to the operator A at this point, then the operator A is Fredholm.

Proof. Above all, we show that the existence of a right local regularizer of the operator A_t at the point $t \in \Gamma$ and its equivalence to the operator A at this point imply that the operator A also admits a local regularization from the right at the point t. Here we do not use that the operator A is of local type.

Let B_r be a right local regularizer of the operator A_t at the point t and a_1 be some function from C_t satisfying the condition $a_1 I - a_1 A_t B_r \in \mathcal{S}$. Due to the equivalence of the operators A and A_t at the point t, there exists a function $a_2 \in C_t$ with $|a_2(A-A_t)| < \|B_r\|^{-1}$. In other words, it is possible to choose an operator $T \in \mathcal{S}$ in such a way that $\|a_2(A-A_t) + T\| < \|B_r\|^{-1}$. But then $\|a_2(A-A_t)B_r + TB_r\| < 1$.

The operator $K = I + a_2(A-A_t)B_r + TB_r$ is invertible, and consequently by Theorem 1.5, the operator $I + a_2(A-A_t)B_r$ is Fredholm.

Let B_0 be some regularizer of the operator $I + a_2(A-A_t)B_r$, and suppose a_t to be a non-negative function of class C_t equal to zero at those points $\tau \in \Gamma$ for which either $a_1(\tau) \neq 1$ or $a_2(\tau) \neq 1$.

Then $a_t = a_t a_1 = a_t a_2$ and, therefore, $a_t I - a_t A B_r B_0 =$ $= a_t I - a_t(I + (A-A_t)B_r)B_0 + a_t(I - A_t B_r)B_0 = a_t(I-(I + a_2(A-A_t)B_r)B_0)$ $+ a_t(a_1 I - a_1 A_t B_r)B_0$. Taking into account the compactness of the operators $I - (I + a_2(A-A_t)B_r)B_0$ and $a_1 I - a_1 A_t B_r$, we deduce that the operator $a_t I - a_t A B_r B_0$ is also compact, i.e., the operator $B_t = B_r B_0$ is a right local regularizer of the operator A at the point t.

Using that A is an operator of local type, we show now how to construct a "global" right regularizer from its right local ones.

The neighbourhoods U_t of the points $t \in \Gamma$ in which the functions a_t are positive form an open covering of the contour Γ. We choose a finite subcovering $\{u_{t_j}\}_{j=1}^{N}$ from it. Then the function $a = \sum_{j=1}^{N} a_{t_j}$ is positive at all points of the contour Γ. Hence the operator $B = (\sum_{j=1}^{N} a_{t_j} B_{t_j})a^{-1}I$ is well-defined. It is just the desired regularizer. Indeed, since A is of local type, we have $Aa_{t_j}I \sim a_{t_j}A$ and, therefore, $Aa_{t_j}B_{t_j} \sim a_{t_j}AB_{t_j}$. But we have $a_{t_j}AB_{t_j} \sim a_{t_j}I$, which implies

$$A(\sum a_{t_j}B_{t_j})a^{-1}I = \sum (Aa_{t_j}B_{t_j})a^{-1}I \sim \sum (a_{t_j}I)a^{-1}I = I.$$

The existence of a left regularizer of the operator A is proved in a similar way (of course, from its existence we conclude that the right regularizer constructed above is also a left one). Owing to Theorem 1.8, there results the Fredholmness of A. $\equiv$

We remark that the proof of the local principle described here can be literally transferred to the case of operators acting in $L_p(X)$, where X is a compact Hausdorff space with measure.
This general case was considered in the fundamental work of the theory of local type operators by SIMONENKO [5], where for the first time the notion of such operators was introduced and the local principle was proved.

1.2. SOME INFORMATION ON BOUNDARY PROPERTIES OF FUNCTIONS ANALYTIC AND MEROMORPHIC IN FINITELY CONNECTED DOMAINS

Let us begin with the consideration of functions analytic in the unit disk $\Delta = \{\zeta : |\zeta| < 1\}$. The proofs of the results on such functions formulated here can be found in Ch. 2 of the book of PRIVALOV [2] and in Ch. 5 of HOFFMAN [1].

<u>DEFINITION 1.12.</u> Functions of the form

$$f^{(e)}(z) = \exp(i\nu + \frac{1}{2\pi} \int_{-\pi}^{\pi} \frac{e^{i\theta} + z}{e^{i\theta} - z} \ln k(\theta)d\theta), \qquad (1.7)$$

where ν is a real constant and k a non-negative function with summable logarithm, are said to be <u>outer</u>.

Before formulating the theorem about properties of outer functions, we recall that we understand by angular boundary values (or limits) of the function f at the points t of a contour the limit $\lim_{z \to t} f(z)$, when the point z tends to t along paths non-tangent to the contour.

THEOREM 1.15. Functions of the form (1.7) are analytic in Δ and have a.e. on the unit circle angular boundary values $f^{(e)}(e^{i\theta})$ fulfilling the condition $|f^{(e)}(e^{i\theta})| = k(\theta)$. Every analytic function not having zeros in Δ with a continuous bounded logarithm is an outer function.

The class of all functions analytic and bounded in Δ is denoted by H_∞. According to Fatou's theorem, the angular boundary values exist a.e. on the circle for any function from H_∞.

DEFINITION 1.13. A function from H_∞, the angular boundary values of which are a.e. equal to one, taken absolutely, is called inner.

In particular, Blaschke products, i.e. functions of the form

$$B(z) = z^l \prod_{k=1}^{\infty} \frac{\overline{a_k}}{|a_k|} \frac{a_k - z}{1 - \overline{a_k} z},$$

are inner functions, where l is a non-negative integer and $\{a_k\}$ is a sequence of points from Δ with $\sum (1 - |a_k|) < \infty$. A Blaschke product vanishes at the points a_k and, in case $l > 0$, at $z = 0$ and only at these points.

An example of an inner function not having zeros in Δ is provided by the function

$$S(z) = \exp(-\frac{1}{2\pi} \int_{-\pi}^{\pi} \frac{e^{i\theta}+z}{e^{i\theta}-z} d\mu(\theta)),$$

where μ is a singular (i.e. concentrated on a set of Lebesgue measure zero) measure on $[-\pi,\pi]$. Such a function is referred to as singular.

THEOREM 1.16. Any inner function $f^{(i)}$ can be represented in the form $f^{(i)} = \lambda BS$, where B is a Blaschke product, S is a singular function and λ is a constant ($|\lambda| = 1$). Such a representation is unique.

DEFINITION 1.14. All functions representable in the form of a quotient of a function from H_∞ to an outer function form the class D.

THEOREM 1.17. A function f not identically zero belongs to the class D if and only if it can be represented in the form of a product of an outer by an inner function. Such a representation is unique. Outer functions and only they belong to the class D together with their reciprocal ones.

It is evident that functions of class D have angular boundary values a.e. on the unit circle.

DEFINITION 1.15. The function f analytic in Δ belongs to the class H_p, $0 < p < \infty$, if

$$\sup_{r<1} \int_{-\pi}^{\pi} |f(re^{i\theta})|^p \, d\theta < \infty . \tag{1.8}$$

We note that, for any function f analytic in Δ, $M_{p,f}(r) = \int_{-\pi}^{\pi} |f(re^{i\theta})|^p \, d\theta$ is a non-decreasing function of r. Consequently, condition (1.8) is equivalent to $\lim_{r\to 1-0} M_{p,f}(r) < \infty$ or, what is the same, $\sup_k M_{p,f}(r_k) < \infty$, where $\{r_k\}$ is an arbitrary sequence of positive scalars tending to one from below.

The classes H_p with $0 < p \leq \infty$ are called Hardy classes. The relation between Hardy classes and the class D is expressed by the following theorem.

THEOREM 1.18 (P.Ya. POLUBARINOVA-KOCHINA). A function f defined in Δ belongs to the class H_p $(0 < p \leq \infty)$ if and only if $f \in D$ and the function $f^{(b)}$ made up of its angular boundary values on the circle belongs to the class L_p.

COROLLARY 1.5 (V.I. SMIRNOV). If $f \in H_p$ and $f^{(b)} \in L_{p_1}$ for some $p_1 > p$, then $f \in H_{p_1}$.

Let $\mathcal{D}$ be a simply connected domain in the extended complex plane, the boundary of which is a closed rectifiable Jordan curve Γ.

DEFINITION 1.16. We shall say that a function f analytic in $\mathcal{D}$ is an element of the class $E_p(\mathcal{D})$, $0 < p < \infty$, if it is possible to find an expanding sequence of domains $\mathcal{D}_k$ with rectifiable boundaries Γ_k such that:

1) $\mathcal{D}_k \cup \Gamma_k \subseteq \mathcal{D}$,

2) $\bigcup_k \mathcal{D}_k = \mathcal{D}$,

3) $\sup_k \int_{\Gamma_k} |f(t)|^p |dt| < \infty$.

The classes $E_p(\mathcal{D})$ are called Smirnov classes. This definition of Smirnov classes $E_p(\mathcal{D})$ was suggested by M.V. KELDYSH and M.A. LAVRENT'EV. In SMIRNOV's original definition it was required that the curves Γ_k are images of a circle under a conformal mapping of the disk Δ onto the domain $\mathcal{D}$. The definitions of KELDYSH/LAVRENT'EV and SMIRNOV are equivalent (concerning this topic, see DANILYUK [6] p. 91, GOLUZIN [1] p. 423 and PRIVALOV [2] p. 203). Consequently, the following criterion of the membership of functions to Smirnov classes holds.

THEOREM 1.19. The function φ defined in a simply connected domain $\mathcal{D}$ belongs to the class $E_p(\mathcal{D})$ if and only if the function $\psi = \varphi \circ \omega \sqrt[p]{\omega'}$ belongs to the Hardy class H_p . Here by ω a function is denoted, which realizes the conformal mapping of the disk Δ onto the domain $\mathcal{D}$, and $\sqrt[p]{\omega'}$ is an arbitrary branch of the pth root of ω' continuous in Δ .

Below we shall be concerned with functions analytic, generally speaking, in a multiply connected domain $\mathcal{D}^+$. It is assumed that this domain is obtained from a bounded simply connected domain $\mathcal{D}^+(0)$ by removal of pairwise disjoint simply connected domains $\mathcal{D}^+(j)$ $(j = 1,\dots,m-1)$ lying inside it. The boundaries of the domains $\mathcal{D}^+(j)$ $(j = 0,\dots,m-1)$ are supposed to be Jordan rectifiable curves. We denote by $\mathcal{D}^-(j)$ the complement to $\mathcal{D}^+(j) \cup \Gamma(j)$. The contour $\Gamma = \bigcup_{j=0}^{m-1} \Gamma(j)$ is assumed to be oriented in such a manner that at an anticlockwise revolution along it the domain $\mathcal{D}^+$ is on the left.

LEMMA 1.1. Every function f analytic in the domain $\mathcal{D}^+$ admits a unique representation

$$f(z) = \sum_{j=0}^{m-1} f_j(z) \, , \quad z \in \mathcal{D}^+ \, , \tag{1.9}$$

where the function f_0 is analytic in $\mathcal{D}^+(0)$, and the functions f_j $(j = 1,\ldots,m-1)$ are analytic in $\mathcal{D}^-(j)$ and vanish at infinity.

Proof. Set $f_j(z) = \frac{1}{2\pi i} \int_{\gamma_{j,z}} f(\tau) \frac{d\tau}{\tau - z}$, where the contour $\gamma_{j,z}$ $(\subset \mathcal{D}^+)$, which is oriented in the same way as $\Gamma(j)$, is chosen in such a manner that it borders together with $\Gamma(j)$ a doubly connected domain lying in $\mathcal{D}^+$ and not containing the point z. Clearly, $\gamma_{j,z}$ is not unique, but, in view of Cauchy's integral theorem, the value $f_j(z)$ does not depend on the choice of $\gamma_{j,z}$. The equation (1.9) is nothing else than Cauchy's integral formula written in a shortened form.

It is sufficient to prove the uniqueness of the representation (1.9) for the function $f(z) \equiv 0$. If $\sum_{j=0}^{m-1} f_j(z) = 0$, then, for any $k = 1,\ldots,m-1$,

$$f_k(z) = - \sum_{j \neq k} f_j(z) \, , \quad z \in \mathcal{D}^+ \, .$$

The left-hand side of the equality obtained is analytic in $\mathcal{D}^-(k)$, and the right one in $\mathcal{D}^+ \cup \overline{\mathcal{D}^+(k)}$. Therefore, the function f_k is analytic in $\mathcal{D}^-(k) \cup \mathcal{D}^+ \cup \mathcal{D}^+(k)$, i.e. in the entire complex plane. Since, in addition, $f_k(\infty) = 0$, we can conclude that $f_k \equiv 0$, by Liouville's theorem.

But then we have $f_0(z) = - \sum_{k=1}^{m-1} f_k(z) = 0$ for $z \in \mathcal{D}^+$. Thus, the function f_0 is also identically equal to zero in $\mathcal{D}^+(0)$. This proves the uniqueness of the representation in the form (1.9) for $f(z) \equiv 0$. Hence Lemma 1.1 has been proved. $\equiv$

In what follows, we need a generalization of the Smirnov classes to the case of multiply connected domains.

DEFINITION 1.17 (KHAVINSON [1]). The function f analytic in the m-connected domain $\mathcal{D}^+$ belongs to class $E_p(\mathcal{D}^+)$ $(0 < p < \infty)$ if and only if there exists an expanding sequence of m-connected domains $\mathcal{D}_k$ satisfying conditions 1) - 3) of Definition 1.16 and, moreover, the requirement

4) the lengths $|\Gamma_k|$ of the contours Γ_k are equibounded.

THEOREM 1.20 (Decomposition theorem). The function f analytic in the m-connected domain $\mathcal{D}^+$ belongs to the class $E_p(\mathcal{D}^+)$ if and only if the functions f_j from the representation (1.9) fulfil the conditions $f_o \in E_p(\mathcal{D}^+(0))$, $f_j \in E_p(\mathcal{D}^-(j))$, $j = 1,\dots,m-1$.

Proof. Let $f \in E_p(\mathcal{D}^+)$. We represent the boundary Γ_k of the domain $\mathcal{D}_k$ occurring in Definition 1.17 in the form $\Gamma_k = \bigcup_{j=o}^{m-1} \Gamma_k(j)$, where $\Gamma_k(j)$ is a closed curve, which borders together with $\Gamma(j)$ a doubly connected domain lying in $\mathcal{D}^+$. According to equation (1.9) and Minkowski's integral inequality, the following relation holds:

$$\int_{\Gamma_k(j)} |f_j(\tau)|^p |d\tau| \le \Big(\Big(\sum_{l \neq j} \int_{\Gamma_k(j)} |f_l(\tau)|^p |d\tau|\Big)^{\frac{1}{\max(1,p)}} + \Big(\int_{\Gamma_k(j)} |f(\tau)|^p |d\tau|\Big)^{\frac{1}{\max(1,p)}}\Big)^{\max(1,p)} . \qquad (1.10)$$

Denoting by M_l the upper bound of the absolute value of the function f_l in the domain $\mathcal{D}_l$, on the strength of the maximum modulus principle, we can claim that it is also an upper bound for $|f(\tau)|$ for $\tau \in \Gamma_k(j)$ $(j \neq l)$. Therefore,

$$\int_{\Gamma_k(j)} |f_l(\tau)|^p |d\tau| \le M_l^p |\Gamma_k(j)| < M_l^p |\Gamma_k| .$$

Moreover, $\int_{\Gamma_k(j)} |f(\tau)|^p |d\tau| \le \int_{\Gamma_k} |f(\tau)|^p |d\tau|$. As the function f belongs to the class $E_p(\mathcal{D}^+)$, we deduce from condition 4) of Definition 1.17 that the right-hand side of the inequality (1.10) is bounded from above by a quantity not depending upon k . This fact as well as Definition 1.16 imply that the functions f_j belong to the Smirnov classes in appropriate domains.

Vice versa, let the conditions $f_o \in E_p(\mathcal{D}^+(0))$, $f_j \in E_p(\mathcal{D}^-(j))$,

$j = 1,\dots,m-1$ be fulfilled. We introduce curves $\Gamma_k(j)$ which are images of a circle with radius r_k under a conformal mapping of the unit disk on the domain $\mathcal{D}^+(j)$ for $j = 0$, and on $\mathcal{D}^-(j)$ for $j = 1,\dots,m-1$. Here $\{r_k\}$ is an arbitrary increasing sequence of positive numbers tending to one. Then (at least beginning with some k) the contour $\Gamma = \bigcup_{j=0}^{m-1} \Gamma_k(j)$ is the boundary of a m-connected domain $\mathcal{D}_k$ satisfying conditions 1) and 2) of Definition 1.17. Condition 4) of this definition is fulfilled, since the derivative of the function ω describing a conformal mapping of the disk on a domain with rectifiable boundary, always belongs to the class H_1 (see PRIVALOV [2] p. 173, DANILYUK [6] p. 88, GOLUZIN [1]). Now condition 3) can be readily verified in the same manner as it was done above for the function f_j. Consequently $f \in E_p(\mathcal{D}^+)$, which completes the proof of the theorem. ≡

In TUMARKIN, KHAVINSON [1] it was shown that the omission of condition 4) in Definition 1.17 does not lead to an extension of the notion of classes E_p in finitely connected domains (for a simply connected domain this fact is a consequence of Theorem 1.19 and the relation $\omega' \in H_1$). In other words, Definition 1.16 can be used for arbitrary finitely connected domains[1].

This allowed for considering Smirnov classes in Jordan domains with non-rectifiable boundary. For such domains the decomposition theorem does not hold as it was clarified in TUMARKIN, KHAVINSON [2,3].

The class of all functions analytic and bounded in some domain $\mathcal{D}$ will be denoted by $E_\infty(\mathcal{D})$.

In what follows, we shall simply write E_p instead of $E_p(\mathcal{D}^+)$, whereas E_p^- denotes the class of functions defined on the set $\mathcal{D}^- = \mathcal{D}^-(0) \cup (\bigcup_{j=1}^{m-1} \mathcal{D}^+(j))$, the restriction of which on $\mathcal{D}^-(0)$ ($\mathcal{D}^+(j)$ respectively) is a function of class $E_p(\mathcal{D}^-(0))$ ($E_p(\mathcal{D}^+(j))$). The class $\mathring{E}_p^-$ consists of functions from E_p^- vanishing at infinity.

It is clear from the proof of Theorem 1.20 that the system of domains $\mathcal{D}_k$

---

1) Note that in the monograph DUREN [1] the problem of description of multiply connected domains for which Definitions 1.16 and 1.17 are equivalent is referred to as an open one.

occurring in Definition 1.17 can be chosen as one and the same for all functions of a Smirnov class. Hence, applying Minkowski's and Hölder's integral inequalities in a standard manner, we can easily derive the following result.

THEOREM 1.21. 1) The classes E_p^+, E_p^- and $\overset{\circ}{E}{}_p^-$ are lineals,

2) for any $p_1, p_2 \in (0,\infty]$, $p_1 < p_2$, the inclusions $E_{p_1}^+ \supseteq E_{p_2}^+$, $E_{p_1}^- \supseteq E_{p_2}^-$ hold,

3) if $f \in E_{p_1}^+ \ (E_{p_1}^-)$, $g \in E_{p_2}^+ \ (E_{p_2}^-)$, then $fg \in E_p^+ \ (E_p^-)$, where $p = (p_1^{-1} + p_2^{-1})^{-1}$.

More profound properties of functions from Smirnov classes are gathered below in Theorems 1.22, 1.24, 1.25. For the case of a simply connected domain $\mathcal{D}^+$, the arguments of these theorems can be found in the monographs GOLUZIN [1], DANILYUK [6], PRIVALOV [2]. With the help of the decomposition theorem 1.20, these theorems can be transferred to the more general case of a finitely connected domain in an obvious manner[1], which justifies the references to GOLUZIN [1], DANILYUK [6], PRIVALOV [2] before the formulations of Theorems 1.22, 1.24, 1.25.

THEOREM 1.22 (see PRIVALOV [2] p. 204, DANILYUK [6] p. 91, GOLUZIN [1] p. 422). Every function f of the class $E_p^+ \ (E_p^-)$, $p > 0$, has limits $f^+(t) \ (f^-(t))$ along all non-tangent paths at almost all points $t \in \Gamma$. The function $f^+ \ (f^-)$, which is defined on Γ by these limits, belongs to the class L_p.

DEFINITION 1.18. The function f defined a.e. on Γ we assume to belong to the class $L_p^+ \ (L_p^-, \overset{\circ}{L}{}_p^-)$, $0 < p \leq \infty$, if there exists a function of class $E_p^+ \ (E_p^-, \overset{\circ}{E}{}_p^-)$ the boundary values of which coincide with f at almost all $t \in \Gamma$.

We denote by $M_p^{\pm} \ (= L_p^{\pm} + R)$ the class of functions f representable in the form $f = f_1 + f_2$, where $f_1 \in L_p^{\pm}$ and $f_2 \in R$.

---

1) Concerning this topic, see TUMARKIN, KHAVINSON [3], where Theorem 1.24 is formulated (in particular, on p. 73) for the case of a finitely connected domain.

The classes $E_p^{\pm} + R$, $L_p^{\pm} + C$ are introduced in a similar way. Sometimes it will be convenient to use the notations $L^{\pm} = \bigcup_{p>0} L_p^{\pm}$, $M^{\pm} = \bigcup_{p>0} M_p^{\pm}$.

Beginning with Chapter 2, we shall, as a rule, not deal with scalar functions from classes L_p , $L_p^{\pm}$, $M_p^{\pm}$ etc., but with n-dimensional vector functions and $n \times n$ matrix functions, the elements of which belong to these classes. We denote by X^n the space of n-dimensional vector functions, the components of which belong to some function class X (in the case $X = L_p$, this notation was already used in Section 1.1). It would be natural to denote the class of $n \times n$ matrix functions with elements from X by $X^{n \times n}$. However, we renounce this notation, since it is too troublesome. Instead, the class of such matrix functions is simply denoted by X . From the context it will be always clear whether we speak about a scalar or a matrix function.

Vector functions of the class $(E_p^{\pm} + R)^n$ can be reproduced from their limits on Γ in a unique manner. Let us show this. If the vector functions f_1 and f_2 from the class $(E_p^{\pm} + R)^n$ have angular limits coinciding a.e. on Γ , then their difference $f = f_1 - f_2$ is also a vector function from $(E_p^{\pm} + R)^n$ and has angular limits equal to zero a.e. on Γ . But in view of the following uniqueness theorem, this is possible only for $f = 0$.

<u>THEOREM 1.23 (PRIVALOV [2] p. 292, GOLUZIN [1] p. 413).</u> A function meromorphic in a domain with rectifiable boundary, which has angular limits equal to zero on a set of positive linear measure, is identically equal to zero.

Taking into account the injectivity of the correspondence between classes $E_p^{\pm} + R$ and $M_p^{\pm}$ (respectively between $E_p^{\pm}$ and $L_p^{\pm}$), we sometimes shall identify vector functions from $(E_p^{\pm} + R)^n$ with vector functions from $(M_p^{\pm})^n$ formed by their limits.

In particular, if $f \in (M_p^{\pm})^n$, under $f(z)$, $z \in \mathcal{D}^{\pm}$ we shall simply understand the value at the point z of that vector function of class $(E_p^{\pm} + R)^n$, the boundary value of which is f . In connection with this, note that the identification of functions of class H_p with corresponding functions of class L_p^+ on the circle is in general use. For the latter class even the notation H_p (or H_p^+) is preserved (see HOFFMAN [1]). For the sake of uniformity, the class L_p^- on the circle is denoted by H_p^-. Clearly, all the assertions of Theorem 1.21 may be transferred from the classes $E_p^{\pm}$ to classes $(E_p^{\pm} + R)^n$, $(L_p^{\pm})^n$ and $(M_p^{\pm})^n$, if the product fg is defined componentwise.

In the classic case of functions analytic in a domain and continuous up to the boundary, the values in the domain are expressed via boundary ones with the aid of Cauchy's integral. This method cannot be transferred directly to the case of classes $E_p^{\pm}$ for $p < 1$, because a boundary function is not summable. However, for functions from the classes $E_1^{\pm}$, Cauchy's integral remains to be a natural means of continuation from a contour to a domain.

<u>THEOREM 1.24 (PRIVALOV [2] pp. 202, 206, DANILYUK [6] p. 133, GOLUZIN [1] pp. 420–423).</u> The vector function f defined on Γ a.e. belongs to $(L_1^+)^n$ ($(\overset{\circ}{L}{}_1^-)^n$, respectively) iff $f \in L_1^n$ and

$\int_\Gamma f(\tau) \frac{d\tau}{\tau - z} = 0$ for $z \in \mathcal{D}^-$ ($\mathcal{D}^+$) or, equivalently,

$$\int_\Gamma f(\tau)\tau^k d\tau = 0 \tag{1.11}$$

for all non-negative (negative) integers k.

If this condition is fulfilled, then the function Φ from $(E_1^+)^n$ ($(\overset{\circ}{E}{}_1^-)^n$), whose boundary value is f, can be determined by the formula

$$\Phi(z) = \frac{\varepsilon}{2\pi i} \int_\Gamma f(\tau) \frac{d\tau}{\tau - z}, \quad z \in \mathcal{D}^+ \ (\mathcal{D}^-),$$

where $\varepsilon = 1$ if $z \in \mathcal{D}^+$ and $\varepsilon = -1$ if $z \in \mathcal{D}^-$.

In accordance with Theorems 1.24 and 1.22, any vector function f of the class $(L_p^+)^n$ ($(\overset{\circ}{L}{}_p^-)^n$), $p \geq 1$ lies in the subspace of L_p^n consisting of all functions satisfying (1.11) and, for $p = 1$, coincides with it. Corollary 1.5 shows that for the circle the coincidence holds for

$p > 1$, too. To characterize the situation in general, we introduce another definition.

DEFINITION 1.19 (see PRIVALOV [2] p. 250, DANILYUK [6] p. 90).

A simply connected domain $\mathcal{D}$ is called Smirnov, if the derivative of the function ω , which maps Δ conformally on $\mathcal{D}$, is an outer function.

A closed Jordan curve Γ will be called Smirnov (curve of class $\mathcal{C}$), if each of the domains in which Γ decomposes the extended complex plane is Smirnov.

The class of Smirnov curves is so wide that it is rather difficult to construct examples of curves not belonging to this class. For the first time, such an example was built by M.V. KELDYSH and M.A. LAVRENT'EV. Their construction, with some simplifications, is described in PRIVALOV [2] pp. 229-250. An analytic description of all non-Smirnov curves the interested reader can find in DUREN, SHAPIRO, SHIELDS [1]. Various sufficient conditions of membership of curves to the Smirnov class are established in PRIVALOV [2] pp. 250-257. In particular, one can find there LAVRENT'EV's result that a curve is Smirnov as soon as

$$\operatorname{ess\,sup}\left\{\frac{s(t_1,t_2)}{|t_1 - t_2|} : t_1, t_2 \in \Gamma\right\} < \infty . \tag{1.12}$$

Here $s(t_1,t_2)$ is the distance between points t_1 and t_2 measured along Γ .

THEOREM 1.25 (PRIVALOV [2] pp. 264-266, DANILYUK [6] pp. 92-95).

If the contour Γ consists only of Smirnov curves, then, for $0 < p_1 < p_2 \leq \infty$, the equalities $(L^+_{p_2})^n = (L^+_{p_1})^n \cap L^n_{p_2}$ and $(L^-_{p_2})^n = (L^-_{p_1})^n \cap L^n_{p_2}$ are valid. Especially, in case that $p \geq 1$ we have $f \in (L^+_p)^n$ $((\overset{\circ}{L}{}^-_p)^n$ respectively) if and only if $f \in L^n_p$ and condition (1.11) is fulfilled. If at least one component of the contour Γ is not a Smirnov curve, then there exists a function $f \in L_\infty$ such that $f \in L^+_1$ $(\overset{\circ}{L}{}^-_1)$, but $f \notin L^+_p$ $(\overset{\circ}{L}{}^-_p)$ for all $p > 1$.

Due to Theorem 1.25, in the case of a contour Γ consisting of Smirnov curves, the lineals $(L^+_p)^n$ and $(L^-_p)^n$ are subspaces in L^n_p . In fact, this assertion is true for every rectifiable contour Γ , however, the corresponding argument requires some preliminary constructions, which we are now going to deal with.

We shall identify the space L_q^n with the dual to L_p^n $(1 \leq p < \infty$, $q = p/(p-1))$ via the bilinear form

$$\langle f,g\rangle = \sum_{j=1}^{n} \int_{\Gamma} f_j(t)g_j(t)dt , \tag{1.13}$$

where $f = \sum_{j=1}^{n} f_j e_j \in L_p^n$, $g = \sum_{j=1}^{n} g_j e_j \in L_q^n$.

It goes without saying that in this case L_p^n and L_q^n have to be regarded as real Banach spaces. The norm in L_p^n consistent with the form (1.13) (i.e. satisfying the condition $\|f\|_{L_p^n} = \sup\{\langle f,g\rangle : g \in L_q^n,$ $\|g\|_{L_q^n} = 1\})$ looks as

$$\|f\|_{L_p^n} = \Big(\int_{\Gamma} \Big(\sum_{j=1}^{n} |f_j(t)|^2\Big)^{p/2} |dt|\Big)^{1/p} ,$$

where $f = \sum_{j=1}^{n} f_j e_j$ and e_j denotes the j th unit vector.

<u>THEOREM 1.26.</u> The orthogonal (in the sense of the form (1.13)) complement to $(L_p^+)^n$ $((\mathring{L}_p^-)^n)$ coincides with $(L_q^+)^n$ $((\mathring{L}_q^-)^n)$.

<u>Proof.</u> Obviously, it is sufficient to prove the statement for $n = 1$. If $f \in L_q^+$, $g \in L_p^+$, then we deduce $fg \in L_1^+$ and $\langle f,g\rangle = 0$ by Theorem 1.24. The same theorem enables us to conclude that $\langle f,g\rangle = 0$, if $f \in \mathring{L}_q^-$, $g \in \mathring{L}_p^-$, since in this case the function $tf(t)g(t)$ belongs to the class $\mathring{L}_1^-$.

Conversely, let the function $f \in L_q$ be such that $\langle f,g\rangle = 0$ for all $g \in L_p^+$ $(\mathring{L}_p^-)$. Setting $g(t) = t^k$, where k is a non-negative (negative) integer, we obtain $\int_{\Gamma} f(t)t^k dt = 0$. According to Theorem 1.24, this means that $f \in L_1^+$ $(\mathring{L}_1^-)$.

The following considerations are first made for the case of a simply connected domain $\mathcal{D}^+$. Suppose that the function ω realizes a conformal mapping of the unit disk onto $\mathcal{D}^+$. After substituting $\tau = \omega(\zeta)$ the condition $\int_{\Gamma} f(\tau)g(\tau)d\tau = 0$ can be rewritten as

$$\int_{\mathbb{T}} f(\omega(\zeta)) \sqrt[q]{\omega'(\zeta)}\, g(\omega(\zeta)) \sqrt[p]{\omega'(\zeta)}\, d\zeta = 0 \qquad (1.14)$$

(for $p = 1$, i.e. $q = \infty$, we put $\sqrt[q]{\omega'} = 1$) where $\mathbb{T}$ is the unit circle.

The function $\varphi = (f \circ \omega)\sqrt[q]{\omega'}$ belongs to the class L_q $(\subset L_1)$ simultaneously with f. If g ranges over the whole class L_p^+, then in view of Theorem 1.19, $(g \circ \omega)\sqrt[p]{\omega'}$ ranges over the whole class H_p. By what was proved above, relation (1.14) implies $\varphi \in H_1$. From this relation and the condition $\varphi \in L_q$, in virtue of Corollary 1.5, one can conclude $\varphi \in H_q$. Applying Theorem 1.19 once more, we observe $f \in L_q^+$.

Considering a conformal mapping of the disk onto the domain $\mathcal{D}^-$, it can be shown in a similar way that the orthogonal complement to $\mathring{L}_p^-$ consists of functions of the class $\mathring{L}_q^-$.

Now we return to the general case of a m-connected domain $\mathcal{D}^+$. A function $f \in L_q$ orthogonal to L_p^+ lies in L_1^+ as was already shown. Applying the decomposition theorem 1.20 to that function from E_1^+ the boundary value of which is f, we get the representation $f = \sum_{j=0}^{m-1} f_j$, where f_0 (f_j) is the boundary value on Γ of a function from $E_1(\mathcal{D}^+(0))$ $(\mathring{E}_1^-(\mathcal{D}^-(j))$, $j = 1,\ldots,m-1)$. Let us choose as g the boundary value on Γ of a function from $E_q(\mathcal{D}^+(0))$. Then, of course, $g \in L_q^+$ and thus

$$\int_\Gamma f(t)g(t)dt = \sum_{j=0}^{m-1} \int_\Gamma f_j(t)g(t)dt = 0 .$$

Directly by definition we check that, for $j = 1,\ldots,m-1$, the functions f_jg belong to the class L_1^+. As a result of this we obtain that in the last sum all summands are equal to zero, with the exception of the first one. Hence,

$$\int_\Gamma f_0(t)g(t) = 0 . \qquad (1.15)$$

Furthermore, the product f_0g can be continued to a function analytic in the closure of the domains $\mathcal{D}^+(j)$ $(j = 1,\ldots,m-1)$. Owing to

Cauchy's classical integral theorem, $\int_{\Gamma(j)} f_o(t)g(t)dt = 0 \quad (j=1,\dots,m-1)$.
Thus, condition (1.15) can be rewritten as

$$\int_{\Gamma(0)} f_o(\tau)g(\tau)d\tau = 0 .$$

In summary, it has been shown that the boundary value on $\Gamma(0)$ of the function f_o analytic in $\mathcal{D}^+(0)$ belongs to the orthogonal complement of the set of boundary values of functions from $E_p(\mathcal{D}^+(0))$. Due to the result established above for simply connected domains, f_o is the boundary value of a function from $E_q(\mathcal{D}^+(0))$.
Taking as g the boundary value on Γ of functions from the class $E_p(\mathcal{D}^-(j))$ and acting by the scheme described, we can state that f_j is in fact the boundary value of a function from $E_q(\mathcal{D}^-(j))$.
By Theorem 1.20, this just means that $f \in L_q^+$. Thus, also in case of a multiply connected domain the orthogonal complement to L_p^+ is the class L_q^+. Analogously, the assertion concerning the classes $\mathring{L}_p^-$, $\mathring{L}_q^-$ allow to be transferred to the case of multiply connected domains. ≡

<u>COROLLARY 1.6.</u> For any $p \in [1,\infty]$, the sets $(L_p^+)^n$, $(\mathring{L}_q^-)^n$ are subspaces in L_p^n.

Indeed, for $p \in (1,\infty]$, $(L_p^+)^n$ $((\mathring{L}_p^-)^n)$ is the orthogonal complement to some lineal and is therefore a subspace. The closedness of the classes $(L_1^+)^n$ and $(\mathring{L}_1^-)^n$ was already mentioned.

<u>THEOREM 1.27.</u> Let the vector functions $\Phi^+ \in (M_1^+)^n$ and $\Phi^- \in (M_1^-)^n$ coincide a.e. on some arc $\gamma \subseteq \Gamma$. Then each of these vector functions is the analytical continuation of the other through γ.

<u>Proof.</u> By definition of the classes $(M_1^\pm)^n$, there exist representations $\Phi^+ = \varphi^+ + r_1$, $\Phi^- = \varphi^- + r_2$ with $\varphi^\pm \in (L_1^\pm)^n$ and $r_j \in R^n$. In its turn, $r_j = r_j^+ - r_j^-$, where $r_j^\pm$ is a rational vector function with poles concentrated in $\bar{\mathcal{D}}^\mp$, $j = 1,2$. The coincidence of the functions Φ^+ and Φ^- a.e. on γ, obviously, implies that the vector functions $\Psi^+ = \varphi^+ + r_1^+ - r_2^+$ and $\Psi^- = \varphi^- + r_1^- - r_2^-$ belonging to the classes $(L_1^+)^n$ and $(L_1^-)^n$, respectively, coincide a.e. on γ. Changing the vector functions $r_2^\mp$ by one and the same constant, one

may always obtain $\Psi^- \in (\overset{\circ}{L}{}_1^-)^n$. In view of Theorem 1.24, for $z \in \mathcal{D}^+$, the following equalities hold:

$$\Psi^+(z) = \frac{1}{2\pi i} \int_{\Gamma\setminus\gamma} \Psi^+(\tau) \frac{d\tau}{\tau-z} + \frac{1}{2\pi i} \int_{\gamma} \Psi^+(\tau) \frac{d\tau}{\tau-z} ,$$

$$0 = \frac{1}{2\pi i} \int_{\Gamma\setminus\gamma} \Psi^-(\tau) \frac{d\tau}{\tau-z} + \frac{1}{2\pi i} \int_{\gamma} \Psi^-(\tau) \frac{d\tau}{\tau-z} .$$

Having regard to the relation $\int_\gamma \Psi^+(\tau) \frac{d\tau}{\tau-z} = \int_\gamma \Psi^-(\tau) \frac{d\tau}{\tau-z}$, we come to the representation

$$\Psi^+(z) = \frac{1}{2\pi i} \int_{\Gamma\setminus\gamma} \Psi^+(\tau) \frac{d\tau}{\tau-z} - \frac{1}{2\pi i} \int_{\Gamma\setminus\gamma} \Psi^-(\tau) \frac{d\tau}{\tau-z} . \tag{1.16}$$

Consequently, the vector function Ψ^+ is analytically continuable from the domain $\mathcal{D}^+$ through the arc γ . But then, in some neighbourhood of γ , the vector function $\Psi^+ - r_1^- + r_2^+$ coinciding a.e. on γ with Φ^+ (and Φ^-) is also analytic. Thus Theorem 1.27 is proved. ≡

If, in particular, $\gamma = \Gamma$, then we can conclude from formula (1.16) that $\Psi^+ = 0$, i.e. $\Phi^+ \in R^n$. Taking into account that a rational function that has no poles is a constant, we get the following uniqueness theorem.

<u>THEOREM 1.28.</u> The intersection of the classes $(M_1^+)^n$ and $(M_1^-)^n$ consists precisely of vector functions of class R^n , the intersection of the classes $(L_1^+)^n$ and $(L_1^-)^n$ consists of constants, and the intersection of $(L_1^+)^n$ and $(\overset{\circ}{L}{}_1^-)^n$ contains the zero function only.

Finally, we want to insert some remarks on the notion of the index of a function.

<u>DEFINITION 1.20.</u> Suppose f is a function continuous and nonvanishing on Γ . Then the integer

$$\operatorname{ind} f = \frac{1}{2\pi} \{\arg f(t)\}_\Gamma$$

is called the (Cauchy) <u>index</u> of f . Here $\{\arg f(t)\}_\Gamma$ denotes the sum of the argument increments of the function f , when the variable t ranges over all closed curves $\Gamma_{(j)}$ which form the contour Γ , once in each case in positive direction.

If the function f admits a continuation to a function F , which is continuous in $\mathcal{D}^+ \cup \Gamma$ and analytic in $\mathcal{D}^+$ with the exception of a finite number of poles , then, due to the argument principle,

$$\text{ind } f = N - P \, , \tag{1.17}$$

where N and P are the numbers of zeros and poles of the function F in $\mathcal{D}^+$ calculated with regard to their multiplicity.

Starting from this result, we take equation (1.17) as the definition of the index for functions f belonging to the class M^+ together with their inverses. This definition is correct, since the condition $f \in M^+$ guarantees the finiteness of the number of poles, and the condition $f^{-1} \in M^+$ the finiteness of zeros of the meromorphic continuation of f from the contour Γ into $\mathcal{D}^+$.

The index of invertible elements of class M^+ defined in this way will be denoted by the symbol ind_+ .

Analogously, for functions f satisfying the condition $f^{\pm 1} \in M^-$ we introduce the index via the formula

$$\text{ind}_- f = P - N \, , \tag{1.17'}$$

where P and N are the numbers of poles and zeros in $\mathcal{D}^-$ of f .

THEOREM 1.29.

1) If i is one of the functions ind_+, ind_-, ind defined on the invertible elements of M^+, M^-, C, respectively, then

$$i(f_1 f_2) = i(f_1) + i(f_2),$$

2) $\text{ind}_+ f \geq 0$ if $f \in L^+$, $\text{ind}_- f \leq 0$ if $f \in L^-$,

3) for a rational function f the value $\text{ind}_- f - \text{ind}_+ f$ coincides with the difference of the number of poles and zeros of this function that lies on Γ taking into account their multiplicity,

4) if $f^{\pm 1} \in M_1^+$ and $f^{\pm 1} \in M_1^-$, then $\text{ind}_+ f = \text{ind}_- f$ (= ind f) .

Proof. Assertions 1) and 2) result directly from the definition. Assertion 3) follows from the fact that $\text{ind}_- f - \text{ind}_+ f$ is the difference of the number of poles and zeros (with regard to their multiplicity) of the function f in $\mathcal{D}^+ \cup \mathcal{D}^-$, i.e. in the complement to the contour Γ with respect to the complete plane, and the corresponding difference in the complete plane is equal to zero. Finally, under the condition $f \in M_1^+ \cap M_1^-$, the function f is rational by Theorem 1.28 and does not have poles on Γ. Moreover, since $f^{-1} \in M_1^{\pm}$, the function f does not have any zeros on Γ either. According to assertion 3), we have $\operatorname{ind}_- f = \operatorname{ind}_+ f$, which proves the theorem. ≡

Immediately from the definition of classes M we deduce that, for each matrix function $G \in M^+$ (M^-), one can choose a scalar function $f \in R$ not vanishing on Γ and such that

$$fG \in L^+ \quad (L^-) . \tag{1.18}$$

Applying assertion 2) of Theorem 1.29 to the function $\det(fG)$, we find by the equality $\operatorname{ind}_{\pm}(\det fG) = n \cdot \operatorname{ind} f + \operatorname{ind}_{\pm} \det G$ that the value $\operatorname{ind} f$ is bounded from below, if $G \in M^+$, and from above, if $G \in M^-$. The corresponding exact bound of the value $\operatorname{ind} f$, when f ranges over the set of all non-degenerating functions from R satisfying condition (1.18), we shall denote by $\mathcal{P}_+(G)$ $(-\mathcal{P}_-(G))$. Clearly, $\mathcal{P}_{\pm}(G)$ is less or equal than the total number of poles of the entries of G in $\mathcal{D}^{\pm}$.

1.3. ON THE OPERATOR OF SINGULAR INTEGRATION IN SPACES OF SUMMABLE FUNCTIONS

In the previous section it was shown that the classes $(L_1^+)^n$ and $(\overset{\circ}{L}{}_1^-)^n$ are subspaces in L_1^n, the only common element of which is 0. This allows for correctly defining the operator S of the lineal $\mathcal{L}_1^n = (L_1^+ \dotplus \overset{\circ}{L}{}_1^-)^n$ as follows: if $\varphi^+ \in (L_1^+)^n$, $\varphi^- \in (L_1^-)^n$, then $S(\varphi^+ + \varphi^-) = \varphi^+ - \varphi^-$.

Furthermore, we define operators in $\mathcal{L}_1^n$ by $P = \frac{1}{2}(I + S)$ and $Q = \frac{1}{2}(I - S)$. Obviously, P projects $\mathcal{L}_1^n$ onto $(L_1^+)^n$ along $(\overset{\circ}{L}{}_1^-)^n$, and Q projects $\mathcal{L}_1^n$ onto $(\overset{\circ}{L}{}_1^-)^n$ along $(L_1^+)^n$.

In view of Theorem 1.24, we can write that with $\varphi = \varphi^+ + \varphi^-$ $(\varphi^+ \in (L_1^+)^n$, $\varphi^- \in (\overset{\circ}{L}{}_1^-)^n)$ for the analytic continuations $\Phi^{\pm}$ of the vector functions $\varphi^{\pm}$ in $\mathcal{D}^{\pm}$ the following formulae are valid:

$$\Phi^+(z) = \frac{1}{2\pi i} \int_\Gamma \varphi(\tau) \frac{d\tau}{\tau - z}, \quad z \in \mathcal{D}^+,$$
$$\Phi^-(z) = -\frac{1}{2\pi i} \int_\Gamma \varphi(\tau) \frac{d\tau}{\tau - z}, \quad z \in \mathcal{D}^-. \tag{1.19}$$

But for angular limits on Γ of a Cauchy type integral with summable density φ, the following formula holds a.e. on Γ (see PRIVALOV [2] p. 194, DANILYUK [6] p. 125):

$$\lim_{z \to t} \frac{1}{2\pi i} \int_\Gamma \varphi(\tau) \frac{d\tau}{\tau - z} = \pm \frac{1}{2} \varphi(t) + \frac{1}{2\pi i} \int_\Gamma \varphi(\tau) \frac{d\tau}{\tau - t},$$

where the sign is chosen depending on whether the point z lies in $\mathcal{D}^+$ or $\mathcal{D}^-$, and the integral is to be understood in the sense of the principal value.

Passing in formulae (1.19) to the limit for $z \to t$ along non-tangent paths, we get the following result.

THEOREM 1.30. If $\varphi \in \mathcal{L}_1^n$ a.e. on Γ, then the formulae

$$(P\varphi)(t) = \frac{1}{2} \varphi(t) + \frac{1}{2\pi i} \int_\Gamma \varphi(\tau) \frac{d\tau}{\tau - t},$$

$$(Q\varphi)(t) = \frac{1}{2} \varphi(t) - \frac{1}{2\pi i} \int_\Gamma \varphi(\tau) \frac{d\tau}{\tau - t}$$

are valid.

These formulae were obtained for the first time by Yu.V. SOKHOTSKIǏ and are called after him[1]). Subtracting the second Sokhotskiǐ formula from the first one, we find that the relation

$$(S\varphi)(t) = \frac{1}{\pi i} \int_\Gamma \varphi(\tau) \frac{d\tau}{\tau - t}$$

holds a.e. on Γ.

In view of the latter equality, the operator S is called the operator of singular integration along the contour Γ.

It is evident that the conditions $\varphi \in L_p^n$ and $S\varphi \in L_p^n$ $(p \geq 1)$ are fulfilled simultaneously iff $\varphi = \varphi^+ + \varphi^-$, where $\varphi^+ \in L_p^n \cap (L_1^+)^n$ and $\varphi^- \in L_p^n \cap (\mathcal{L}_1^-)^n$. Therefore, the natural domain of the operators S, P and Q in the space L_p^n coincides with the lineal

1) In the English literature these formulae are usually called after PLEMELJ.

$(L_p \cap L_1^+)^n \dotplus (L_p \cap \overset{\circ}{L}{}_1^-)^n$. According to Theorems 1.24 and 1.25, in case of a Smirnov contour Γ this lineal agrees with $\mathcal{L}_p^n = (L_p^+ \dotplus \overset{\circ}{L}{}_p^-)^n$. If Γ is not Smirnov and $p > 1$ then it contains $\mathcal{L}_p^n$ as a proper part. However, for our purposes it is convenient to consider the operators S , P and Q in the spaces L_p^n , assuming their domains of definition to be the lineal $\mathcal{L}_p^n$, irrespective of the nature of the contour Γ . Henceforth we shall do so.

<u>THEOREM 1.31.</u>

1) The operators S, P and Q studied in L_p^n $(1 \leq p \leq \infty)$ are closed and, for $p < \infty$, densely defined,
2) the operators S (P, Q) and $-S$ (Q, P) considered in the spaces L_p^n and L_q^n $(1 < p < \infty,\ q = p/(p-1))$, respectively are adjoint to each other in the sense of Definition 1.6,
3) the operator S is involutory, and P as well as Q are idempotent: $S^2 = I|_{\mathcal{L}_p^n}$, $P^2 = P$, $Q^2 = Q$,
4) for the function $\varphi \in \mathcal{L}_{2p}$ the equality

$$(S\varphi)^2 = 2S(\varphi S\varphi) - \varphi^2 \tag{1.20}$$

is true.

<u>Proof.</u> Since $P = \frac{1}{2}(I + S)$ and $Q = \frac{1}{2}(I - S)$, it suffices to verify the assertions concerning the operator S .

1) Let $\{\varphi_k\}$ be a sequence from the domain of the operator S considered in L_p^n , $\varphi_k \to \varphi$ and $S\varphi_k \to \Psi$. Then $\varphi_k \in \mathcal{L}_p^n$, i.e. $\varphi_k = \varphi_k^+ + \varphi_k^-$ $(\varphi_k^+ \in (L_p^+)^n$, $\varphi_k^- \in (\overset{\circ}{L}{}_p^-)^n)$, and $S\varphi_k = \varphi_k^+ - \varphi_k^-$. Consequently, each of the sequences $\{\varphi_k^+ + \varphi_k^-\}$ and $\{\varphi_k^+ - \varphi_k^-\}$ and, thus, also each of the sequences $\{\varphi_k^+\}$ and $\{\varphi_k^-\}$ converges in L_p^n . Denoting the limits of the latter two sequences by φ^+ and φ^- respectively, we obtain $\varphi = \varphi^+ + \varphi^-$ and $\Psi = \varphi^+ - \varphi^-$. In view of Corollary 1.6, we have $\varphi^+ \in (L_p^+)^n$, $\varphi^- \in (\overset{\circ}{L}{}_p^-)^n$. Hence $\varphi \in \mathcal{L}_p^n$ and $S\varphi = \Psi$, which proves the closedness of the operator S .

   For $p < \infty$ the lineal R^n being located in $\mathcal{L}_p^n$ is dense in L_p^n , therefore, for $p < \infty$, the operator S is densely defined.

2) Suppose $f \in \mathcal{L}_p^n$, $g \in \mathcal{L}_q^n$ and $f = f^+ + f^-$, $g = g^+ + g^-$, where $f^+ \in (L_p^+)^n$, $g^+ \in (L_q^+)^n$, $f^- \in (\overset{\circ}{L}{}_p^-)^n$, $g^- \in (\overset{\circ}{L}{}_q^-)^n$, then we obtain, taking into account Theorem 1.26,

$$\langle Sf,g\rangle = \langle f^+-f^-, g^++g^-\rangle = \langle f^+,g^+\rangle - \langle f^-,g^+\rangle +$$

$$\langle f^+,g^-\rangle - \langle f^-,g^-\rangle = \langle f^+,g^-\rangle - \langle f^-,g^+\rangle$$

and

$$\langle f, -Sg\rangle = \langle f^++f^-, g^--g^+\rangle = \langle f^+,g^-\rangle + \langle f^-,g^-\rangle -$$

$$\langle f^+,g^+\rangle - \langle f^-,g^+\rangle = \langle f^+,g^-\rangle - \langle f^-,g^+\rangle .$$

Thus $\langle Sf,g\rangle = \langle f,-Sg\rangle$ for arbitrary f, g from the corresponding domains of definition.

Assertion 3) is obvious.

Now we are going to prove assertion 4). If $\varphi = \varphi^+ + \varphi^-$ with $\varphi^+ \in L_{2p}^+$, $\varphi^- \in L_{2p}^-$, then $S\varphi = \varphi^+ - \varphi^-$, $\varphi S\varphi = (\varphi^+)^2 - (\varphi^-)^2$. But in virtue of assertion 3) from Theorem 1.21, $(\varphi^+)^2 \in L_p^+$, $(\varphi^-)^2 \in \overset{\circ}{L}{}_p^-$. Consequently, the function $\varphi S\varphi$ belongs to the domain of S, where $S(\varphi S\varphi) = (\varphi^+)^2 + (\varphi^-)^2$. Hence
$2S(\varphi S\varphi) - \varphi^2 = 2(\varphi^+)^2 + 2(\varphi^-)^2 - (\varphi^+)^2 - 2\varphi^+\varphi^- - (\varphi^-)^2 = (\varphi^+-\varphi^-)^2 = (S\varphi)^2$. Theorem 1.31 has completely been proved. ≡

Now we turn to the question on the boundedness of the operator S in the spaces L_p^n, $1 < p < \infty$. Above all, we note that the operator S is defined componentwise. In other words, the vector function $f = \sum_{j=1}^{n} f_j e_j$ belongs to the domain of S iff the scalar functions f_j belong to this domain. If this condition is fulfilled, then we have

$$Sf = \sum_{j=1}^{n} (Sf_j)e_j .$$

In particular, the operator S is bounded in L_p^n for any natural n if and only if it is bounded in L_p.

Further, since the operator S is densely defined and closed, it is bounded if and only if its domain of definition is the whole space. Consequently, the boundedness of the operator S in L_p, $1 < p < \infty$, is equivalent to the fact that the space L_p is the direct sum of its subspaces L_p^+ and $\overset{\circ}{L}{}_p^-$.

THEOREM 1.32. In the case of a fixed contour Γ the operator S is either bounded in all L_p, $1 < p < \infty$, or is not bounded in any one of these spaces.

Proof. Let the operator S be bounded in some space L_p. Due to equation (1.20), we can conclude that, for $\varphi \in \mathcal{L}_{2p}$,

$$\|(S\varphi)^2\|_{L_p} \leq 2\|S(\varphi S\varphi)\|_{L_p} + \|\varphi\|^2_{L_p} .$$

But

$$\|(S\varphi)^2\|_{L_p} = \|S\varphi\|^2_{L_{2p}} , \quad \|\varphi^2\|_{L_p} = \|\varphi\|^2_{L_{2p}}$$

and

$$\|S(\varphi S\varphi)\|_{L_p} \leq \|S\|_{L_p} \cdot \|\varphi(S\varphi)\|_{L_p} \leq \|S\|_{L_p} \cdot \|\varphi\|_{L_{2p}} \|S\varphi\|_{L_{2p}} .$$

Consequently,

$$\|S\varphi\|^2_{L_{2p}} \leq 2\|S\|_{L_p} \|\varphi\|_{L_{2p}} \cdot \|S\varphi\|_{L_{2p}} + \|\varphi\|^2_{L_{2p}} .$$

Solving this quadratic inequality relative to $\|S\varphi\|_{L_{2p}}$, we get

$$\|S\varphi\|_{L_{2p}} \leq (\|S\|_{L_p} + \sqrt{1+\|S\|^2_{L_p}}) \cdot \|\varphi\|_{L_{2p}} .$$

The last inequality means that the operator S regarded in L_{2p} is bounded on its natural domain. Moreover, the following estimation is valid:

$$\|S\|_{L_{2p}} \leq \|S\|_{L_p} + \sqrt{1+\|S\|^2_{L_p}} .$$

Repeating this reasoning, we deduce that the operator S is bounded in the spaces L_r with $r = 2^k p$ $(k = 0,1,\ldots)$. With the help of the interpolation theorem 1.13 we conclude that the operator S is bounded in all spaces L_r with $p \leq r < \infty$.

Now we utilize assertion 2) of Theorem 1.31. Taking into account that S is densely defined, this assertion implies that the operator S is bounded in L_p iff it is bounded in L_q.

In this way, the operator S is bounded in spaces L_r not only for

$r \in [p,\infty)$, but also for $r \in (1,p/(p-1)]$. If $p \leq 2$, then there remains nothing to show. If $p > 2$, it is still necessary to establish the boundedness of S in L_r for $r \in (q,p)$, using the boundedness of S in L_p and L_q and again applying Theorem 1.13. ≡

Theorem 1.32 was proved in PAATASHVILI [1] (see also COTLAR [1]), where relation (1.20) occurs for the first time. The proof described here is borrowed from KHVEDELIDZE [4].

Acting as in KHVEDELIDZE [4], we introduce the following

<u>DEFINITION 1.21.</u> We say that the contour Γ belongs to the <u>class</u> $\mathcal{R}$ if the operator S of singular integration along Γ is bounded in the spaces L_p, $1 < p < \infty$.

<u>THEOREM 1.33 (KHAVIN [1]).</u> A contour of class $\mathcal{R}$ consists only of Smirnov curves.

<u>Proof.</u> Let $\Gamma \in \mathcal{R}$, and let the function f $(\in L_1^+)$ be summable with some power $p > 1$. Then, due to the boundedness of S in L_p , one has $L_p = L_p^+ \dot{+} \mathring{L}_p^-$, thus $f = f^+ + f^-$, where $f^+ \in L_p^+$, $f^- \in \mathring{L}_p^-$. Hence $f - f^+ = f^-$. The left-hand side of the equality obtained is a function of class L_1^+ and the right-hand side of class $\mathring{L}_1^-$. Owing to Theorem 1.28, we therefore conclude that $f = f^+$, i.e. $f \in L_p^+$. Hence, for a contour of class $\mathcal{R}$ the correctness of the equality $L_1^+ \cap L_p = L_p^+$ has been shown. According to Theorem 1.25, we can conclude that each component of the contour Γ is a Smirnov curve. ≡

Note that Theorem 1.33 cannot be converted: In that very article of KHAVIN [1] there was constructed a Smirnov curve Γ such that not all functions continuous on Γ are elements of the lineal $\mathcal{L}_1$. By PAATASHVILI, KHUSKIVADZE [1] another necessary condition for the membership of a curve Γ to the class $\mathcal{R}$ was discovered. It consists in the requirement

$$\operatorname*{ess\,sup}_{\xi\in\Gamma}\ \operatorname*{ess\,sup}_{\varrho>0}\ \varrho^{-1} l_\xi(\varrho) < \infty ,$$

where $l_\xi(\varrho)$ is the length of that part of Γ , which lies in the disk with centre at ξ and radius ϱ . In this paper it was suspected that this condition is sufficient, too.

Obviously, the contour Γ belongs to the class $\mathcal{R}$ if and only if every of its components has this property.

THEOREM 1.34. Each of the following conditions is sufficient for the curve Γ to belong to class $\mathcal{R}$:

1) Γ is a curve of bounded rotation,

2) Γ is a piecewise smooth curve.

Recall that Γ is called a curve of bounded rotation (or a Radon curve), if the function $t'(s)$ is of bounded variation, where s is a natural parameter on Γ.

Assertion 1) of Theorem 1.34 for curves without recurrence points is proved in DANILYUK, SHELEPOV [1] (a detailed explanation can be found in the monograph of DANILYUK [6] pp. 133-139); to Radon curves with recurrence points it was generalized in GORDADZE [6]. The membership of all smooth curves to the class $\mathcal{R}$ was shown in CALDERON [1]. The last assertion can be transferred to the case of piecewise smooth curves with the aid of results from GORDADZE [3], where it is established that each contour consisting of smooth arcs of class $\mathcal{R}$ belongs also to the class $\mathcal{R}$.

Let us remark that assertion 2) of Theorem 1.34 preceded chronologically some weaker sufficient conditions of membership to class $\mathcal{R}$. These conditions are special cases of the criterion 2) (Lyapunov or piecewise Lyapunov[1]) contours etc.) They can be found (with names of their authors and corresponding references) in KHVEDELIDZE [4], p. 62ff. After the article of CALDERON [1] there appeared papers, in which further weakenings of sufficient conditions for the membership of curves to class $\mathcal{R}$ were derived (e.g. KOKILASHVILI, PAATASHVILI [3]). Nevertheless, sufficient and necessary conditions for $\Gamma \in \mathcal{R}$ which can be checked effectively have not been found yet.

Here we introduce one subclass of class $\mathcal{R}$ which we shall deal with,

---

1) Recall that the curve $t=t(s)$ is said to be a Lyapunov curve, if the function $t'(s)$ satisfies a Hölder condition.

investigating the factorization of piecewise continuous functions.

DEFINITION 1.22 (KHVEDELIDZE [4]). To the class $\widetilde{\mathcal{R}}$ we shall ascribe contours $\Gamma \in \mathcal{R}$, which are representable as the union of a finite number of arcs satisfying condition (1.12), do not have pairwise common inner points, and are smooth in a neighbourhood of their common points t_j.

Let us clarify that condition (1.12) may or may not be fulfilled for the whole contour $\Gamma\ (\in \widetilde{\mathcal{R}})$, since the points t_j can be recurrence points for Γ.

Each piecewise smooth contour belongs to class $\widetilde{\mathcal{R}}$.

We have to examine operators of the type $AQFPB$ and $APFQB$ in the space L_p^n, where A, B and F are $n \times n$ matrix functions. In the general case, the domain of definition of these operators consists of vector functions $\varphi \in L_p^n$ satisfying

$$\begin{aligned} &\varphi_1 = B\varphi \in \mathcal{L}_1^n, \quad \varphi_2 = FP\varphi_1\ (FQ\varphi_1) \in \mathcal{L}_1^n, \\ &\varphi_3 = AQ\varphi_2\ (AP\varphi_2) \in L_p^n. \end{aligned} \tag{1.21}$$

THEOREM 1.35. Let A and B be $n \times n$ matrix functions of classes L_p and L_q respectively, and let one of the following conditions be fulfilled:

1) $F \in M_\infty^+\ (M_\infty^-)$,
2) $F \in M^+\ (M^-)$, and the contour Γ consists of Smirnov curves.

Then the operator $T = AQFPB\ (APFQB)$ is finite dimensional and bounded in L_p^n. To the domain of definition of this operator belong, under condition 1), all vector functions $\varphi \in L_p^n$ satisfying relation (1.21), and in case of condition 2), only those of them for which $FP\varphi_1\ (FQ\varphi_1) \in L_1^n$.

Proof. Consider the case of operator $T = AQFPB$. The matrix function F can be represented as $F = F_0 + F_1$, where F_0 is a rational matrix function with poles concentrated in $\mathcal{D}^+$, and $F_1 \in L_\infty^+$ under condition 1), $F_1 \in L^+$ under condition 2). Regard vector functions $\varphi \in L_p^n$ satisfying condition (1.21), which is necessary for φ to belong to the domain of the operator T. For such vector functions the relation $PB\varphi \in (L_1^+)^n$ is valid. Assuming condition 1) to be fulfilled, we have in this case $F_1PB\varphi \in (L_1^+)^n$ and $QF_1PB\varphi = 0$. If condition 2) is satisfied, generally speaking, we can only claim that $F_1PB\varphi \in (L^+)^n$. If we, in addition, suppose that $F_1PB\varphi \in (L_1^+)^n$, we get $F_1PB\varphi = FPB\varphi - F_oPB\varphi \in L_1^n$. According to Theorem 1.25, $F_1PB\varphi \in (L^+ \cap L_1)^n = (L_1^+)^n$. Therefore, under such suppositions we obtain $QF_1PB\varphi = 0$. It remains to verify the assertion of the theorem for the operator AQF_oPB. First, suppose that $F_o = f_{-k}H_{ij}$, where $f_{-k}(t) = (t-z_o)^{-k}$, $z_o \in \mathcal{D}^+$, k is an integer, H_{ij} is a matrix the (i,j) th entry of which is equal to one and the others are identically equal to zero. Then, for each vector function $\psi \in (\mathcal{L}_1^+)^n$,

$$QFP\psi = QF_o\psi^+ = Q((t-z_o)^{-k}\psi_j^+)e_i = \sum_{l=o}^{k} c_l f_{l-k} e_i ,$$

where c_l are the coefficients of the Taylor series expansion of the function ψ_j^+ $(= P\psi_j)$ at the point z_o. Using the formulae

$$c_l = \frac{1}{2\pi i} \int_\Gamma \psi_j^+(\tau) \frac{d\tau}{(\tau-z_o)^{l+1}} = \frac{1}{2\pi i} \int_\Gamma \psi_j(\tau) \frac{d\tau}{(\tau-z_o)^{l+1}} =$$

$$= \frac{1}{2\pi i} \langle \psi, f_{-l-1} e_j \rangle ,$$

we get

$$QF_oP\psi = \sum_{l=o}^{k-1} \frac{1}{2\pi i} \langle \psi, f_{-l-1} e_j \rangle f_{l-k} e_i .$$

Denoting by A_i and B_j the i th column and the j th row of the matrix function A and B, respectively, we obtain for vector functions φ satisfying relation (1.21) that $AQF_oPB\varphi = T_o\varphi$, where

$$T_o\varphi = \sum_{l=o}^{k-1} \frac{1}{2\pi i} \langle \varphi, f_{-l-1} B_j' \rangle f_{l-k} A_i .$$

Since, for an arbitrary integer s, $f_s B_j' \in L_q^n$ and $f_s A_i \in L_p^n$, then the operator T_o defined by the last equation is bounded, defined on the whole space L_p^n and maps it into the linear hull of the vector functions $f_s A_i$ $(s = -k,\dots,-1)$. Consequently, this operator is finite dimensional. Moreover, $f_s \in \overset{\circ}{L}_\infty^-$ for $s < 0$. Thus, the image of the operator T_o lies in $(\overset{\circ}{L}_p^-)^n$ as soon as $A \in L_p^-$.

In summary, for $F_o = f_{-k}H_{ij}$ the operator AQF_oPB is the restriction of the operator T_o on the lineal of vector functions φ satisfying condition (1.21). From this fact, for the case $F_o = f_{-k}H_{ij}$, the assertion to be proved results.

Taking into account that each rational matrix function with poles concentrated in $\mathfrak{D}^+$ can be represented as the linear combination of special type matrix functions considered above, we find that in the general case the operator AQF_oPB is also finite dimensional and bounded in L_p^n, defined on all vector functions $\varphi \in L_p^n$ satisfying condition (1.21) and, for $A \in L_p^-$, the image of this operator lies in $(\overset{\circ}{L}_p^-)^n$. Theorem 1.35 is completely proved. ≡

CHAPTER 2. GENERAL PROPERTIES OF FACTORIZATION

In Section 2.1 the notion of factorization in L_p , which is fundamental to the present book, will be introduced. As a rule, in the current literature the factorization problem is considered in connection with the study of the Riemann boundary value problem. The definition of factorization of a matrix function proposed below is more general than those previously introduced for the aims of solvability theory for the problem considered.

In the present chapter we study only those properties of factorization which can be formulated and proved regardless of the Riemann boundary value problem. In the next chapter we shall discuss the question concerning the role which is played by the factorability in L_p of the matrix coefficient of a Riemann problem for the solvability theory of this problem.

We do not make any a priori suppositions about the matrix function G except the requirement of measurability. However, it turns out that, for fixed p , the diagonal factor of factorization is defined uniquely (which justifies the notion of partial p-indices), and the left and right factor are defined with the same degree of arbitrariness that holds in the classical case. This will be demonstrated in Section 2.2. In the same section the monotonic dependence of partial p-indices on p is established as well as the fact that if different values of p correspond to equal total p-indices, then we can say that the corresponding families of factorizations also coincide. In connection with the study of the factorization dependence on p , in Section 2.3 the notion of the domain of factorability will be introduced, and its explicit description in case of the power function will be given. The latter result plays an important role in Chapter 5, where we shall study the question of factorability of arbitrary piecewise continuous functions.

Section 2.4 includes the factorability criterion for meromorphic (in particular, rational) matrix functions as well as estimates for their

partial indices. These results are basic to all the following theorems on the existence of a factorization including the theorem on the factorability of continuous matrix functions in Chapter 5. As a consequence of the factorability criterion of meromorphic matrix functions, the factorability of matrix functions, which can be represented as the product of meromorphic functions, will be demonstrated in the same section. In turn, with the help of the latter result, we shall prove the theorem on the preservation of the factorability property under multiplication from the left (right) by matrix functions belonging together with their inverses to the class M_∞^+ (M_∞^-). The study of the connection between factorizations of one and the same matrix function G in different spaces L_p will be led to the very end.
Finally, Section 2.5 contains comments on further reading.

2.1. THE DEFINITION OF FACTORIZATION

DEFINITION 2.1. A factorization in L_p $(1 < p < \infty)$ of a matrix function G relative to the contour Γ is, by definition, a representation

$$G(t) = G_+(t)\Lambda(t)\, G_-(t) , \tag{2.1}$$

where $G_+ \in L_p^+$, $G_- \in L_q^-$, $G_+^{-1} \in L_q^+$, $G_-^{-1} \in L_p^-$ $(q = p/(p-1))$,

$$\Lambda(t) = \mathrm{diag}[t^{\varkappa_1},\dots,t^{\varkappa_n}] , \tag{2.2}$$

and $\varkappa_1 \geq \dots \geq \varkappa_n$ are integers.

As will be shown in Section 2.2, the integers $\varkappa_j$ are uniquely defined by G and p. We shall call them partial p-indices of G. Their sum $\varkappa = \varkappa_1 + \dots + \varkappa_n$ will be called its total p-index.

If it is convenient to emphasize groups of coinciding partial p-indices, we shall write the diagonal factor in the form

$$\Lambda(t) = \mathrm{diag}[t^{\tilde\varkappa_1} I_{l_1},\dots,t^{\tilde\varkappa_k} I_{l_k}] ,$$

where $\tilde\varkappa_1 > \dots > \tilde\varkappa_k$. The number l_j is said to be the multiplicity of the partial p-index $\tilde\varkappa_j$ $(j=1,\dots,k)$.

The partial p-indices and the total p-index will be simply called partial and total indices, if it is clear from the context what a value of the parameter is meant.

In the scalar case ($n = 1$) there exists only one partial p-index, which, consequently, coincides with the total one. In the following, we simply call it p-index (index).

Immediately from the definition we deduce that for the matrix function G to be factorable in some L_p it is necessary that G and its inverse be summable. In Sections 2.3 and 2.4 we shall obtain factorability criteria for some special classes of matrix functions. From these criteria it will be clear that the mentioned necessary condition is far from be sufficient. A necessary and sufficient factorability condition applicable to arbitrary measurable matrix functions fails to be known at present.

Besides the factorization (2.1) one can consider representations of the matrix function G in the form

$$G(t) = G_-(t)\Lambda(t)G_+(t) , \qquad (2.1')$$

where $G_- \in L_p^-$, $G_+ \in L_q^+$, $G_-^{-1} \in L_q^-$, $G_+^{-1} \in L_p^+$, and the matrix function Λ is defined by formula (2.2).

If it is necessary to consider both representations (2.1) and (2.1') simultaneously, we call the first of them a left, and the second a right factorization. In doing so, the partial indices will be called left, unlike the right ones, which occur in the representation (2.1'). The following example shows that the tuples of left and right indices, generally speaking, do not coincide.

<u>Example.</u> Let $G(t) = \begin{bmatrix} t & 1 \\ 0 & t^{-1} \end{bmatrix}$. The equations

$$G(t) = \begin{bmatrix} 1 & 0 \\ 0 & 1 \end{bmatrix} \begin{bmatrix} t & 0 \\ 0 & t^{-1} \end{bmatrix} \begin{bmatrix} 1 & t^{-1} \\ 0 & 1 \end{bmatrix}$$

and

$$G(t) = \begin{bmatrix} 1 & 0 \\ t^{-1} & -1 \end{bmatrix} \begin{bmatrix} 1 & 0 \\ 0 & 1 \end{bmatrix} \begin{bmatrix} t & 1 \\ 1 & 0 \end{bmatrix}$$

provide a left and right factorization of the matrix function G, respectively. Consequently, its left indices are equal to ± 1, and the right ones are equal to zero.

Clearly, from a left factorization in L_p of G one may obtain a right factorization in L_q of G', transposing both parts of (2.1), and a right factorization of G^{-1} in L_p, taking the inverse to both parts of (2.1).

Analogously, from a right factorization of the matrix function G one can get a left factorization of G' and G^{-1}. In particular, the tuple of left (right) p-indices of G coincides with the tuple of right (left) q-indices of G' and differs only by the sign from the tuple of right (left) p-indices of G^{-1}. Therefore, all the assertions about a right factorization can be obtained by restating the corresponding assertions for a left one.

Thanks to the latter remark, in the following we focus our attention on the study of a left factorization and left (partial) indices.

2.2. PROPERTIES OF FACTORIZATION FACTORS

Leaving for a moment the questions of existence we clarify in the present section that degree of arbitrariness with which the factorization factors $G_\pm$ and Λ are defined. First of all, we study the middle factor.

THEOREM 2.1. Let the matrix function G admit a factorization in L_{p_1} and L_{p_2} $(p_1 \le p_2)$. Then $\varkappa_i^{(1)} \ge \varkappa_i^{(2)}$, $i = 1,\dots,n$, where $\varkappa_i^{(j)}$ $(j = 1,2)$ denotes the p_j-indices of G.

Proof. Starting from equation $G_+^{(1)}\Lambda^{(1)}G_-^{(1)} = G_+^{(2)}\Lambda^{(2)}G_-^{(2)}$ [1], we find

$$\Lambda^{(1)}H_- = H_+\Lambda^{(2)}, \tag{2.3}$$

[1] Here and further on, the superscript specifies the factorization we deal with.

where

$$H_- = G_-^{(1)}(G_-^{(2)})^{-1}\ , \quad H_+ = (G_+^{(1)})^{-1}G_+^{(2)}\ . \tag{2.4}$$

If we write equation (2.3) componentwise, it takes the form

$$t^{\varkappa_i^{(1)}-\varkappa_j^{(2)}}\ h_{ij}^-(t) = h_{ij}^+(t)\ , \quad i,j = 1,\dots,n. \tag{2.3'}$$

Since $G_-^{(1)} \in L_{q_1}^-$ and $G_-^{(2)} \in L_{p_2}^-$, then $H_- \in L_p^-$, where $p = (p_2^{-1} + q_1^{-1})^{-1} = (1+p_2^{-1} - p_1^{-1})^{-1} \geq 1$. Similarly $H_+ \in L_p^+$. Applying Theorem 1.28 to equality (2.3') we deduce that h_{ij}^+ is a polynomial of degree less than or equal to $\varkappa_i^{(1)} - \varkappa_j^{(2)}$, if $\varkappa_i^{(1)} \geq \varkappa_j^{(2)}$, and identically equal to zero otherwise.

In spite of the assertion of the theorem, assume, for some r , $\varkappa_r^{(1)} < \varkappa_r^{(2)}$. Then, for $j \leq r \leq i$, $\varkappa_i^{(1)} \leq \varkappa_r^{(1)}$, $\varkappa_j^{(2)} \geq \varkappa_r^{(2)}$ and hence $\varkappa_i^{(1)} < \varkappa_j^{(2)}$. Consequently, $h_{ij}^+ = 0$ for $i = r,\dots,n;$ $j = 1,\dots,r$, and each minor of rth order lying in the first r columns of the matrix function H_+ is equal to zero. Due to Laplace's theorem, in this case the determinant of H_+ is equal to zero a.e. on Γ , which contradicts the non-singularity of the factorization factors. Thus Theorem 2.1 has been proved. ≡

<u>COROLLARY 2.1.</u> The factor Λ is uniquely defined by the matrix function G and the parameter p .

Now we turn to the factorization factors $G_\pm$. According to Theorem 2.1, for increasing p , the total index does not increase, thus $\varkappa^{(1)} \geq \varkappa^{(2)}$. From formulae (2.3) and (2.4) obtained in the proof of Theorem 2.1 we conclude that

$$G_+^{(2)} = G_+^{(1)}H_+\ , \quad G_-^{(2)} = \Lambda^{(2)-1}H_+^{-1}\ \Lambda^{(1)}G_-^{(1)}\ , \tag{2.5}$$

where H_+ is a matrix function, the (i,j) th entry of which is a polynomial of degree not higher than $\varkappa_i^{(1)} - \varkappa_j^{(2)}$ if $\varkappa_i^{(1)} \geq \varkappa_j^{(2)}$, and identically zero otherwise.

Formulae (2.5) mean, in particular, that

$$\det H_+(z) = \frac{\det G_+^{(2)}(z)}{\det G_+^{(1)}(z)}, \quad z \in \mathfrak{D}^+$$

and

$$\det H_+(z) = z^{\varkappa^{(1)}-\varkappa^{(2)}} \cdot \frac{\det G_-^{(1)}(z)}{\det G_-^{(2)}(z)}, \quad z \in \mathfrak{D}^-,$$

so that $\det H_+$ does not vanish in $\mathfrak{D}^+$ and at finite points of $\mathfrak{D}^-$. At the point $z = \infty$ it has a pole of order $\varkappa^{(1)} - \varkappa^{(2)}$. As $\det H_+$ is a polynomial, we deduce that its degree is $\varkappa^{(1)} - \varkappa^{(2)}$, and the zeros (if they exist, i.e. in the case $\varkappa^{(1)} \neq \varkappa^{(2)}$) are located on Γ. A somewhat more exact result for the case $\varkappa^{(1)} \neq \varkappa^{(2)}$ will be obtained in Section 2.4. Here we study the case $\varkappa^{(1)} = \varkappa^{(2)}$.

THEOREM 2.2. Let the matrix function G be factorable in the spaces L_{p_1} and L_{p_2} $(p_1 \leq p_2)$ with one and the same total index. Then

a) the transition from a factorization in L_{p_1} to a factorization in L_{p_2} is accomplished according to formulae (2.5), where

1) $\Lambda^{(1)}(t) = \Lambda^{(2)}(t)$ $(= \Lambda(t) = \mathrm{diag}[t^{\tilde{\varkappa}_1} I_{l_1}, \ldots, t^{\tilde{\varkappa}_k} I_{l_k}])$;

2) if $H_+ = (H_{ij})_{i,j=1}^{k}$ provides a splitting of the matrix H into blocks for which the order of the j th diagonal block is equal to l_j, then the diagonal blocks are constant matrices, and the blocks outside the diagonal are equal to zero, if $i > j$, and consist of polynomials of degree not exceeding $\tilde{\varkappa}_i - \tilde{\varkappa}_j$, if $i < j$;

b) the representation (2.1) with

$$G_+ = G_+^{(1)} H_+, \quad G_- = \Lambda^{-1} H_+^{-1} \Lambda\, G_-^{(1)}, \tag{2.6}$$

where H is an arbitrary matrix function satisfying condition 2), is a factorization of G in the space L_{p_1}.

Proof.

a) Due to Theorem 2.1, $\varkappa_i^{(1)} \geq \varkappa_i^{(2)}$. By assumption we have $\sum_{i=1}^{n} \varkappa_i^{(1)} = \sum_{i=1}^{n} \varkappa_i^{(2)}$. Hence $\varkappa_i^{(1)} = \varkappa_i^{(2)}$ $(= \varkappa_i)$ for $i = 1, \ldots, n$, i.e., condition 1) is fulfilled. Condition 2) results from the properties of the matrix function H_+ established above and the equation $\det H_+ = \prod_{j=1}^{n} \det H_{jj}$, which holds in view of Laplace's theorem.

b) If H_+ is a matrix function fulfilling condition 2), then $H_+^{\pm 1} \in L_\infty^+$, $(\Lambda^{-1} H_+ \Lambda)^{\pm 1} \in L_\infty^-$, so that the relations $G_+^{(1)} \in L_{p_1}^+$, $G_-^{(1)} \in L_{q_1}^-$, $G_+^{(1)-1} \in L_{q_1}^+$, $G_-^{(1)-1} \in L_{p_1}^-$ imply that $G_+ \in L_{p_1}^+$, $G_+^{-1} \in L_{q_1}^+$, $G_- \in L_{q_1}^-$, $G_-^{-1} \in L_{p_1}^-$, where the matrix functions $G_\pm$ are defined via formulae (2.6). Thus, the representation $G_+ \Lambda G_-$ is a factorization of G in L_{p_1}, which proves Theorem 2.2. ≡

If $p_1 = p_2$, the condition of equality of partial indices is fulfilled in view of Corollary 2.1. Therefore, Theorem 2.2 yields the description of all factorizations in the space L_p, if at least one of them is known. If all partial indices are pairwise equal, then H_+ is a constant non-degenerate matrix, and the factorization is uniquely defined by giving one of the matrices $G_\pm$ at some point of $\mathcal{D}^\pm$; commonly, the value $G_-(\infty)$ is fixed.

COROLLARY 2.2. A factorization of the matrix function G in one of the spaces L_p is a factorization in all L_p-spaces in which G admits a factorization with the same total index.

Indeed, if G is factorable in L_{p_1} and L_{p_2} $(p_1 \leq p_2)$ with the same total index, then, as it is clear from the proof of Theorem 2.2, there exists a representation $G = G_+^{(1)} \Lambda G_-^{(1)}$ which is a factorization of G simultaneously in L_{p_1} and L_{p_2}. But then all factorizations of the matrix function G both in L_{p_1} and in L_{p_2} can be obtained by formulae (2.6). In this case, on the matrix function H_+ there are imposed one and the same restrictions. Thus, the sets of factorizations of G in L_{p_1} and L_{p_2} coincide.

THEOREM 2.3 (Interpolation theorem). If the matrix function G admits a factorization with one and the same total index in L_{p_1} and L_{p_2}, then it admits a factorization (with the same total index) in all L_p, $p \in [p_1, p_2]$.

Proof. First of all, let us note that if $p_1 \le p \le p_2$, then $q_2 \le q \le q_1$. If (2.1) is a factorization of G in L_{p_1} , then $G_+^{-1} \in L_{q_1}^+$, $G_- \in L_{q_1}^-$ and, therefore, $G_+^{-1} \in L_q^+$, $G_- \in L_q^-$. According to Corollary 2.2, the representation (2.1) is also a factorization of G in L_{p_2} . Consequently $G_+ \in L_{p_2}^+$, $G_-^{-1} \in L_{p_2}^-$ and, finally, $G_+ \in L_p^+$, $G_-^{-1} \in L_p^-$. Theorem 2.3 is proved. ≡

The following result often proves to be useful in the study of properties of the factorization factors $G_\pm$.

<u>THEOREM 2.4.</u> Let the matrix functions G_1 and G_2 admit a factorization in the space L_p and coincide (a.e.) on some open arc $\gamma \subseteq \Gamma$. Then the factorization factors $G_\pm^{(j)}$ $(j = 1,2)$ are connected via the relations

$$G_+^{(1)}(t) = G_+^{(2)}(t)V(t) \ , \ G_-^{(1)}(t) = W(t)G_-^{(2)}(t), \quad t \in \gamma \ , \qquad (2.7)$$

where V and W are matrix functions analytic and non-degenerating on γ .

Proof. Arguing as in Theorem 2.1, we get the equation (2.3), however, not on the whole contour Γ , but only on the arc γ .
Matrix functions $H_\pm$ determined by formulae (2.4) belong together with their inverses to the classes $L_1^\pm$, since in the case under study $p_1 = p_2$. With the help of Theorem 1.27, we deduce from here that the matrix functions $\Lambda^{(1)}H_-$ and $H_+\Lambda^{(2)}$ and their inverses are analytic on the arc γ . Consequently, either of the matrix functions H_+ and H_- is analytic and non-degenerate on γ . From here and from formula (2.4) relations (2.7) follow, which proves Theorem 2.4. ≡

Formulae (2.7) allow us to make conclusions about the smoothness and integrability of the matrix functions $G_\pm^{(1)}$ and their inverses, if the corresponding properties of $G_\pm^{(2)}$ are known. Theorem 2.4 shows that the behaviour of the factorization factors $G_\pm$ is determined by the local behaviour of the matrix function G itself.

2.3. THE DOMAIN OF FACTORABILITY

DEFINITION 2.2. The set $\mathfrak{F}(G)$ of those values of the parameter $p \in (1,\infty)$ for which the matrix function G can be factored in L_p is said to be the domain of factorability of G.

All results of the present chapter obtained till now, in principle, were related to general laws of the structure of $\mathfrak{F}(G)$. Now they can be formulated more compactly: $\mathfrak{F}(G)$ is the union of disjoint connected (may be, empty) sets $\mathfrak{F}_k(G) = \{p : G$ admits a factorization in L_p with total index $k\}$. If $k_1 < k_2$, then the set $\mathfrak{F}_{k_1}(G)$ is located on the real line more to the right than $\mathfrak{F}_{k_2}(G)$. If $p_0 \in \mathfrak{F}_k(G)$, then a factorization of G in L_{p_0} is a factorization of G in all spaces L_p with $p \in \mathfrak{F}_k(G)$. The partial p-indices of the matrix function G are non-increasing functions with the domain of definition $\mathfrak{F}(G)$ constant on any of the sets $\mathfrak{F}_k(G)$.

For the sake of illustration of the notions introduced, we consider the factorization problem for the power function. As not to complicate matters, we assume the contour Γ to be a simple closed curve dividing the plane into an inner domain $\mathcal{D}^+$ and an outer $\mathcal{D}^-$. Let us take a fixed branch $\psi(z)$ of the power function z^α defined on the plane with a cut connecting the point 0 with ∞ and intersecting Γ at the only point t_0. The parts of the cut connecting 0 with t_0 and t_0 with ∞ will be denoted by l_0 and l_∞, respectively. Clearly, only the case of a non-integer α is of interest.

Suppose that

1) Γ is a curve of the Smirnov class;

2) $$\sup \left\{\frac{s(t,t_0)}{|t - t_0|} : t \in \Gamma\right\} < \infty, \tag{2.8}$$

where $s(t,t_0)$ is the distance between the points t and t_0 along the curve Γ;

3) there is a bounded and continuous function $\arg(t - t_0)$ on $\Gamma \setminus \{t_0\}$.

Conditions 2) and 3) are of local character; they are fulfilled, for example, if a certain neighbourhood of the point t_0 on Γ is a piecewise smooth curve without recurrence points.

The function $\psi(t)$ is continuous on the contour Γ with the exception of the point t_0. At the point t_0 it has limits from the left and right (recall that the contour is oriented anticlockwise) satisfying the relation

$$\psi(t_0 - 0) = e^{2\pi i\alpha}\,\psi(t_0 + 0)\ .$$

Thus, the function ψ is piecewise continuous. In Chapter 5 we shall see that the factorization problem for an arbitrary piecewise continuous matrix function (under some additional restrictions on the contour) can be reduced to the factorization of power functions.

<u>THEOREM 2.5.</u> The function $\psi(t) = t^{\alpha}$ can be factored in L_p if and only if $\{\operatorname{Re}\alpha\} \neq 1 - \frac{1}{p}$.[1)]

<u>Proof.</u> Denote by $\psi_+(z)$ that (uniquely determined) branch of the power function $(z - t_0)^{\alpha}$ defined on the plane with the cut l_∞ and by $\psi_-(z)$ that branch of $(1 - t_0/z)^{-\alpha}$ defined on the plane with the cut l_0 such that $\psi_-(\infty) = 1$ and

$$\psi(t) = \psi_+(t)\psi_-(t)\ , \quad t \in \Gamma\setminus\{t_0\}\ . \tag{2.9}$$

For any integer k we have

$$\psi(t) = \psi_{+,k}(t)t^k\,\psi_{-,k}(t)\ , \quad t \in \Gamma\setminus\{t_0\}\ , \tag{2.10}$$

where $\psi_{+,k}(z) = \psi_+(z)(z-t_0)^{-k}\ ,\quad \psi_{-,k}(z) = \psi_-(z)(1-t_0/z)^k$.

Evidently, the functions $\psi_{+,k}^{\pm 1}$ and $\psi_{-,k}^{\pm 1}$ are analytic in the domains $\mathcal{D}^+$ and $\mathcal{D}^-$, respectively, where $\psi_{+,k}(z) = (z-t_0)^{\operatorname{Re}\alpha-k}(z-t_0)^{i\operatorname{Im}\alpha}$, under a suitable choice of the branches of the factors defined on the plane with cut l_∞. Denoting $(z-t_0)^{k-\operatorname{Re}\alpha}$ by $\varphi(z)$, we get $|\varphi(z)| = |z-t_0|^{k-\operatorname{Re}\alpha}$. Hence, for $\operatorname{Re}\alpha \geq k$, there is $\varphi^{-1} \in L_\infty^+$. Furthermore, we denote by $t_1(s)$ and $t_2(s)$ $(0 \leq s \leq d$, where $2d$ is the length of the curve $\Gamma)$ the begin and the end of the anticlock-

---

1) By $\{x\}$ we denote the fractional part of x : $\{x\} = x - [x]$.

wise oriented arc of the contour Γ with the length $2s$ the centre of which is the point t_o. Then, for real λ,

$$\int_\Gamma |t-t_o|^\lambda |dt| = \int_o^d |t_1(s)-t_o|^\lambda ds + \int_o^d |t_2(s)-t_o|^\lambda ds .$$

Due to condition (2.8), the integrals on the right-hand side of the last equation converge only simultaneously with $\int_o^d s^\lambda ds$, i.e., if and only if $\lambda > -1$. Hence $\varphi \in L_p$ if $p < 1/(\mathrm{Re}\,\alpha - k)$ and $\varphi \notin L_{1/(\mathrm{Re}\alpha-k)}$. Moreover, $\varphi \notin L^+_{1/(\mathrm{Re}\alpha-k)}$. Now we intend to show that $\varphi \in L^+_p$, if $p < 1/(\mathrm{Re}\alpha-k)$. For this purpose, we consider the function $\chi = (\varphi \circ \omega)\sqrt{\omega'}$, where ω conformally maps the unit disk Δ onto the domain $\mathcal{D}^+$. In view of condition 3) imposed upon Γ as well as Theorem 1.15, the function $\varphi \circ \omega$ is outer. With regard to condition 1) imposed on Γ, the function $\sqrt[p]{\omega'}$ is also outer. Together with $\varphi \circ \omega$ and $\sqrt[p]{\omega'}$, their product χ is outer, too. But then, in view of Theorem 1.18, from the relation $\chi \in L_p$ (which is equivalent to the condition $\varphi \in L_p$) we deduce that $\chi \in H_p$. From here and from Theorem 1.19 the assertion to be proved results.

In summary, if $\mathrm{Re}\,\alpha \geq k$, then $\varphi \in L^+_p$ for $p < 1/(\mathrm{Re}\alpha-k)$, and $\varphi^{-1} \in L^+_\infty$. Since $|(z-t_o)^{i\mathrm{Im}\alpha}| = \exp(-\mathrm{Im}\alpha \arg(t-t_o))$, condition 3) guarantees the membership of the function $(z-t_o)^{i\mathrm{Im}\alpha}$ together with its inverse to the class L^+_∞. Consequently, the statements proved above concerning the function φ can be transferred to the function $\psi^{-1}_{+,k}$ without any changes. Analogously, we can argue that $\psi^{-1}_{-,k} \in L^-_{1/(\mathrm{Re}\alpha-k)}$, $\psi^{-1}_{-,k} \in L^-_p$ for $p < 1/(\mathrm{Re}\alpha-k)$.

Thus, if $k \leq \mathrm{Re}\,\alpha$, the representation (2.10) is a factorization of the function $\psi(t)$ in L_p if and only if $q < 1/(\mathrm{Re}\,\alpha - k)$, i.e. $k > \mathrm{Re}\,\alpha + 1/p - 1$. But this is only possible if there exists an integer in the interval $(\mathrm{Re}\,\alpha + 1/p - 1, \mathrm{Re}\,\alpha]$, i.e. if $\{\mathrm{Re}\,\alpha\} < 1 - 1/p$. In this case, the p-index of the function ψ coincides with k and is equal to $[\mathrm{Re}\,\alpha]$.

Supposing that $k > \mathrm{Re}\,\alpha$, we observe that $\psi^{-1}_{+,k} \in L^+_\infty$, $\psi_{-,k} \in L^-_\infty$, $\psi_{+,k} \in L^+_p$, $\psi^{-1}_{-,k} \in L^-_p$ for $p < 1/(k - \mathrm{Re}\,\alpha)$, $\psi_{+,k} \notin L^+_{1/(k-\mathrm{Re}\,\alpha)}$,

$\psi^{-1}_{-,k} \notin L^{-}_{1/(k - \mathrm{Re}\,\alpha)}$. Therefore, the representation (2.10) is a factorization of ψ in L_p if and only if $p < 1/(k - \mathrm{Re}\,\alpha)$, i.e. $k < \mathrm{Re}\,\alpha + 1/p$. But this is only possible when $\{\mathrm{Re}\,\alpha\} > 1 - 1/p$ and $k = [\mathrm{Re}\,\alpha] + 1$.

Summarizing, for $p \in (1, 1/(1 - \{\mathrm{Re}\,\alpha\}))$, the function $\psi(t) = t^{\alpha}$ can be factored in L_p, and its index is equal to $[\mathrm{Re}\,\alpha] + 1$. For $p \in (1/(1- \{\mathrm{Re}\,\alpha\}), \infty)$, it is factorable in L_p with the index $[\mathrm{Re}\,\alpha]$.

It remains to show that the function ψ is not factorable in L_{p_0}, where $p_0 = 1/(1 - \{\mathrm{Re}\,\alpha\})$. Assuming the contrary, on the strength of the p-index monotonicity proved in Theorem 2.1, we find that the p_0-index of the function ψ lies between $[\mathrm{Re}\,\alpha]$ and $[\mathrm{Re}\,\alpha] + 1$ and, therefore, coincides either with the first or with the second of these numbers. According to Corollary 2.2, a factorization of the function ψ for $p > p_0$ would be also suitable for $p = p_0$ in the first case. In the second case, the role of a factorization in L_{p_0} could play any factorization of ψ for $p < p_0$. However, as was shown above, neither of these holds.

Thus, the factorability domain of the function $\psi(t) = t^{\alpha}$, for non-integer α, consists of two components: $\mathfrak{F}_{[\mathrm{Re}\,\alpha]+1} = (1, 1/(1-\{\mathrm{Re}\,\alpha\}))$ and $\mathfrak{F}_{[\mathrm{Re}\,\alpha]} = (1/(1-\{\mathrm{Re}\,\alpha\}), \infty)$, which proves Theorem 2.5. ≡

Consider, in particular, the function $\psi(e^{i\theta}) = e^{i\theta/2}$ defined on the unit circle $= \{e^{i\theta} : -\pi < \theta < \pi\}$. This is the function $t^{1/2}$ with a discontinuity at the point -1. The representation $t^{1/2} = (t + 1)^{-1/2} t (t^{-1}+1)^{1/2}$ provides its factorization for $p \in (1,2)$, and the representation $t^{1/2} = (t + 1)^{1/2}(t^{-1} + 1)^{-1/2}$ yields a factorization for $p \in (2, \infty)$. For $p = 2$, the function $t^{1/2}$ fails to be factorable.

An example of a matrix function, all components of the factorability domain of which are non-empty, will be given in Section 5.4.

2.4. FACTORIZATION OF MEROMORPHIC MATRIX FUNCTIONS AND THEIR PRODUCTS

In this section we are concerned with the question of the connection between the factorization of a matrix function G and its representations in the form

$$G = A_+ A_- , \qquad (2.11)$$

where

$$A_+ \in M_p^+ , \quad A_+^{-1} \in M_q^+ , \quad A_- \in M_q^- , \quad A_-^{-1} \in M_p^- . \qquad (2.12)$$

Formally, the representation (2.11) is more general than the factorization (2.1). Indeed, from a factorization (2.1) of the matrix function G one can obtain the equation (2.11), setting, for example, $A_+ = G_+ \Lambda$, $A_- = G_-$. However, it turns out that the existence of a representation (2.11) implies the factorability of the matrix function G in L_p. Moreover, a factorization of G in L_p can be built effectively, as soon as its representation (2.11) is known. To verify this statement, we shall prove the factorability of A_+ and A_- individually. Then it will be checked that the factorability will remain on multiplication.

Incidentally, estimates for the partial p-indices of the matrix functions A_+, A_- and G will be obtained. Furthermore, in the case $G \in M_p^+$ (M_q^-), the necessity of the condition $G^{-1} \in M_q^+$ (M_p^-) for the factorability of G will be proved.

The following result plays an important role in the proof.

LEMMA 2.1. Let the matrix function B be analytic in the domain $\mathcal{O}$ and $\det B(z_0) = 0$ for some $z_0 \in \mathcal{O}$. Then

$$B = CU , \qquad (2.13)$$

where the matrix function C is analytic in $\mathcal{O}$ and has the same zeros of the determinant with the same multiplicities as B except z_0, where the multiplicity of the zero of $\det C$ is smaller than that of $\det B$ by one;

$$U(z) = \begin{bmatrix} 1 & & & \lambda_1 & & & \\ & \ddots & & \vdots & & & \\ & & 1 & \lambda_{k-1} & & & \\ & & & z-z_o & & & \\ & & & & 1 & & \\ & & & & & \ddots & \\ & & & & & & 1 \end{bmatrix}, \qquad (2.14)$$

where $\lambda_1,\dots,\lambda_{k-1}$ are certain numbers standing in the k th column of U .

Proof. For $z \neq z_o$, the matrix $U(z)$ is invertible, and the inverse is of the form

$$V(z) = \begin{bmatrix} 1 & & & -\lambda_1/(z-z_o) & & & \\ & \ddots & & \vdots & & & \\ & & 1 & -\lambda_{k-1}/(z-z_o) & & & \\ & & & 1/(z-z_o) & & & \\ & & & & 1 & & \\ & & & & & \ddots & \\ & & & & & & 1 \end{bmatrix}.$$

Set $C(z) = B(z)V(z)$ for $z \neq z_o$. As $\det C(z) = (\det B(z))/(z-z_o)$ and the matrix function V is analytic in the plane except the point z_o, it remains only to make sure that the matrix function C is analytic at the point z_o . Denoting the j th columns of C and B by c_j and b_j , $(j = 1,\dots,n)$, respectively, we obtain that $c_j = b_j$ for $j \neq k$, $c_k(z) = \frac{1}{z-z_o}\left(b_k(z) - \sum_{i=1}^{k-1} \lambda_i b_i(z)\right)$.

Hence, it is clear that for the matrix function C to be analytic at the point z_o , it is necessary and sufficient that column $b_k(z_o)$ be a linear combination of the columns $b_i(z_o)$ $(i = 1,\dots,k-1)$ and the scalars λ_i be the coefficients of this linear combination. The existence of such k and $\{\lambda_i\}_{i=1}^{k-1}$ results from the condition $\det B(z_o) = 0$, which proves the lemma. ≡

Remark. In the proof of Lemma 2.1 we choose a column of the matrix $B(z_o)$ which is the linear combination of the previous ones. One can also choose a column being the linear combination of the next ones and, owing to this, obtain a representation (2.13) with a right factor of the kind

$$\begin{bmatrix} 1 & & & & & & \\ & \ddots & & & & & \\ & & 1 & & & & \\ & & & z-z_0 & & & \\ & & & \lambda_{k+1} & 1 & & \\ & & & \vdots & & \ddots & \\ & & & \lambda_n & & & 1 \end{bmatrix} . \qquad (2.14')$$

Considering the rows of $B(z_0)$, one may "split off" a left factor of the kind (2.14) or (2.14').

THEOREM 2.6. The matrix function $G \in M_p^+$ can be factored in L_p if and only if $G^{-1} \in M_q^+$. If this condition is fulfilled, then the partial indices of G are included between $-\mathcal{P}_+(G)$ [1] and $\operatorname{ind}_+\det G + (n-1)\,\mathcal{P}_+(G)$, the total index is equal to $\operatorname{ind}_+\det G$, and the right factorization factor G_- is a rational matrix function.

Proof. If the matrix function $G(\in M_p^+)$ is factorable in L_p, then, proceeding from equation (2.1), we observe

$$G_+^{-1}\, G = \Lambda G_- \quad (= F) . \qquad (2.15)$$

Since $G_+^{-1}G \in M_1^+$ and $\Lambda G_- \in M_q^-$, Theorem 1.28 is applicable, thanks to which the matrix function F defined via equation (2.15) is rational and has no poles on Γ. Moreover, this matrix function is non-degenerating on Γ, since otherwise the matrix function F^{-1} would have poles on Γ, which contradicts the relation $F^{-1} = G_-^{-1}\Lambda^{-1} \in M_p^-$.

Consequently, $F^{-1} \in M_\infty^+$ and, therefore, $G^{-1}\ (= F^{-1}G_+^{-1}) \in M_q^+$.

The necessity of the condition $G^{-1} \in M_q^+$ for the factorability of $G \in M_p^+$ has been proved.

Now we are going to verify its sufficiency.

We shall study representation of G in the form

---

1) $\mathcal{P}_\pm(G)$ denotes the number of poles (counting multiplicities) of G in $\mathcal{D}_\pm$

$$G = G_+^{(j)} \Lambda^{(j)} G_-^{(j)} , \qquad (2.16)$$

where $G_-^{(j)}$ are rational matrix functions with poles and zeros of the determinant concentrated in $\mathfrak{D}^+$, and $\Lambda^{(j)}$ are matrix functions of the type (2.2), i.e.,

$$\Lambda^{(j)}(t) = \mathrm{diag}[t^{\varkappa_1^{(j)}}, \ldots, t^{\varkappa_n^{(j)}}] , \quad \varkappa_1^{(j)} \geq \ldots \geq \varkappa_n^{(j)} ,$$

$G_+^{(j)} \in L_p^+$, $(G_+^{(j)})^{-1} \in M_q^+$, and the number of zeros of $\det G_+^{(j)}$ calculated with regard to their multiplicity is equal to j.

Obviously, such representations exist. We shall obtain one of them by introducing the scalar rational function f of index $\mathfrak{p}_+(G)$ the zeros and poles of which are concentrated in $\mathfrak{D}^+$, where we suppose that $fG \in L_p^+$. Furthermore, we set

$$G_+^{(N)} = fG , \quad \Lambda^{(N)}(t) = t^{-\mathfrak{p}_+(G)} I , \quad G_-^{(N)}(z) = \frac{z^{\mathfrak{p}_+(G)}}{f(z)} I , \qquad (2.17)$$

where N is the number of zeros of the function $\det(fG)$ in $\mathfrak{D}^+$. In virtue of the fact that this function has no poles in $\mathfrak{D}^+$, N may be calculated as follows:

$$N = \mathrm{ind}_+ \det (fG) = n \,\mathrm{ind}\, f + \mathrm{ind}_+ \det G = n\, \mathfrak{p}_+(G) + \mathrm{ind}_+ \det G .$$

Now we describe the procedure of "zero separation" of $\det G_+^{(j)}$, which permits, for a given representation (2.16) of the matrix function G, a representation of precisely this type but with an index j smaller by one. Starting from the representation (2.16) of G defined by formulae (2.17) for $j = N$ and applying this procedure N times, we get a representation (2.16) of G for $j = 0$, i.e., a factorization of this matrix function in L_p.

The procedure mentioned consists in the following: For a given matrix function $G_+^{(j)}$ from the representation (2.16) and an arbitrarily chosen zero z_o of the function $\det G_+^{(j)}$ in $\mathfrak{D}^+$, we construct a matrix function U as indicated in Lemma 2.1.

By $\varkappa_1^{(j-1)}, \ldots, \varkappa_n^{(j-1)}$ we denote the tuple obtained by ordering the sequence $\varkappa_1^{(j)},\ldots,\varkappa_{k-1}^{(j)}$, $\varkappa_k^{(j)}+1$, $\varkappa_{k+1}^{(j)},\ldots,\varkappa_n^{(j)}$ in a non-increasing manner. It is clear that such an ordering can be carried out with the help of one transposition permuting the k th element of the sequence with the l th one $(l \leq k)$. By $T^{(j)}$ we denote the matrix obtained from the unit matrix after permutation of the columns with the same numbers. Then, setting $\Lambda^{(j-1)}(t) = \operatorname{diag}[t^{\varkappa_1^{(j-1)}},\ldots,t^{\varkappa_n^{(j-1)}}] =$ $= T^{(j)} \operatorname{diag}[t^{\varkappa_1^{(j)}},\ldots,t^{\varkappa_k^{(j)}+1},\ldots,t^{\varkappa_n^{(j)}}]T^{(j)}$, $G_+^{(j-1)} = G_+^{(j)}U^{-1}T^{(j)}$, $G_-^{(j-1)} = T^{(j)}W^{(j)}G_-^{(j)}$, where

$$W^{(j)}(z) = T^{(j)}(\Lambda^{(j-1)}(z))^{-1}T^{(j)}U(z)\Lambda^{(j)}(z) =$$

$$\begin{bmatrix} 1 & & & \lambda_1 z^{\varkappa_k^{(j)}-\varkappa_1^{(j)}} & & & \\ & \ddots & & \vdots & & & \\ & & 1 & \lambda_{k-1} z^{\varkappa_k^{(j)}-\varkappa_{k-1}^{(j)}} & & & \\ & & & 1-\frac{z_0}{z} & & & \\ & & & & 1 & & \\ & & & & & \ddots & \\ & & & & & & 1 \end{bmatrix},$$

we obtain the desired representation of the matrix function G.

In fact, the matrix function $G_-^{(j-1)}$ is rational with poles and zeros of its determinant concentrated in $\mathcal{D}^+$, since $G_-^{(j)}$ and $W^{(j)}$ possess these properties. The matrix function $G_+^{(j-1)}$ is analytic in $\mathcal{D}^+$ and, according to Lemma 2.1, its determinant has one zero less than the determinant of $G_+^{(j)}$. At the same time, $G_+^{(j-1)} \in M_p^+$, because $G_+^{(j)} \in L_p^+$ and $U^{-1} \in M_\infty^+$ .Consequently, $G_+^{(j-1)} \in L_p^+$. The validity of the condition $G_+^{(j-1)} \in M_q^+$ is evident, and to verify the equation

$$G = G_+^{(j-1)}\Lambda^{(j-1)}G_-^{(j-1)} \qquad (2.18)$$

itself is a straightforward matter.

Thus, a procedure for changing the representation (2.16) by (2.18) has been constructed. At the same time, under the condition $G^{-1} \in M_q^+$, the existence of a factorization of $G\ (\in M_p^+)$ in L_p with a rational factor G_- has been proved.In view of Theorem 2.2, the factor G_- is rational for every factorization of G, but not only for the one built by means of the procedure described above.

It remains to prove the estimates for the partial and the formula for the total index of the matrix function G. To this end, note that the diagonal factor $\Lambda^{(0)}$ ($= \Lambda$) from the factorization of G is obtained from the matrix function $\Lambda^{(N)}$ as the result of N operations, each of which increases one of the exponents by one and leaves the others (up to a permutation) without changes. As the initial values of all exponents are equal to $-\mathcal{P}_+(G)$, we can claim that the partial indices of G are included between $-\mathcal{P}_+(G)$ and $-\mathcal{P}_+(G) + N = -\mathcal{P}_+(G) + n\mathcal{P}_+(G) + \mathrm{ind}_+\det G = \mathrm{ind}_+\det G + (n-1)\mathcal{P}_+(G)$, and its total index is equal to $-n\mathcal{P}_+(G) + N = \mathrm{ind}_+\det G$. Thus Theorem 2.6 has been proved. $\equiv$

The following statement about the matrix function $G \in M_q^-$ can be proved in a similar way.

THEOREM 2.7. The matrix function $G \in M_q^-$ is factorable in L_p if and only if $G^{-1} \in M_p^-$. If this condition is fulfilled, then the partial indices of G are included between $-(n-1)\mathcal{P}_-(G) + \mathrm{ind}_-\det G$ and $\mathcal{P}_-(G)$, the total index is equal to $\mathrm{ind}_-\det G$, and the left factorization factor G_+ is a rational matrix function.

Remark that the property of the right (left) factorization factor established in Theorem 2.6 (2.7) is characteristic. In fact, if the matrix function G can be factored in L_p and the factor G_- (G_+) is rational, then, obviously, $G \in M_p^+$ (M_q^-).

The estimates for the partial indices from Theorems 2.6 and 2.7 are exact, although they are not necessarily attained. This can be easily seen by simplest examples of diagonal matrix functions.

In the case of a Smirnov contour Γ, Theorems 2.6 and 2.7 can be made more precise.

THEOREM 2.8. A matrix function G of class M^+ (M^-) defined on a Smirnov contour Γ can be factored in L_p if and only if $G \in L_p$ (L_q) and $G^{-1} \in M_q^+$ (M_p^-). In this case the formulae for the total index and the estimates for the partial indices formulated in Theorems 2.6 and 2.7 continue to be valid.

Proof. If the matrix function $G \in M^+$ belongs to the class L_p, then according to Theorem 1.25, $G \in M_p^+$. Consequently, under these conditions, Theorem 2.6 may be applied to G. It remains only to prove that the requirement $G \in L_p$ is necessary for the factorability in L_p of the matrix function G of class M^+. For this purpose, as in the proof of Theorem 2.6, we proceed from the factorization (2.1) of G to equation (2.15), the left-hand side $(G_+^{-1}G)$ of which belongs to the class M^+, and the right-hand side (ΛG_-) to M_q^-. But in case of a Smirnov contour $M^+ \cap M_q^- = (M^+ \cap L_q) \cap M_q^- = M_q^+ \cap M_q^- = R$. Therefore, the matrix function $F = G_+^{-1}G$ is rational. Hence $G\ (= G_+F) \in M_p^+ \subset L_p$. The case $G \in M^-$ can be treated analogously. ≡

With regard to Theorem 2.8, in case of a Smirnov contour, the total index of $G \in M^{\pm}$ is one and the same for all possible factorizations. Consequently, the factorability domain of such matrix functions consists of no more than one component, which (in view of the general properties of the factorability domain) is a connected subset of the ray $(1, \infty)$.

It appears that every connected subset of the ray $(1, \infty)$ may be the factorability domain of a matrix function $G \in M^{\pm}$. Appropriate examples can be built even in case of scalar functions defined on the unit circle. To this end, we state a corollary from Theorem 2.8 related to the indicated special case.

COROLLARY 2.3. The function $f \in D$ can be factored in L_p if and only if in its inner-outer decomposition $f = f^{(i)}f^{(e)}$ the inner function $f^{(i)}$ is a finite Blaschke product and the outer function $f^{(e)}$ satisfies the conditions $f^{(e)} \in L_p$, $(f^{(e)})^{-1} \in L_q$.

Proof. If the function f is factorable in L_p, then, in particular, it is summable and, according to Theorem 1.18, belongs to the class H_1. Consequently, Theorem 2.8 can be applied, in view of which $f \in L_p$

and $f^{-1} \in M_q^+$. The first of the relations obtained is equivalent to $f^{(e)} \in L_p$, and the second implies the finiteness of the Blaschke product B belonging to the factor $f^{(i)}$ as well as the membership of $(f^{(e)})^{-1}$ to the class L_q. Furthermore, the function $\varphi = B^{-1}f$ is an invertible element of H_1 ($\varphi^{-1} = Bf^{-1} \in H_q$, because Bf^{-1} belongs to M_q^+ and does not have poles in the unit disk). Consequently, φ is an outer function. In this way, it has been proved that $f^{(i)} = B$.

Vice versa, if $f = Bf^{(e)}$, where B is a finite Blaschke product and $f^{(e)}$ an outer function with $f^{(e)} \in L_p$, $f^{(e)-1} \in L_q$, then $f \in H_p$, $f^{-1} \in M_q^+$. It remains to utilize the sufficiency part of Theorem 2.6 to prove Corollary 2.3. ≡

Due to Theorem 1.15 and Corollary 2.3, the function $f = f^{(e)}$ defined by formula (1.7) is factorable in L_p if and only if $k \in L_p$ and $k^{-1} \in L_q$. Thus, an example of a function G ($\in H_1$) with a given connected subset σ of the ray $(1, \infty)$ as factorability domain can be built as soon as a non-negative function k defined on the interval $(-\pi, \pi)$ will be found, for which $k \in L_p$, $k^{-1} \in L_q$ for all $p \in \sigma$, $k \notin L_p$ for p lying to the right of σ, and $k^{-1} \notin L_q$ for p to the left of σ. A function k satisfying the conditions enumerated can be found, for instance, among the continuous and non-vanishing functions on $(-\pi, \pi)$ converging to zero if $\theta \to -\pi$ and to infinity if $\theta \to \pi$.

Actually, let $p_1 = \inf\{p : p \in \sigma\}$, $p_2 = \sup\{p : p \in \sigma\}$.
We impose the following requirements on the asymptotic behaviour of the function k at the right end of the domain of definition:

$$k(\theta) \sim \begin{cases} \ln(\pi - \theta) & , \text{ if } p_2 = \infty , \\ (\pi - \theta)^{-1/p_2} & , \text{ if } p_2 < \infty , \ p_2 \notin \sigma , \\ (\pi - \theta)^{-1/p_2} \ln^{\varepsilon}(\pi - \theta) \ (0 < \varepsilon < 1/p_2), & \text{ if } p_2 \in \sigma . \end{cases}$$

Then it is easy to verify that $k \in L_p$ for $p \in \sigma$, and $k \notin L_p$ for $p \notin \sigma$, $p \geq p_2$. The demands on the function k^{-1} can be satisfied due to the asymptotic behaviour of the function k at the left end chosen in a similar way.

We still mention that, owing to Corollary 2.3, an inner function f different from a finite Blaschke product yields an example of a function belonging to class L_∞ together with its inverse, which cannot be factored in any L_p .

On the strenght of either Theorem 2.6 or 2.7, it is possible to obtain the following result on factorization of rational matrix functions.

THEOREM 2.9. Let X be a rational matrix function. If the poles and the zeros of the determinant of X do not lie on Γ , then it admits a representation

$$X(t) = X_+(t)\Lambda(t)\,X_-(t) \tag{2.19}$$

where X_+ (X_-) is a rational matrix function with poles and zeros of the determinant concentrated in $\mathcal{D}^-$ ($\mathcal{D}^+$) , and Λ is a matrix function of the kind (2.2). This representation provides a factorization of X in all L_p , $1 < p < \infty$. On the other hand, if at least one pole or a zero of the determinant of X is placed on Γ , then the set $F(X)$ is empty.

Proof. If the poles and the zeros of the determinant of X do not lie on Γ , then X and X^{-1} can be viewed as elements of the class M_∞^+ . Hence, by Theorem 2.6, there follows the existence of a representation (2.19) in which the X_--factor is rational and X_+ is analytic and non-degenerate in $\mathcal{D}^+$. Since the matrix function X itself is rational, the factor X_+ is also a rational matrix function. Consequently, $X_+^{\pm 1} \in L_\infty^+$, $X_-^{\pm 1} \in L_\infty^-$, so that the representation (2.19) is a factorization of X in all L_p , $1 < p < \infty$.

We could obtain the same conclusion, considering X as an element of M_∞^- and applying Theorem 2.7. If one pole of the matrix function X lies on Γ , then $X \notin L_1$, and if X degenerates on Γ , then X^{-1} has poles on Γ and, therefore, $X^{-1} \notin L_1$. Thus, the necessary condition that $F(X)$ is non-empty indicated in Theorem 2.6 is violated, which proves Theorem 2.9. $\equiv$

Evidently, the estimates for the partial indices both from Theorem 2.6 and from Theorem 2.7 can be applied to rational matrix functions.

Now we are prepared to prove the factorability of matrix functions of the type (2.11).

THEOREM 2.10. Let the matrix function G be representable in the form (2.11), where $A_\pm$ are matrix functions satisfying conditions (2.12). Then this matrix function can be factored in L_p, and its factorization may be obtained from the representation (2.11) with the help of a finite number of algebraic operations. The total p-index $\varkappa$ of G is equal to $\mathrm{ind}_+ \det A_+ + \mathrm{ind}_- \det A_-$, and the partial p-indices are included between $-P_+(A_+) - (n-1)P_-(A_-) + \mathrm{ind}_- \det A_-$ and $P_-(A_-) + (n-1)P_+(A_+) + \mathrm{ind}_+ \det A_+$.

Proof. According to Theorems 2.6 and 2.7, the matrix functions A_+ and A_- admit a factorization in L_p:

$$A_+ = G_+^{(1)} \Lambda^{(1)} G_-^{(1)}, \tag{2.20}$$

$$A_- = G_+^{(2)} \Lambda^{(2)} G_-^{(2)}, \tag{2.21}$$

where $G_-^{(1)}$ and $G_+^{(2)}$ are rational, have no poles and do not degenerate on Γ. Consequently, the matrix function $X = \Lambda^{(1)} G_-^{(1)} G_+^{(2)} \Lambda^{(2)}$ has the same properties. Therefore, the representation (2.19) holds. Setting $G_+ = G_+^{(1)} X_+$, $G_- = X_- G_-^{(2)}$, we obtain a factorization (2.1) of G in L_p, since $G_+^{(1)} \in L_p^+$, $(G_+^{(1)})^{-1} \in L_q^+$, $G_-^{(2)} \in L_q^-$, $(G_-^{(2)})^{-1} \in L_p^-$ and $X_+^{\pm 1} \in L_\infty^+$, $X_-^{\pm 1} \in L_\infty^-$.

Each of the factorizations (2.20), (2.21) and (2.19) is realized with the help of a finite number of algebraic operations. Hence, a factorization of G can be derived with the same degree of effectiveness. The partial p-indices of the matrix function G coincide with the partial p-indices of X. To get lower bounds of the latter, we represent X in the form

$$X = G_+^{(1)-1} A_+ G_+^{(2)} \Lambda^{(2)}. \tag{2.22}$$

From the non-degeneracy and analyticity of $G_+^{(1)}$ and $G_+^{(2)}$ in $\mathcal{D}^+$ it follows that $P_+(X) \leq P_+(A_+) - \varkappa_n^{(2)}$, where $\varkappa_n^{(2)}$ is the smallest of the partial indices of A_-. According to Theorem 2.7, for the value $\varkappa_n^{(2)}$, the estimate from below

$$\varkappa_n^{(2)} \geq -(n-1)P_-(A_-) + \text{ind}_- \det A_-$$

holds, owing to which we have

$$P_+(X) \leq P_+(A_+) + (n-1)P_-(A_-) - \text{ind}_-\det A_- .$$

Since, by Theorem 2.6, the value $-P_+(X)$ serves as a lower bound for the partial indices of X, the desired lower bound for the partial indices of G has been derived. An upper bound of the partial indices may be established analogously, and the formula for the total index results from equation (2.22), taking into account that $\text{ind}_+\det G_+^{(j)} = 0$ $(j = 1,2)$ and $\text{ind} \det \Lambda^{(2)} = \text{ind}_-\det A_-$. Thus Theorem 2.10 is completely proved. ≡

It is of interest that the factorization factors $G_\pm$ of a matrix function G of the type (2.11) differ from $A_\pm$ by a rational factor. In fact,

$$G_+ = A_+(G_-^{(1)})^{-1}(\Lambda^{(1)})^{-1}X_+ \quad (= A_+G_+^{(2)}\Lambda^{(2)}X_-^{-1}\Lambda) ,$$
$$G_- = X_-(\Lambda^{(2)})^{-1}(G_+^{(2)})^{-1}A_- \quad (=\Lambda^{-1}X_+^{-1}\Lambda^{(1)}G_-^{(1)}A_-) .$$

This implies, in particular, that if some of the matrix functions $A_+^{\pm 1}$, $A_-^{\pm 1}$ is summable with larger exponent that it is caused by relations (2.12), or if it is smooth on some arc $\gamma \subset \Gamma$, then this property is acquired by the corresponding matrix functions $G_+^{\pm 1}$, $G_-^{\pm 1}$.

<u>THEOREM 2.11.</u> Let the matrix function G be factorable in L_p, let $\{\varkappa_j\}_{j=1}^n$ be the set of its partial p-indices, and suppose B_+ (B_-) to be a matrix function belonging to class M_∞^+ (M_∞^-) together with its inverse. Then the matrix function B_+GB_- is also factorable in L_p, its partial p-indices are included between $\varkappa_n - P_+(B_+) - (n-1)P_-(B_-) + \text{ind}_-\det B_-$ and $\varkappa_1 + P_-(B_-) + (n-1)P_+(B_+) + \text{ind}_+\det B_+$, and the total p-index is equal to $\varkappa + \text{ind}_+\det B_+ + \text{ind}_-\det B_-$.

<u>Proof.</u> If (2.1) is a factorization of G in L_p, then, setting e.g.

$$A_+ = B_+G_+ , \quad A_- = \Lambda G_-B_- , \tag{2.23}$$

we obtain a representation of the matrix function $\hat{G} = B_+GB_-$ in the form (2.11). Consequently, $\hat{G}$ satisfies the conditions of Theorem 2.10 and, according to this theorem, it is factorable in L_p. The upper estimate for the partial indices and the formula for the total p-index of $\hat{G}$ can be also deduced from Theorem 2.10, taking into account that $P_+(A_+) = P_+(B_+)$, $P_-(A_-) \leq P_-(B_-) + \varkappa_1$, $\mathrm{ind}_+\det A_+ = \mathrm{ind}_+\det B_+$, and $\mathrm{ind}_-\det A_- = \varkappa + \mathrm{ind}_-\det B_-$. In order to derive the lower estimate for the partial indices stated in the theorem, we have to choose $A_+ = B_+G_+\Lambda$, $A_- = G_-B_-$.

Theorem 2.10 allows us to solve the question of the connection between the factorizations of one and the same matrix function G in different L_p to the very end.

<u>THEOREM 2.12.</u> Let $G = G_+^{(1)}\Lambda^{(1)}G_-^{(1)}$ be a factorization of the matrix function G in L_{p_1}. If the rational matrix function X is such that

$$G_+^{(1)} X \in M_{p_2}^+ , \quad X^{-1}(G_+^{(1)})^{-1} \in M_{q_2}^+ ,$$
$$X^{-1} \Lambda^{(1)}G_-^{(1)} \in M_{q_2}^- , \quad G_-^{(1)}(\Lambda^{(1)})^{-1} X \in M_{p_2}^- , \qquad (2.24)$$

then G can be factored in L_{p_2}, and the total indices $\varkappa^{(1)}$ and $\varkappa^{(2)}$ of G in L_{p_1} and L_{p_2} are connected via the relation

$$\varkappa^{(1)} - \varkappa^{(2)} = N - P , \qquad (2.25)$$

where N (P) is the number of zeros (poles) of the function $\det X$ on Γ.

Conversely, if G is factorable in L_{p_1} and L_{p_2} with the total indices $\varkappa^{(1)}$ and $\varkappa^{(2)}$ respectively, then there exists a triangular polynomial matrix function Y, the degree of whose determinant is equal to $|\varkappa^{(1)} - \varkappa^{(2)}|$ such that relations (2.24) are satisfied for $X = Y$, if $\varkappa^{(1)} \geq \varkappa^{(2)}$, and for $X = Y^{-1}$, if $\varkappa^{(1)} < \varkappa^{(2)}$.

<u>Proof.</u> If the requirements (2.24) are fulfilled, then the matrix function G satisfies the conditions of Theorem 2.10 for $p = p_2$, $A_+ = G_+^{(1)}X$, $A_- = X^{-1}\Lambda^{(1)}G_-^{(1)}$. Consequently, this matrix function

can be factored in L_{p_2} , where its total p_2-index is equal to $\mathrm{ind}_+\det(G_+^{(1)}X) + \mathrm{ind}_-(X^{-1}\Lambda^{(1)}G_-^{(1)}) = \mathrm{ind}_+\det X + \mathrm{ind}_-\det X^{-1} + \mathrm{ind}\ \det\Lambda^{(1)} = \mathrm{ind}_+\det X - \mathrm{ind}_-\det X + \varkappa^{(1)}$.

This implies $\varkappa^{(1)} - \varkappa^{(2)} = \mathrm{ind}_-\det X - \mathrm{ind}_+\det X$. Applying statement 2) of Theorem 1.29 to the function $f = \det X$, we therefore get the relation (2.25).

For the sake of definiteness, in the following it is assumed that $\varkappa^{(1)} \geq \varkappa^{(2)}$.

In order to prove the converse statement, we study the matrix function H_+ occurring in the formulae (2.5) of transfer from a factorization in L_{p_1} to a factorization in L_{p_2} . Splitting off left factors of the type (2.14) as indicated in the remark to Lemma 2.1, we obtain the representation $H_+ = XH_1$, where H_1 is a polynomial matrix function with constant determinant different from zero, and X is the product of $\varkappa^{(1)} - \varkappa^{(2)}$ factors of the kind (2.14). Hence, it is a polynomial upper-triangular matrix and $\deg\det X = \deg\det H = \varkappa^{(1)} - \varkappa^{(2)}$.

As was shown in Section 2.2, all the zeros of $\det H$ and, thus, of the diagonal elements of X lie on the contour Γ .

Since $G_+^{(1)}X = G_+^{(2)}H_1^{-1}$, $X^{-1}\Lambda^{(1)}G_-^{(1)} = H_1\Lambda^{(2)}G_-^{(2)}$ and H_1 is analytic and non-degenerate at all finite points, the relations (2.24) are valid. Separating from H left factors of the type (2.14'), but not of type (2.14), we could obtain a lower-triangular matrix function. Theorem 2.12 has been proved. ≡

From Theorem 1.21 we deduce that a matrix function X satisfying conditions (2.24) belongs to the class $M^+_{p_2q_1/(p_2+q_1)}$ and its inverse X^{-1} to $M^+_{p_1q_2/(p_1+q_2)}$. Hence, for $p_2 \geq p_1$, the entries of X must be summable on Γ , therefore, $P = 0$. Similarly, for $p_2 \leq p_1$, we have $N = 0$. Furthermore, $p_2q_1/(p_2+q_1) > 1/2$ and $p_1q_2/(p_1+q_2) > 1/2$, thus, the elements of the matrix functions $X^{\pm 1}$ are summable with exponent 1/2 . Consequently, their poles lying on Γ must be simple.

If, in particular, Y is the triangular polynomial matrix function occurring in the converse statement, then all roots of its diagonal elements must be simple. The number of different roots of det Y lies, therefore, between $-[-|\varkappa^{(1)} - \varkappa^{(2)}|/n]$ (i.e. the smallest integer greater than or equal to $|\varkappa^{(1)} - \varkappa^{(2)}|/n$) and $|\varkappa^{(1)} - \varkappa^{(2)}|$. For $n = 1$, the first value is precisely equal to $|\varkappa^{(1)} - \varkappa^{(2)}|$.

2.5. COMMENTS

The factorization of the type (2.1) as well as the notions of partial and total indices, under classical assumptions concerning the matrix function G and the contour Γ, where introduced in the paper of MUSKHELISHVILI and VEKUA [1] (see also the monographs N. VEKUA [4], MUSKHELISHVILI [1]).

Factorization in L_p was first studied by PRIVALOV [1], for a review of subsequent results, see KHVEDELIDZE [1]. The right-factorization was presented for the first time in GOHBERG, KREĬN [3].

It would seem that the first detailed study of the relations between the factorization of one and the same matrix function for different values of the parameter p is accomplished in the present book. Theorems 2.1 and 2.3 were formulated in SPITKOVSKIĬ [2]. At the same place the notion of the factorability domain was introduced.

Corollary 2.1 is usually called the invariance theorem for partial indices (see N. VEKUA [4], GAKHOV [3], MUSKHELISHVILI [1], SIMONENKO [4]). Theorem 2.2 on the connections between factorizations with one and the same total index was actually proved in MUSKHELISHVILI, VEKUA [1] (see also GOHBERG [4]).

For the factorization of power functions, see VEKUA [4], GAKHOV [3], GOHBERG, KRUPNIK [4].

The method of "splitting off zeros" was first applied to the factorization problem by GAKHOV [1-3]. The factorization problem for rational matrix functions (Theorem 2.9) was solved by N.P. VEKUA [1] (see also N. VEKUA [4]), and for matrix functions of the type (2.11) with Hölder factors $A_{\pm}$ (up to Γ) by GAKHOV. In a spezial case, the latter problem was studied by N. VEKUA, KVESELAVA [1,2] and SHERMAN [1]. The general

Theorem 2.10 has been proved here, in principle, by the same method as was used by GAKHOV for the result mentioned above.
Note that a somewhat different procedure of constructing a factorization based on the preceding transition to a matrix function with constant determinant and with trigonometric polynomials as elements, has been recently proposed in JONCKHEERE, DELSARTE [1].
This applies, in particular, to rational matrix functions with zero partial indices.

In Ch. 1 of CLANCEY, GOHBERG [3] formulae are presented which express the partial indices of rational matrix functions by spectral data of the matrix polynomial corresponding to it. For more results in this direction we refer to BART, GOHBERG, KAASHOEK [1-3]. The factorization of a matrix function of the kind (2.11), if the contour Γ is the real line and det $A_{\pm}$ are entire functions, was considered in HURD [1] in connection with applications to problems of determining a stationary temperature regime.

Theorems 2.6 - 2.8 and 2.11 are proved in SPITKOVSKIĬ [13] (see also SPITKOVSKIĬ [2]). The estimates of the partial indices established in these theorems generalize results of NIKOLAĬCHUK [1] obtained by him for the classical case by the same method of "separating zeros". The formulae for the total index stated in these theorems are analogues of the classical formula of MUSKHELISHVILI (see VEKUA [4], MUSKHELISHVILI [1]), according to which the total index of a non-degenerate Hölder matrix function is equal to the index of its determinant. For Muskhelishvili's formula and its extensions to other classes, we refer to Chapter 5.

Remark that the formula for the total index and the non-negativity of the partial indices of a matrix function analytic in $\mathcal{D}^+$ (and continuous in $\mathcal{D}^+ \cup \Gamma$) are also presented in the article of CHEBOTAREV and GAKHOV [1].
Theorem 2.12 is published for the first time.

CHAPTER 3. THE CRITERION OF FACTORABILITY. Φ-FACTORIZATION AND ITS BASIC PROPERTIES

In this chapter we shall study relationships between properties of the Riemann boundary value problem for a piecewise analytic vector function and the factorability of its coefficient, i.e. the matrix function G. It is well-known that in the classical case, if we seek for a solution of the Riemann problem with a Hölder matrix G, the factorability of G is equivalent to the Fredholmness of the boundary value problem. The transition to the solution of the Riemann problem with measurable matrix function G in the Smirnov classes $E_p^{\pm}$ leads to a new quality. It turns out that the factorability of G, generally speaking, does not imply the Fredholmness of the corresponding boundary value problem. In fact, the vector-valued Riemann boundary value problem with a factorable matrix function G and the associate problem considered in the classes L_p and L_q, respectively, have finite defect numbers, and the indices of the problems are opposite. However, these problems are, in general, not normally solvable, i.e. their images can be not closed. Nevertheless, these images are in a sense well-situated. More strictly speaking, the following weakened closedness condition is fulfilled: the image of the Riemann boundary value problem (and of the problem associate to it) contains all those rational vector functions belonging to its closure. Moreover, the factorability of the matrix coefficient of a Riemann problem in L_p is equivalent to the property described above, a property intermediate between finiteness of the defect numbers and Fredholmness. This fact is the main result of Sections 3.1 and 3.2. In Section 3.3 it will be further stated that to guarantee the normal solvability and, thus, the Fredholmness of the Riemann problem with factorable coefficient G, certain two operators K and K_1 composed with the help of the factorization factors of G in a certain way as well as the projectors P and Q have to be bounded in L_p^n. In order to make the notions of Fredholmness of a boundary value problem and factorability of its coefficient equivalent, it is natural to confine the definition of factorization given in Ch. 2 by adding the requirement of boundedness of the mentioned operators K and K_1 in the space L_p^n. This special type of factorization ensuring the Fredholmness of the corresponding Riemann problem will be introduced in Section 3.4 and called Φ-factorization. In what follows we shall almost always have to deal with just Φ-factorization.

The results of Sections 3.1 - 3.4 will be formulated and proved under the only suppositions that the contour is rectifiable. The matrix function G is unbounded in general. In Section 3.5, the contour Γ is, in addition, assumed to belong to the class $\mathfrak{R}$, on which, by the very definition, the operator of singular integration S is bounded in the spaces L_p, $1 < p < \infty$. The condition $\Gamma \in \mathfrak{R}$ permits us to consider the closed operator $P + GQ$ [1]) instead of the Riemann problem with matrix coefficient G. This enables us to utilize the powerful apparatus of operator theory for investigation.

Here we shall demonstrate the necessity of the condition $G^{-1} \in L_\infty$ for $P + GQ$ to be a Φ_--operator and the criterion of Φ-factorability resulting from this fact.

Finally, under the condition $G \in L_\infty$, in Section 3.6 we change over from the examination of closed, generally speaking, unbounded operators to the study of bounded ones. It will be established that the problem of Φ-factorability of a matrix function G defined on a composite contour can be reduced to the corresponding problem for its restrictions on the connected components of the contour. Φ-factorability is preserved and of local character if the images of the contours are smooth enough.

In 3.7 comments for further reading will be given.

---

1) Incidentally, note that the equation $(P+GQ)\varphi = f$ is, in principle, - the characteristic system of singular integral equations with Cauchy kernels. In the classical case (if G is a matrix with Hölder elements and Γ a piecewise smooth contour) it was studied in the monographs MUSKHELISHVILI [1] and VEKUA [4].

3.1. ON SOLVABILITY OF THE RIEMANN BOUNDARY VALUE PROBLEM WITH FACTORABLE MATRIX COEFFICIENT

Let G be a matrix function of n th order and g a n-dimensional vector function both defined on Γ. The vector-valued Riemann boundary value problem is stated in the following way:

> Find n-dimensional vector functions φ^+ and φ^- analytic in $\mathcal{D}^+$ and $\mathcal{D}^-$, respectively, satisfying the condition
>
> $$\varphi^+(t) + G(t)\varphi^-(t) = g(t) \qquad (3.1)$$
>
> imposed on their boundary values on the contour Γ.

Problem (3.1) is viewed in the class L_p. This means that the vector function g is defined in L_p^n, and the components of the desired vector functions $\varphi^{\pm}$ must belong to the classes E_p^+ and $\overset{\circ}{E}{}_p^-$ respectively. Equation (3.1) must hold a.e. on the contour Γ, which henceforth (up to Section 3.5) is assumed to be merely rectifiable.

Together with problem (3.1) we consider the problem

$$\psi^-(t) + G'(t)\psi^+(t) = h(t) \qquad (3.2)$$

in the class L_q. Problem (3.2) is called associate [1)] to problem (3.1).

The pair $\{\varphi^+,\varphi^-\}$ is referred to as the solution of problem (3.1). We shall call φ^+ the "+"-component of the solution, and φ^- the "-"-component. Borrowing the terminology from operator theory (for which we have serious reasons, as will become clear later on), the solution set of the homogeneous problem (3.1) is called its kernel, and the set of vector-functions $g(t)$, for which the inhomogeneous problem is solvable, is said to be its image.

Clearly, the kernel and the image of problem (3.1) are lineals. The dimension α_1 of the first of them and the codimension (in L_p^n) β_1 of the closure of the second are called defect numbers of problem (3.1).

---

1) Sometimes the problem $\psi^+ + (G')^{-1}\psi^- = h$ related to (3.2) in an obvious manner is called associate to (3.1).

The difference $\alpha_1 - \beta_1$, which makes sense if at least one of the numbers α_1 and β_1 is finite, is referred to as its <u>index.</u> Problem (3.1) is said to be normally solvable, if its image is closed; it is called Fredholm, if it is normally solvable and has a finite index.

Analogous terms and notations are introduced for problem (3.2). In particular, the defect numbers α_2 and β_2 of problem (3.2) are the dimension of its kernel and the codimension in L_q^n of the closure of its image, respectively.

THEOREM 3.1.

1) The orthogonal complement [1] to the image of problem (3.1) contains the set of "+"-components of the kernel of problem (3.2) and, for $G \in L_\infty$, coincides with it.
2) The orthogonal complement to the image of problem (3.2) contains the set of "−"-components of the kernel of problem (3.1) and, provided that $G \in L_\infty$, coincides with it.
3) In the case of a scalar function G different from zero on a subset of positive measure of each of the components of the contour Γ, at least one of the defect numbers α_1 and α_2 is equal to zero.

Proof.

1) The image of problem (3.1) consists precisely of all vector functions of the form $\varphi^+ + G\varphi^-$, where $\varphi^+ \in (L_p^+)^n$, and the vector function $\varphi^- \in (\overset{\circ}{L}{}_p^-)^n$ is such that $G\varphi^- \in L_p^n$. Therefore, the vector function $\psi \in L_q^n$ is orthogonal to the image of problem (3.1) if and only if it is orthogonal to $(L_p^+)^n$ (in view of Theorem 1.26, this condition means that $\psi = \psi^+ \in (L_q^+)^n$), and

$$\langle G\varphi^- , \psi^+ \rangle = 0 \qquad (3.3)$$

for all $\varphi^- \in (\overset{\circ}{L}{}_p^-)^n$ with $G\varphi^- \in L_p^n$.

If ψ^+ is the "+"-component of a vector function from the kernel

[1] Here orthogonality is understood in the sense of the bilinear form (1.13).

of problem (3.2), then $G'\psi^+ \ (= -\psi^-) \in (\overset{\circ}{L}{}_q^-)^n$, and
$\langle G\varphi^-,\psi^+\rangle = \langle \varphi^-,G'\psi^+\rangle = \langle \varphi^-,-\psi^-\rangle = 0$. Thus, condition (3.3) is fulfilled. Hence, the "+"-components of the kernel of problem (3.2) are orthogonal to the image of problem (3.1).

If $G \in L_\infty$, condition (3.3) simply means that $\langle \varphi',G'\psi^+\rangle = 0$ for all $\varphi^- \in (\overset{\circ}{L}{}_p^-)^n$, i.e. $G'\psi^+ \in (\overset{\circ}{L}{}_q^-)^n$ (here Theorem 1.26 is used again). Consequently, for $G \in L_\infty$, the orthogonal complement to the image of problem (3.1) does not contain vector functions different from "+"-components of the kernel of problem (3.2).
The argument for statement 2) is similar.

3) Let $n = 1$, and let $\{\varphi^+,\varphi^-\}$ and $\{\psi^+,\psi^-\}$ be solutions of the homogeneous problems (3.1) and (3.2), respectively. Consider the function $f = \psi^+ G\varphi^-$. On the one hand, $\psi^+ \in L_q^+$ and $G\varphi^- = -\varphi^+ \in L_p^+$, so that $f \in L_1^+$. On the other hand, $G\psi^+ = -\psi^-$, and $\varphi^- \in \overset{\circ}{L}{}_p^-$, hence $f \in \overset{\circ}{L}{}_1^-$. According to Theorem 1.28, $f(t) = 0$ a.e. on Γ . Denoting by γ_j those subsets of the curves $\Gamma(j)$ of positive linear measure on which $G(t) \not\equiv 0$, we get $\psi^+(t)\varphi^-(t) = 0$ a.e. on γ_j $(j = 0,\ldots,m-1)$. Owing to Theorem 1.23, we thus deduce that either $\psi^+ = 0$ or $\varphi^- = 0$. In the first case we have $\psi^- = -G'\psi^+ = 0$, i.e. the homogeneous problem (3.2) has only the trivial solution and, hence $\alpha_2 = 0$. In the second case we have $\alpha_1 = 0$.
Theorem 3.1 is completely proved. ≡

By Theorem 1.2, the codimension of the closure of the image of problem (3.1) (and 3.2)) coincides with the dimension of its orthogonal complement. On the strength of Theorem 3.1, we conclude from this that the defect numbers of (3.1) and (3.2) are connected by the inequalities

$$\alpha_1 \le \beta_2 \ , \quad \alpha_2 \le \beta_1 \ . \tag{3.4}$$

In virtue of the same Theorem 3.1, for $G \in L_\infty$, inequalities (3.4) turn into equalities. However, in the general case, the equality sign in relations (3.4) fails to be true.

EXAMPLE. Assume the contour Γ to be a circle and $G(t) = t\, f(t)$, where f is an outer function with the properties $f \notin L_p$, $f^{-1} \in L_q$, $f \in L_r$ for some $r > 0$. The kernel of the corresponding problem (3.1) is trivial. Actually, if $\varphi^+ \in L_p^+$, $\varphi^- \in \overset{\circ}{L}{}_p^-$ and $\varphi^+ + G\varphi^- = 0$, then $f^{-1}(t)\varphi^+(t) = - t\, \varphi^-(t)$. The right-hand side of the equation obtained lies in the class L_p^-, and the left one in L^+. Since a circle is a Smirnov curve, in accordance with Theorems 1.25 and 1.28, we conclude that either of the functions $f^{-1}(t)\varphi^+(t)$ and $t\,\varphi^-(t)$ is equal to a certain constant c a.e. on Γ. Hence $\varphi^+(t) = c\, f(t)$. Taking into consideration that $\varphi^+ \in L_p^+$ and $f \notin L_p$, we find that $c = 0$. Consequently, $\varphi^+ = \varphi^- = 0$, i.e., the homogeneous problem (3.1) does not have nontrivial solutions. At the same time, the orthogonal complement of the image of the corresponding problem (3.2) is nontrivial: it contains, in any case, the function f^{-1}, as for $\psi^+ \in L_q^+$, $\psi^- \in \overset{\circ}{L}{}_q^-$, we have $\langle f^{-1}, \psi^- + G\psi^+ \rangle = \langle f^{-1}, G\psi^+ \rangle = \langle f^{-1}, tf\,\psi^+ \rangle = \langle 1, t\,\psi^+ \rangle = 0$. Thus, for the considered scalar function G, $\alpha_1 = 0$ and $\beta_2 > 0$.

Below in this section we will be concerned with matrix functions G which can be factored in L_p. A criterion of solvability and formulae for the defect numbers of problems (3.1) and (3.2) will be obtained, meaning, in particular, that the equality sign in the inequalities (3.4) is attained for all factorable matrix functions G irrespective of whether they are bounded or not.

THEOREM 3.2. Let the matrix function G admit a factorization (2.1) in L_p. Then

1) problem (3.1) with the right-hand side g is solvable if and only if

$$G_+^{-1}g \in \mathcal{L}_1^n\ ^{1)}, \quad \varphi_0^+ = G_+PG_+^{-1}g \in (L_p^+)^n, \qquad (3.5)$$
$$\varphi_0^- = G_-^{-1}\Lambda^{-1}QG_+^{-1}g \in (\overset{\circ}{L}{}_p^-)^n ;$$

1) Concerning the definition of $\mathcal{L}_1^n$ see p.47.

2) if the conditions (3.5) are fulfilled, then the general solution of problem (3.1) is of the form

$$\varphi^+ = \varphi_0^+ + G_+\varrho \quad , \quad \varphi^- = \varphi_0^- + G_-^{-1}\,\Lambda^{-1}\varrho \; , \tag{3.6}$$

where ϱ is a vector function, the j th element of which is a polynomial of degree less than or equal to $\varkappa_j - 1$, if $\varkappa_j > 0$, and equal to zero, if $\varkappa_j \leq 0$.

<u>Proof.</u> Assume that $\{\varphi^+,\varphi^-\}$ is a solution of problem (3.1). Substituting the representation (2.1) of the matrix function G into the boundary condition (3.1), we obtain

$$f^+ + \Lambda f^- = G_+^{-1}g \; , \tag{3.7}$$

where $f^+ = G_+^{-1}\varphi^+$ and $f^- = G_-\varphi^-$. Since $G_+^{-1} \in L_q^+$, $\varphi^+ \in (L_p^+)^n$, we have $f^+ \in (L_1^+)^n$. Analogously, $f^- \in (\mathring{L}_1^-)^n$. Therefore, equation (3.7) means that $G_+^{-1}g \in \mathcal{L}_1^n$, so that $PG_+^{-1}g \; (\in (L_1^+)^n)$ and $QG_+^{-1}g \; (\in (\mathring{L}_1^-)^n)$ are well-defined.

Rewriting equation (3.7) componentwise, we get

$$f_j^+ - P(G_+^{-1}g)_j = -t^{\varkappa_j} f_j^- + Q(G_+^{-1}g)_j \quad (=: \varrho_j) \; , \quad j = 1,\dots,n \; .$$

Applying Theorem 1.28 to the latter equation and taking into account that the functions f_j^- and $Q(G_+^{-1}g)_j$ vanish at infinity, we find that ϱ_j is identically zero, if $\varkappa_j \leq 0$, and a polynomial of degree not higher than $\varkappa_j - 1$ otherwise.

Thus, $f_j^+ = P(G_+^{-1}g)_j + \varrho_j$, $f_j^- = t^{-\varkappa_j} Q(G_+^{-1}g)_j - t^{-\varkappa_j}\varrho_j$.
Returning to the vector functions $\varphi^{\pm}$, we obtain formulae (3.6).
According to the assumption, $\{\varphi^+,\varphi^-\}$ is the solution of problem (3.1), thus $\varphi^+ \in (L_p^+)^n$, $\varphi^- \in (\mathring{L}_p^-)^n$. Since $\varrho \in (L_\infty^+)^n$, $\Lambda^{-1}\varrho \in (\mathring{L}_\infty^-)^n$, the conditions $G_+\varrho \in (L_p^+)^n$ and $G_-^{-1}\,\Lambda^{-1}\varrho \in (\mathring{L}_p^-)^n$ are satisfied and, therefore, $\varphi_0^+ \in (L_p^+)^n$, $\varphi_0^- \in (\mathring{L}_p^-)^n$.

In this way, the necessity of conditions (3.5) for the solvability of problem (3.1) has been proved as well as the fact that every solution of the problem is of the form (3.6).

THEOREM 3.7. Assume G to be a bounded measurable matrix function factorable in L_p. Then problems (3.1) and (3.2) are normally solvable (Fredholm) only simultaneously.

At the end of Section 3.5 we shall construct examples of factorable (of course, unbounded) functions for which the operators $G_+SG_+^{-1}$ and K are bounded, but the operator $G_-SG_-^{-1}$ is unbounded. In doing so, the existence of a matrix function G will be proved for which problem (3.1) is Fredholm, but problem (3.2) fails to be normally solvable.

3.4. LEFT AND RIGHT Φ-FACTORIZATION. A CRITERION OF SIMULTANEOUS FREDHOLMNESS OF THE RIEMANN BOUNDARY VALUE PROBLEM AND THE ASSOCIATE PROBLEM TO IT

It is clear from Theorem 3.6 that among all matrix functions which can be factored in L_p a special role is played by those functions for which the operators (3.16) are bounded in the space L_p^n. Therefore, it is convenient to introduce a shortened notation for factorizations satisfying the additional condition just mentioned.

DEFINITION 3.1. A factorization (2.1) of a certain matrix function G in L_p will be called a Φ-factorization, if the operators $K = G_-^{-1}\Lambda^{-1}QG_+^{-1}$ and $K_1 = G_+PG_+^{-1}$ associated with it are bounded in L_p^n.

Above all, we intend to show that Φ-factorability is a property depending only on the matrix function itself as well as the value of the parameter p, but not on the concrete choice of the factorization.

THEOREM 3.8. Two different factorizations of a matrix function G in L_p are either both Φ-factorizations or neither of them is a such one.

Proof. According to Theorem 3.6, a factorization of G in L_p is a Φ-factorization if and only if problem (3.1) with the matrix coefficient G is normally solvable, i.e. irrespective of which factorization from all possible ones has been chosen. ≡

We may also give a direct proof based upon Theorem 2.2 on the general form of a factorization and assertion 4) of Lemma 3.3. However, it is omitted here.

The domain of Φ-factorability $\Phi(G)$ may be defined in a similar manner as it was done for the factorability domain $\mathfrak{F}(G)$. Obviously $\Phi(G) \subseteq \mathfrak{F}(G)$. More exactly, $\Phi(G) = \bigcup_k \Phi_k(G)$, where $\Phi_k(G)$ $(\subseteq \mathfrak{F}_k(G))$ is the set of such values p for which the matrix function G admits a Φ-factorization in the space L_p with total index k. The next theorem shows that $\Phi_k(G)$ is a connected subset of $\mathfrak{F}_k(G)$.

THEOREM 3.9 (Interpolation theorem). Assume that the matrix function G admits a Φ-factorization with one and the same total index in L_{p_1} and L_{p_2} $(p_1 \leq p_2)$. Then G admits a Φ-factorization in all L_p, where $p \in [p_1, p_2]$.

Proof. Suppose that equation (2.1) provides a Φ-factorization of G in L_p. As in the proof of Corollary 2.2, we establish that (2.1) is a factorization of G in L_p $(p_1 \leq p \leq p_2)$, which is, due to Theorem 3.8, a Φ-factorization of G in L_{p_2}. Thus, the operators (3.16) are bounded in the spaces $L^n_{p_1}$ and $L^n_{p_2}$. By Theorem 1.13, they are bounded in L^n_p, $p \in [p_1, p_2]$, either. Consequently, (2.1) is a Φ-factorization of G in L_p for any $p \in [p_1, p_2]$. Theorem 3.9 is proved. ≡

The class of transformations preserving the factorability of a matrix function, which was considered in Theorem 2.11, preserves, actually, also the Φ-factorability. Now we are going to show this.

THEOREM 3.10. Let G be a matrix function admiting a Φ-factorization in L_p, and suppose B_+ and B_- to be matrix functions belonging together with their inverses to classes M^+_∞ and M^-_∞, respectively. Then the matrix function $\hat{G} = B_+ G B_-$ admits a Φ-factorization in L_p, too.

Proof. On the strength of formulae (2.23) and the remark after Theorem 2.10, we may claim that the factorization factors of G and $\hat{G}$ are connected by the relations

$$\hat{G}_+ = B_+ G_+ X \,, \quad \hat{G}_- = \hat{\Lambda}^{-1} X^{-1} \Lambda \, G_- B_- \,,$$

where X is some rational matrix function not having poles and non-degenerating on Γ. In accordance with this, the operators constructed analogously to formulae (3.16) are, for $\hat{G}$, of the form

$$\hat{K} = \hat{G}_-^{-1} \hat{\Lambda}^{-1} Q \hat{G}_+^{-1} = B_-^{-1} G_-^{-1} \Lambda^{-1} X Q X^{-1} G_+^{-1} B_+^{-1} ,$$

$$\hat{K}_1 = \hat{G}_+ P \hat{G}_+^{-1} = B_+ G_+ X P X^{-1} G_+^{-1} B_+^{-1} .$$

Due to assertion 4) of Lemma 3.3, the boundedness of the operators $K = G_-^{-1} \Lambda^{-1} Q G_+^{-1}$ and $K_1 = G_+ P G_+^{-1}$ implies the boundedness of operators $G_-^{-1} \Lambda^{-1} X Q X^{-1} G_+^{-1}$ and $G_+ X P X^{-1} G_+^{-1}$. Multiplying the first of them from the left by B_-^{-1} and the second one by B_+, and both from the right by B_+^{-1}, in view of the condition $B_{\pm}^{\pm 1} \in L_\infty$, we do not go beyond the class of bounded operators. Moreover, it is easy to see, that we just obtain the operators $\hat{K}$ and $\hat{K}_1$, which proves Theorem 3.10. ≡

Above problem (3.1) was viewed as the main subject and problem (3.2) as an auxiliary one. As a matter of fact, they are completely equivalent. If we assume

$$\varphi^-(t) + G(t)\varphi^+(t) = g(t) \qquad (3.18)$$

to be the initial problem and $\psi^+(t) + G'(t)\psi^-(t) = h(t)$ the problem constructed by it, then the right-factorization of G will be the natural means of investigation. On the condition that G admits a right factorization in L_q, the normal solvability (and Fredholmness) of problem (3.18) in L_q^n is equivalent to the boundedness of the operators $G_- Q G_-^{-1}$ and $G_+^{-1} \Lambda^{-1} P G_-^{-1}$ in L_q^n (after obvious transformations this result is a direct consequence of Theorem 3.6').

In accordance with this, representation (2.1') of the matrix function G is said to be its right Φ-factorization (in L_q), as soon as the operators $G_- Q G_-^{-1}$ and $G_+ \Lambda^{-1} P G_-^{-1}$ are bounded in L_q^n. For the sake of uniformity, a Φ-factorization of the form (2.1), when it occures besides a right Φ-factorization, is referred to as a left Φ-factorization. With the help of the notions introduced we formulate the crucial result of the present chapter.

THEOREM 3.11. The following assertions are equivalent:

1) problems (3.1) and (3.2) are Fredholm and their indices are opposite,

2) the matrix function G is Φ-factorable from the left in L_p, and the matrix function G' is Φ-factorable from the right in L_q.

Proof. Suppose assertion 1) to be fulfilled. Then, in particular, the indices of Problems (3.1) and (3.2) are finite and opposite, and the images of these problems involve all vector functions from R^n belonging to their closure. According to Theorem 3.5, G is factorable in L_p (and, thus, G' is factorable from the right in L_q). Since, in addition, problems (3.1) and (3.2) are normally solvable, with regard to Theorems 3.6 and 3.6', we deduce that a (left) factorization of G in L_p and a (right) factorization of G' in L_q are Φ-factorizations. Consequently, 1) implies 2).

Vice versa, if 2) applies, then, thanks to Theorems 3.6 and 3.6' respectively, problems (3.1) and (3.2) are Fredholm and their indices are opposite by Theorem 3.5. Thus Theorem 3.11 is proved. ≡

In case $G \in L_\infty$, the indices of problems (3.1) and (3.2) are automatically opposite, in virtue of Theorem 3.1. Moreover, by restating Theorem 3.7, with regard to the definition of a right Φ-factorization, we obtain the following result.

THEOREM 3.7'. The matrix function $G \in L_\infty$ admits a left Φ-factorization in L_p if and only if the matrix function G' admits a right Φ-factorization.

Therefore, for bounded matrix functions, Theorem 3.11 admits the following simplification.

COROLLARY 3.4. If the matrix function G belongs to the class L_∞, then for the simultaneous Fredholmness of problems (3.1) and (3.2), it is necessary and sufficient for G to be Φ-factorable in L_p.

3.5. Φ-FACTORIZATION OF UNBOUNDED MATRIX FUNCTIONS ON CONTOURS OF CLASS $\mathfrak{R}$

Beginning with this section, we shall assume that the contour Γ belongs to class $\mathfrak{R}$. As noted in Section 1.1, under this condition, the spaces L_p^n $(1 < p < \infty)$ split into the direct sum of the subspaces $(L_p^+)^n$ and $(\mathring{L}_p^-)^n$, and P and Q are bounded projections related to the mentioned decomposition.

It is quite natural to emphasize the contours of class $\mathfrak{R}$ for the study of questions associated with Φ-factorization, as can be seen from the following theorem.

THEOREM 3.12. Let $G = A_+A_-$, where $A_+^{\pm 1} \in M_\infty^+$, $A_-^{\pm 1} \in M_\infty^-$. The condition $\Gamma \in \mathfrak{R}$ is necessary for the Φ-factorability of G in L_p for at least one value $p \in (1,\infty)$ and sufficient for its Φ-factorability for all such p.

Proof. According to Theorem 2.10 and the remark after it, the matrix function G admits a factorization (2.1) in which $G_+^{\pm 1} \in L_\infty^+$, $G_-^{\pm 1} \in L_\infty^-$. Therefore, the condition of boundedness of the operator $K = G_-^{-1}\Lambda^{-1}QG_+^{-1}$, which is necessary and sufficient (due to Corollary 3.3) for the Φ-factorability of G, is equivalent to the boundedness of the operator $Q = \frac{1}{2}(I-S)$. $\equiv$

In this way, condition $\Gamma \in \mathfrak{R}$ is necessary for the Φ-factorability of rational and even constant non-degenerate matrix functions.

To the measurable matrix function G we assign the operator of multiplication by this matrix function in the space L_p^n, denoting it by the same letter G. We shall consider this operator on the lineal $\mathrm{dom}_p G$ of such vector functions $\varphi \in L_p^n$ for which $G\varphi \in L_p^n$. Apparently, the operator of multiplication defined in this fashion is closed. It is bounded if and only if $G \in L_\infty$. In the latter case, the norm of the operator of multiplication does not depend upon p for the norm in L_p^n introduced by us. Moreover, it can be calculated via the formula

$$\|G\| = \operatorname{ess\,sup}_{t\in\Gamma} \|G(t)\|^{1)} . \qquad (3.19)$$

In what follows, the operators $P + GQ$ and $Q + G'P$ defined on the lineals $(L_p^+)^n \dotplus ((\overset{\circ}{L}{}_p^-)^n \cap \operatorname{dom}_p G)$ and $((L_q^+)^n \cap \operatorname{dom}_q G') \dotplus (\overset{\circ}{L}{}_q^-)^n$, respectively, will play an important role. Obviously, these operators are closed. Problems (3.1) and (3.2) may be rewritten in the form $(P+GQ)\varphi = g$ and $(Q+G'P)\psi = h$, where the vector functions $\varphi \in L_p^n$ and $\psi \in L_q^n$ are taken from the domain of definition of the corresponding operator. Consequently, the image of the operator $P + GQ$ coincides with the image of problem (3.1), and $\varphi \in \ker(P+GQ)$ if and only if the pair $\{P\varphi, Q\varphi\}$ is the solution of the homogeneous problem (3.1). Therefore, the defect numbers of the operator $P + GQ$ coincide with the corresponding defect numbers of problem (3.1).
Besides, this operator is normally solvable, Fredholm etc. iff problem (3.1) possesses the corresponding property. The operator $Q + G'P$ is associated with problem (3.2) in an analogous manner.

Summarizing, in case $\Gamma \in \mathcal{R}$, Theorems 3.5 – 3.7 and 3.9 – 3.11 may be interpreted as conditions (necessary or sufficient) of finiteness of the index and Fredholmness of the operators $P + GQ$ and $Q + G'P$. In the present section we shall obtain a necessary condition for $P + GQ$ and $Q + G'P$ to be Φ_--operators (of course, it will be also necessary for the Φ-factorability of the matrix function G). To this end, we need the following auxiliary result.

<u>LEMMA 3.4.</u> Let G be a measurable matrix function, such that G^{-1} does not belong to L_∞. Then, for all $\varepsilon > 0$, we may find a matrix function $G_1 \in L_\infty$ such that $\|G_1\| < \varepsilon$ and the matrix function $G - G_1$ degenerates on a set of positive measure.

---

1) We identify the matrix **A** with the operator of multiplication by **A** in the space $\mathbb{C}^n$ supplied with the scalar product $(x,y) = \sum x_j \bar{y}_j$, where $x = \sum x_j e_j$, $y = \sum y_j e_j$, and the norm $\|x\| = (x,x)^{1/2}$. In this case, the norm of **A** is the largest eigenvalue of its left modulus $(AA^*)^{1/2}$ (or, equivalently, of its right modulus $(A^*A)^{1/2}$).

Proof. Unboundedness of the matrix function G^{-1} means that, for all $\varepsilon > 0$, the set $\Gamma_\varepsilon = \{t \in \Gamma : \inf_{\|x\|=1} \|G(t)x\| < \varepsilon/3, \ x \in \mathbb{C}^n\}$ has positive measure. We represent the matrix function G as the limit of an almost everywhere converging sequence of matrix functions $G^{(k)}$ the elements of which are step functions. Due to Egorov's theorem, there exists a set Δ_ε of positive measure lying in Γ_ε on which the sequence $\{G^{(k)}\}$ converges uniformly. Now we choose a number k so large that $\|G^{(k)}(t)-G(t)\| < \varepsilon/3$ for all $t \in \Delta_\varepsilon$. Then $\inf_{\|x\|=1} \|G^{(k)}(t)x\| < 2\varepsilon/3$ for all $t \in \Delta_\varepsilon$. Therefore, we may choose a vector function $x(t)$ defined on Δ_ε with $\|x(t)\| = 1$ such that $\|G^{(k)}(t)x(t)\| < 2\varepsilon/3$. Since the elements of the matrix function $G^{(k)}$ are step functions, the elements of $x(t)$ can also be chosen as step functions. In doing so, we have $\|G(t)x(t)\| \leq \|G^{(k)}(t)x(t)\| + \|G(t)-G^{(k)}(t)\| < \varepsilon$. As G_1 we choose a matrix function equal to zero outside Δ_ε and defined by the equation $G_1(t)y = (y,x(t))G(t)x(t)$ on Δ_ε. The matrix function G_1 is measurable, because G and x are measurable.

Apparently, $\|G_1(t)\| = \begin{cases} 0, & t \notin \Delta_\varepsilon, \\ \|G(t)x(t)\|, & t \in \Delta_\varepsilon \end{cases}$ and, therefore, $\|G_1\| < \varepsilon$.
Further, $G_1(t)x(t) = G(t)x(t)$ for $t \in \Delta_\varepsilon$, so that the matrix function $G - G_1$ degenerates on the set Δ_ε. The proof of Lemma 3.4 is complete. ≡

Notice that in case $n = 1$ the assertion of the lemma is evident.

THEOREM 3.13. If at least one of the operators $P + GQ$ and $Q + G'P$ is a Φ_--operator, then $G^{-1} \in L_\infty$.

Proof. For the sake of definiteness, we consider the operator $P + GQ$. If it is a Φ_--operator, then by Theorem 1.4, there exists an $\varepsilon > 0$ such that the operator $P + G_1Q$ is also a Φ_--operator, as soon as $\|G-G_1\| < \varepsilon$. If, in addition, $G^{-1} \notin L_\infty$, then in view of Lemma 3.4, we may assume that G_1 degenerates on a set of positive measure.
Since the image of the operator $P + GQ$ is a subspace of finite codimension in L_p, it has a nontrivial intersection with any infinite-

dimensional lineal contained in L_p^n, in particular, with the lineals M_j $(j = 1,\dots,n)$ of those vector functions the j th component of which is a polynomial of t^{-1} vanishing at infinity, and the remaining components are identically equal to zero. In other words, there are vector functions $\varphi_j \in (L_p^+)^n$, $\varphi_j^- \in (\overset{\circ}{L}{}_p^-)^n$ and $h_j \in M_j$, $h_j(t) \not\equiv 0$ such that $\varphi_j^+ + G_1\varphi_j^- = h_j$. Composing the matrix functions Φ_+, Φ_- and H from the vector functions φ_j^+, φ_j^- and h_j respectively, taken as columns, we obtain $\Phi_+ + G_1\Phi_- = H$ or $H - \Phi_+ = G_1\Phi_-$, from which we conclude that $\det(H - \Phi_+) = \det(G_1\Phi_-)$ (a.e. on Γ). The left-hand side of the latter equation is a function of class $M_{p/n}^+$ having a pole at zero of at least n th order. Therefore, it is not identically zero. At the same time, its angular limits on Γ are equal to zero at almost all points at which the matrix function G_1 degenerates, i.e. on a set of positive measure. This contradicts Theorem 1.23, therefore, the supposition $G^{-1} \notin L_\infty$ is not consistent with the requirement that $P + GQ$ is a Φ_--operator. Theorem 3.13 is proved. ≡

<u>COROLLARY 3.5.</u> For the Φ-factorability of the matrix function G in some L_p, it is necessary that $G^{-1} \in L_\infty$ and $G \in L_1$.

<u>Proof.</u> The condition $G \in L_1$ is necessary for factorability, all the more, it is necessary for Φ-factorability. According to Theorem 3.6, Φ-factorability of G implies Fredholmness of the operator $P + GQ$, from which the inclusion $G^{-1} \in L_\infty$ follows, by Theorem 3.13. ≡

<u>THEOREM 3.14.</u> A left factorization of the matrix function G in L_p is its Φ-factorization if and only if $G^{-1} \in L_\infty$ and the operator $G_+SG_+^{-1}$ is bounded in the space L_p^n. A right factorization of G in L_p is its Φ-factorization iff $G^{-1} \in L_\infty$ and the operator $G_-SG_-^{-1}$ is bounded in L_p^n.

<u>Proof.</u> We intend to prove the first part of the theorem. The necessity of the condition $G^{-1} \in L_\infty$ for Φ-factorability of G is precisely Corollary 3.5; the boundedness of the operator $G_+SG_+^{-1}$ is equivalent to the boundedness of operator $G_+PG_+^{-1}$. The latter demand is involved

in the definition of Φ-factorization. Conversely, if $G^{-1} \in L_\infty$ and the operator $G_+SG_+^{-1}$ (and, thus, also $G_+QG_+^{-1}$) is bounded, then the operator $G^{-1}G_+QG_+^{-1}$, a part of which is the first operator K from (3.16), is bounded. The argument concerning the second assertion is similar, thus the proof of Theorem 3.14 is complete. ≡

<u>COROLLARY 3.6.</u> The matrix function $G \in L_\infty$ is left (right) Φ-factorable in L_p if and only if the matrix function G^{-1} is right (left) Φ-factorable in L_p.

<u>Proof.</u> Assume the representation (2.1) to be a left Φ-factorization of $G \in L_\infty$ in L_p. By Theorem 3.14, $G^{-1} \in L_\infty$ and the operator $G_+SG_+^{-1}$ is bounded in L_p^n. In this case, the operator $G^{-1}(G_+SG_+^{-1})G = G_-^{-1}\Lambda^{-1}S\Lambda G_-$ and, consequently, also the operator $G_-^{-1}SG_-$ differing from it by a finite-dimensional bounded summand are bounded. Due to Theorem 3.14, the representation $G^{-1} = G_-^{-1}\Lambda^{-1}G_+^{-1}$ is a right Φ-factorization of G^{-1} in L_p. Hence, the right Φ-factorability of G^{-1} follows from the left Φ-factorability of the matrix function $G \in L_\infty$. The remaining assertions can be verified analogously. ≡

<u>COROLLARY 3.7.</u> The factorization

$$G(t) = G_+(t)t^{\varkappa}G_-(t) \tag{3.20}$$

of the scalar function G in L_p is a Φ-factorization if and only if $G^{-1} \in L_\infty$ and the operator S is bounded in the space L_p with weight $|G_+|^p$.

This result follows from Theorem 3.14 and the fact (verifiable by the very definition) that the boundedness of the operator fSf^{-1} in L_p is equivalent to the boundedness of S in the space L_p with weight $|f|^p$.

For a fairly wide class of contours (including, in particular, piecewise Lyapunov and smooth curves), a necessary and sufficient condition for the operator S to be bounded in the space L_p with weight ϱ is the following one:

$$\sup\{|\gamma|^{-1}(\int_\gamma \varrho(t)|dt|)^{1/p}(\int_\gamma \varrho(t)^{-q/p}|dt|)^{1/q}\} < \infty . \tag{3.21}$$

The supremum in (3.21) is taken over all arcs $\gamma \subseteq \Gamma$, and $|\gamma|$ denotes the length of arc γ.

In this way, the factorization (3.20) in L_p of the scalar function G defined on a contour from the class under study is its Φ-factorization if and only if $G^{-1} \in L_\infty$ and

$$\sup\{|\gamma|^{-1}(\int_\gamma |G_+(t)|^p|dt|)^{1/p}(\int_\gamma |G_+(t)|^{-q}|dt|)^{1/q}\} < \infty .$$

With the help of Theorem 3.14 one may obtain necessary and sufficient conditions of Φ-factorability of matrix functions from classes $M^{\pm}$.

THEOREM 3.15. The matrix function $G \in M^+$ is left Φ-factorable in L_p if and only if $G^{-1} \in M^+_\infty$ and the operator GSG^{-1} is defined on the whole space L^n_p. It is right Φ-factorable in L_p iff $G^{-1} \in M^+_\infty$ and $G \in L_q$.

Proof. Since the contour Γ belongs to class $\mathfrak{R}$ and is, specifically, a Smirnov curve, Theorem 2.8 can be applied, thanks to which the conditions $G \in L_p$, $G^{-1} \in M^+_q$ are necessary for G to be left factorable in L_p. According to Theorem 3.14, for the Φ-factorability of G it is necessary that $G^{-1} \in L_\infty$. Consequently, if the considered matrix function G ($\in M^+$) is Φ-factorable in L_p, then $G^{-1} \in L_\infty \cap M^+_q = M^+_\infty$.

Furthermore, if the relations $G \in L_p$, $G^{-1} \in M^+_\infty$ are fulfilled, then the matrix function $G \in M^+$ admits a factorization (2.1) with a rational matrix function G_- with poles and the zeros of the determinant only in $\mathfrak{D}^+$. In accordance with assertion 4) of Lemma 3.3, the operators $GSG^{-1} = G_+(\Lambda G_-)S(\Lambda G_-)^{-1}$ and $G_+SG_+^{-1}$ are bounded only simultaneously. Therefore, the condition of boundedness (or, equivalently, the condition to be defined everywhere) of the operator GSG^{-1} is also necessary for the Φ-factorability of G. Conversely, if this condition is satisfied, then by statement 1) of Lemma 3.3, we have $G \in L_p$. If, in addition, the condition $G^{-1} \in M^+_\infty$ applies, then, because of Theorem 2.8, G can be factored in L_p and the operator $G_+SG_+^{-1}$ is bounded (in view of the boundedness of GSG^{-1}).

The necessity of conditions $G \in L_q$ and $G^{-1} \in L_\infty$ for G to be right Φ-factorable in L_p can be verified analogously. However, they prove to be also sufficient, as the factorization factor G_- is a rational matrix function not having poles and non-degenerating on Γ and the operator $G_- S G_-^{-1}$ is bounded in L_p^n along with S. The proof of Theorem 3.15 is complete. $\equiv$

The argument for the following result differs only in detail from that given above, hence, it is omitted.

<u>THEOREM 3.15'.</u> The matrix function $G \in M^-$ is left Φ-factorable in L_p if and only if $G \in L_q$, $G^{-1} \in M_\infty^-$; it is right Φ-factorable in L_p iff $G^{-1} \in M_\infty^-$ and the operator GSG^{-1} is bounded in L_p^n .

Consider the outer function defined via formula (1.7). Let us assume that $k \in L_p$, $k^{-1} \in L_\infty$ and condition (3.21) is not fulfilled for any $p \in (1,\infty)$ substituting $\varrho = k^p$. Then, in view of Theorem 3.15, the function $f^{(e)}$ fails to be left Φ-factorable in L_p for any $p \in (1,\infty)$, although it is right Φ-factorable in L_p for all $p \in (1,\infty)$. Problem (3.1) with $G = f^{(e)}$ is not Fredholm, although the associate problem (3.2) is Fredholm.

It remains to specify a function k having the desired properties. Set $k(\theta) = s$ for $2^{1-2s}-1 \leq \frac{\theta}{\pi} < 2^{2-2s}-1$ and $k(\theta) = 1$ for $2^{2-2s}-1 \leq \frac{\theta}{\pi} < 2^{3-2s}-1$ $(s = 1,2,\ldots)$. Then $\int_{-\pi}^{\pi} k(\theta)^p d\theta =$ $\pi\left(\sum_{s=1}^{\infty} \frac{s^p}{2^{2s-1}} + \frac{4}{3}\right)$, so that $k \in L_p$ for all $p < \infty$, and $|k(\theta)| \geq 1$, thus $k^{-1} \in L_\infty$. Now we choose the interval $\gamma = \{e^{i\theta} : 2^{1-2s}-1 \leq \frac{\theta}{\pi} < 2^{3-2s}-1\}$. Then

$$\int_\gamma k(\theta)^p d\theta = \pi(2^{2-2s} + s^p \cdot 2^{1-2s}) , \quad \int_\gamma k(\theta)^{-q} d\theta = \pi(2^{2-2s} + s^{-q} \cdot 2^{1-2s})$$

and

$$\frac{1}{|\gamma|}\left(\int_\gamma k(\theta)^p d\theta\right)^{1/p}\left(\int_\gamma k(\theta)^{-q} d\theta\right)^{1/q} =$$

$$= \frac{2^{2s-1}}{3} \cdot (2^{2-2s} + s^p \cdot 2^{1-2s})^{1/p}(2^{2-2s} + s^{-q} \cdot 2^{1-2s})^{1/q} =$$

$$= \frac{1}{3}(2+s^p)^{1/p}(2+s^{-q})^{1/q} .$$

Passing to the limit $s \to \infty$, the expression obtained increases infinitely. Consequently, condition (3.21) is not fulfilled.

Utilizing Theorem 3.15', one may quote examples in which problem (3.1) is Fredholm, but the associate problem (3.2) is not Fredholm. Choosing a function $k \in L_\infty$ in such a way that $k^{-1} \notin L_\infty$, but $k^{-1} \in L_p$ for all $p < \infty$, we obtain, by Corollary 2.3, a function $f^{(e)}$ which is factorable in all spaces L_p, $p \in (1,\infty)$, but in none of them it is Φ-factorable, because the necessary condition for Φ-factorability $(f^{(e)})^{-1} \in L_\infty$ fails to be fulfilled.

3.6. Φ-FACTORIZATION OF BOUNDED MEASURABLE MATRIX FUNCTIONS ON CONTOURS OF CLASS $\mathfrak{R}$

Under the conditions $G \in L_\infty$ and $\Gamma \in \mathfrak{R}$, the operators $P + GQ$ and $Q + G'P$ are, obviously, defined and bounded on the whole of the space L_p^n, $1 < p < \infty$. We now introduce auxiliary operators

$$T_Q(G) = QG|_{\text{im } Q} \quad \text{and} \quad T_P(G') = PG'|_{\text{im } P}, \qquad (3.22)$$

which will be considered in the spaces $(\overset{\circ}{L}{}_p^-)^n$ and $(L_q^+)^n$, respectively. These operators are also bounded and defined everywhere.

LEMMA 3.5.

1) The operator $P + GQ$ $(Q + G'P)$ considered in the space L_p^n (L_q^n) is normally solvable if and only if the operator $T_Q(G)$ $(T_P(G'))$ is normally solvable, and the corresponding defect numbers of these operators coincide;
2) the operators $T_Q(G)$ and $T_P(G')$ are normally solvable only simultaneously.

Proof. It is easily shown that $\text{im}\,(P + GQ) = (L_p^+)^n \dotplus \text{im}\, T_Q(G)$ and that $\varphi \in \ker\,(P + GQ)$ if and only if $\varphi^- = Q\varphi \in \ker T_Q(G)$, $P\varphi = -PG\varphi^-$. Assertion 1) for the pair of operators $P + GQ$ and $T_Q(G)$ follows from here immediately. Incidentally, this assertion may be proved by Corollary 1.2, representing the space L_p^n in the form of the direct sum $(\overset{\circ}{L}{}_p^-)^n \dotplus (L_p^+)^n$. In fact, for such a decomposition, the blocks of the operator $A = P + GQ$ in the representation (1.1) are $A_{11} = T_Q(G)$,

$A_{12} = 0$, $A_{22} = I|_{(L_p^+)^n}$, $A_{21} = PG|_{(\overset{\circ}{L}{}_p^-)^n}$ and the assumptions of Corollary 1.2 are satisfied, where m (= dim A_{21} ker A_{11} / (im $A_{22} \cap A_{21}$ ker A_{11})) = 0 .
The part of assertion 1) related to the operators $Q + G'P$ and $T_P(G')$ can be proved analogously. This is left to the reader.

Proceeding to the proof of assertion 2), first of all, note that the images of operators $T_Q(G)$ and $T_P(G')$ coincide with the images of operators QGQ (considered in L_p^n) and $PG'P$ (considered in L_q^n), respectively. Besides, the operators QGQ and $PG'P$ are adjoint to each other: if $f \in L_p^n$, $g \in L_q^n$, then by Theorem 1.31,
$\langle QGQf,g \rangle = \langle QGQf, Pg \rangle = \langle GQf, Pg \rangle = \langle Qf, G'Pg \rangle =$
$\langle Qf, PG'Pg \rangle = \langle f, PG'Pg \rangle$.

According to Theorem 1.7, the images of operators QGQ (in L_p^n) and $PG'P$ (in L_q^n) are closed only simultaneously. Consequently, the operators $T_Q(G)$ and $T_P(G')$ have the same property, which proves Lemma 3.5. ≡

THEOREM 3.16. The matrix function $G \in L_\infty$ defined on a contour of class $\mathfrak{R}$ is Φ-factorable in L_p if and only if the operator $P + GQ$ is Fredholm in the space L_p^n . In this case, the dimension of the kernel of operator $P + GQ$ coincides with the sum of positive partial indices of G , the codimension of the image is opposite to the sum of negative partial indices, and the index is equal to the total index of G . The operator $K + K_1$ (where K and K_1 are defined by formulae (3.16)) is a two-sided regularizer for the operator $P + GQ$.

Proof. Comparing assertions 1) and 2) of Lemma 3.5, we find that the images of the operators $P + GQ$ (in L_p^n) and $Q + G'P$ (in L_q^n) are closed only simultaneously. Due to Theorem 3.1, the indices of these operators are also finite only simultaneously. Consequently, the Fredholmness of one of these operators is equivalent to the Fredholmness of the other. With regard to Corollary 3.4, this enables us to derive the criterion of Φ-factorability mentioned in the theorem. Taking Theorem 3.1 into consideration, the formulae for the defect numbers of the operator $P + GQ$ are obtained from relations (3.8).

Now we evaluate the product $K + K_1$ by $P + GQ$:

$$(K+K_1)(P+GQ) = (G_-^{-1} \Lambda^{-1} QG_+^{-1} + G_+PG_+^{-1})(P + G_+\Lambda G_-Q)$$

$$= G_-^{-1} \Lambda^{-1} QG_+^{-1}P + G_+PG_+^{-1}P + G_-^{-1} \Lambda^{-1} Q \Lambda G_-Q + G_+P\Lambda G_-Q$$

$$= G_-^{-1} \Lambda^{-1} QG_+^{-1}P + P - G_+QG_+^{-1}P + Q - G_-^{-1} \Lambda^{-1} P\Lambda G_-Q$$

$$+ G_+P\Lambda G_-Q = I + (G_-^{-1} \Lambda^{-1} - G_+)(QG_+^{-1}P - P\Lambda G_-Q) .$$

Taking into account that $QG_+^{-1}P = 0$ and that the operator $(G_-^{-1} \Lambda^{-1} - G_+)P\Lambda G_-Q$ is finite dimensional by Theorem 1.35, we recognize that $K + K_1$ is a left regularizer for $P + GQ$. Now we may either analogously check that the product $(P + GQ)(K + K_1)$ also differs from I by a finite-dimensional summand, or we may use that general fact that a left regularizer of a Fredholm operator is also its right regularizer. ≡

In the formulation of Theorem 3.16 the operator $P + GQ$ can be replaced by $T_Q(G)$; then the restriction of the operator K on $\operatorname{im} Q$ serves as a regularizer of this operator.

<u>COROLLARY 3.8.</u> The matrix function $G \in L_\infty$ is Φ-factorable in L_p with a non-negative (non-positive, zero) tuple of partial indices if and only if the operator $P + GQ$ is Fredholm and right-invertible (left-invertible, two-sided invertible) in the space L_p^n. In case of Φ-factorability of G with a non-positive (zero) tuple of partial indices, the operator $K + K_1$ is left (two-sided) inverse to the operator $P + GQ$. If the partial indices are non-negative, the operator $G_-^{-1}Q \Lambda^{-1}G_+^{-1} + K_1$, which differs from $K + K_1$ by a finite-dimensional summand, serves as a right inverse.

<u>Proof.</u> We have merely to verify the statement concerning the operators inverse to $P + GQ$. If all partial indices are non-positive, then in the transformations of the product $(K + K_1)(P + GQ)$ carried out above,

we have $\Lambda \in L_\infty^-$. Thus, the operator $P\Lambda G_-Q$ and along with it also $(G_-^{-1}\Lambda^{-1} - G_+)P\Lambda G_-Q$ is equal to zero. Consequently, in case of non-positive partial indices, we have $(K + K_1)(P + GQ) = I$, i.e., the operator $K + K_1$ is a left inverse for $P + GQ$. The remaining cases can be treated similarly. ≡

Now we intend to state several results without proof related to right Φ-factorization.

THEOREM 3.16'. The matrix function G admits a right Φ-factorization in L_p iff the operator $Q + GP$ is Fredholm in the space L_p^n . In this case the kernel dimension of $Q + GP$ is opposite to the sum of negative right indices of G , and the codimension of the image is equal to the sum of positive right indices.

COROLLARY 3.8'. The matrix function $G \in L$ admits a right Φ-factorization in L_p with a non-negative (non-positive, zero) tuple of partial indices if and only if the operator $Q + GP$ is Fredholm and left-invertible (right-invertible, two-sided invertible).

In Theorem 3.16' and Corollary 3.8' the operator $Q + GP$ can be replaced by $T_P(G)$.

As an example of some importance we take up the power functions regarded in Section 2.3. Suppose that the contour Γ belongs to class $\mathfrak{R}$ and condition (2.8) is fulfilled. The branch $\arg(t - t_0)$ continuous on $\Gamma \setminus \{t_0\}$ is assumed to be bounded (here t_0 is the chosen point of discontinuity of the function $\psi(t) = t^\alpha$). Then, in particular, the assumptions on the contour Γ stipulated in Theorem 2.5 are satisfied. The statement of this theorem can now be made more precise.

THEOREM 3.17. The domain of Φ-factorability of the function $\psi(t) = t^\alpha$ consists of one component filling the whole ray $(1,\infty)$ if $\operatorname{Re}\alpha$ is an integer, and consists of the two components $(1, \frac{1}{1-\{\operatorname{Re}\alpha\}})$ and $(\frac{1}{1-\{\operatorname{Re}\alpha\}}, \infty)$ otherwise.

Proof. Let $p \in (\frac{1}{1-\{\operatorname{Re}\alpha\}}, \infty)$. In Theorem 2.5 it was proved that the representation (2.10), where $k = [\operatorname{Re}\alpha]$ and $\psi_{+,k}$ is a fixed branch of the function $(z - t_o)^{\alpha-k}$, $t_o \in \Gamma$, is a factorization of ψ in L_p. Since $\psi^{-1} \in L_\infty$ and in view of Corollary 3.7, it suffices to prove the boundedness of the operator S in the space L_p with weight

$|\psi_{+,k}(t)|^p = |t-t_o|^{p\{\mathrm{Re}\ \alpha\}}\exp(-p\ \mathrm{Im}\ \alpha\ \arg(t-t_o))$.

The boundedness condition for the function $\arg(t - t_o)$ allows us to substitute the considered weight by $|t - t_o|^{p\{\mathrm{Re}\ \alpha\}}$. The boundedness of the operator S on a contour of class $\mathfrak{R}$ in the space L_p with such a weight results from a theorem of B.V. KHVEDELIDZE (see. e.g. NIKOLAĬCHUK [1] , p.81).

If $\{\mathrm{Re}\ \alpha\} = 0$, then there is nothing to prove. Otherwise, we still have to examine the case $p \in (1, \frac{1}{1-\{\mathrm{Re}\ \alpha\}})$, which can be done analogously. The non-existence of a factorization and, thus, also of a Φ-factorization for $p = \frac{1}{1 - \{\mathrm{Re}\ \alpha\}}$ $(\{\mathrm{Re}\ \alpha\} \neq 0)$ was shown in Theorem 2.5. ≡

Remark. Theorem 3.17 remains true even without assumption (2.8). In fact, in verifying the existence of a factorization in L_p for the function t^α with $p \neq (1 - \{\mathrm{Re}\ \alpha\})^{-1}$ (see Theorem 2.5) and the boundedness of the operator S in the space L_p with appropriate weight, we did not take advantage of condition (2.8). This condition plays an essential role only in the argument concerning the non-existence of a factorization in L_p with $p = \frac{1}{1 - \{\mathrm{Re}\ \alpha\}}$. However, it appears that the non-existence of a Φ-factorization may be proved without using condition (2.8), too. The corresponding proof for arbitrary piecewise continuous functions will be given in Section 5.3. Therefore, it is omitted here.

Besides the operator $P + GQ$, we sometimes have to study operators of more general type, namely $AP + BQ$, where $A, B \in L_\infty$. In order to derive a criterion of Fredholmness for such an operator, we need the following auxiliary proposition.

LEMMA 3.6. If $G \in L_\infty$, then, for all $\varepsilon > 0$, there exists a matrix function $\hat{G} \in L_\infty$ such that $\|G - \hat{G}\| < \varepsilon$, $\hat{G}^{-1} \in L_\infty$.

Proof. The lemma will be proved by induction on the order k of G. For $k = 1$, the assertion to be proved is obvious. Assuming that it is true for $k = n-1$, we consider a $n \times n$ matrix function $G \in L_\infty$. Taking $\varepsilon > 0$, we choose a matrix function $(\hat{G}_{ij})_{i,j=1}^{n-1}$ differing from $(G_{ij})_{i,j=1}^{n-1}$ by less than $\varepsilon/2$ with respect to the norm, the determinant $a(t)$ of which satisfies the relation $a^{-1} \in L_\infty$ (by the induction hypothesis, this can always be done).

Setting $\hat{G}_{in} = G_{in}$, $\hat{G}_{nj} = G_{nj}$ $(i,j = 1,\dots,n-1)$, $\hat{G}_{nn}(t) = \varepsilon(t) + G_{nn}(t)$, where $|\varepsilon(t)| = \varepsilon/2$, $\arg \varepsilon(t) = \arg(\det G(t)/a(t))$, we have

$$|\det \hat{G}(t)| = |\det G(t)+a(t)\varepsilon(t)| = |\det G(t)|+|a(t)\varepsilon(t)| \geq \frac{\varepsilon}{2}|a(t)|.$$

Since $a^{-1} \in L_\infty$, the matrix function $\hat{G}$ also belongs to class L_∞. At the same time, $\|G - \hat{G}\| < \varepsilon$, which proves Lemma 3.6. ≡

THEOREM 3.18. If A, $B \in L_\infty$ and $AP + BQ$ is a Φ_+- or Φ_--operator, then A^{-1}, $B^{-1} \in L_\infty$.

Proof. Let $AP + BQ$ be a Φ_--operator. Due to Theorem 1.4, there exists an $\varepsilon > 0$ such that if $\|A - A_1\| < \varepsilon$, then $A_1P + BQ$ is also a Φ_--operator. In accordance with Lemma 3.6, we may assume that $A_1^{-1} \in L_\infty$. But then, along with operator $A_1P + BQ$, the operator $A_1^{-1}(A_1P + BQ) = P + A_1^{-1}BQ$ is also Fredholm. In view of Theorem 3.13, the matrix function $B^{-1}A_1$ inverse to $A_1^{-1}B$ belongs to the class L_∞. Therefore, B^{-1} $(= B^{-1}A_1A_1^{-1})$ also belongs to L_∞.

Furthermore, for $B \in L_\infty$, $Q + B^{-1}AP$ is a Φ_--operator together with $AP + BQ$. Using Theorem 3.13 again (that part related to operators of the kind $Q + GP$), we observe that $A^{-1}B \in L_\infty$. From this we get $A^{-1} \in L_\infty$.

If $AP + BQ$ is a Φ_+-operator, then arguing as above, we conclude that $P + A_1^{-1}BQ$ is a Φ_+-operator. In virtue of Lemma 3.5 and Theorem 3.1, $Q + (A_1^{-1}B)'P$ is then a Φ_--operator.

The further argument is trivial. ≡

THEOREM 3.19. Let A and B be $n \times n$ matrix function of class L_∞. The operator $AP + BQ$ is Fredholm in L_p^n if and only if $A^{-1} \in L_\infty$ and the matrix function $G = A^{-1}B$ is Φ-factorable in L_p. In this case the kernel dimension of the operator $AP + BQ$ is equal to the sum of positive partial indices of G, and the codimension of the image is opposite to the sum of negative partial indices.

Proof. The necessity of the condition $A^{-1} \in L_\infty$ follows from the preceding theorem. If this condition applies, then in view of the equality $AP + BQ = A(P + GQ)$, the operators $AP + BQ$ and $P + GQ$ are Fredholm only simultaneously and have the same defect numbers. To complete the proof, we have to apply Theorem 3.16. ≡

Notice that the following condition will also be necessary and sufficient for the Fredholmness of the operator $AP + BQ$ in the space L_p^n: $B^{-1} \in L_\infty$ and the matrix function $B^{-1}A$ is Φ-factorable from the right in L_p.

Theorems 3.16 and 3.16' serve as a chain link, with the help of which results and methods of the theory of linear bounded operators can be used for the study of problems related to Φ-factorization. In particular, the stability property of Fredholmness under small perturbations (Theorem 1.4) leads to the next result.

THEOREM 3.20. If the matrix function $G \in L_\infty$ is Φ-factorable from the left (right) in L_p, then, for some $\varepsilon > 0$, the matrix function F is also Φ-factorable from the left (right) in L_p, as soon as $\|F - G\| < \varepsilon$, where the total indices of F and G coincide.

THEOREM 3.21.

1) Let Γ and L be two simple smooth closed contours, and assume β to be a diffeomorphism of the contour L onto Γ. The matrix function $G \in L_\infty$ defined on Γ is Φ-factorable from the left (right) in L_p iff $G_0 = G \circ \beta$ defined on L is Φ-factorable in L_p from the left (right), if β remains the orientation, and from the right (left), if β changes the orientation. In the first case the total index of G_0 is equal to the total index

of G, and in the second case it is opposite to it.

2) Assume Γ to be a closed contour consisting of m connected curves Γ_j, and let G be a measurable bounded matrix function defined on Γ. By G_j, $j = 0,\dots,m-1$ we denote the matrix function defined on Γ_j and coinciding on it with G. The matrix function G is Φ-factorable from the left (right) in L_p if and only if the matrix function G_0 is Φ-factorable from the left (right) in L_p and G_j, $j = 1,\dots,m-1$, are Φ-factorable from the right (left) in L_p. In this case the total index of G is equal to the left (right) total index of G_0 minus the right (left) total indices of G_j, $j = 1,\dots,m-1$.

<u>Proof.</u> 1) We introduce the operator W mapping from $L_p^n(\Gamma)$ onto $L_p^n(L)$ by the rule $(W\varphi)(t) = \varphi(\beta(t))$. Clearly, W provides an isomorphism of the spaces $L_p^n(L)$ and $L_p^n(\Gamma)$. Therefore, $P + GQ$ is Fredholm iff the operator $W(P + GQ)W^{-1} = WPW^{-1} + WGQW^{-1} = WPW^{-1} + (WGW^{-1})(WQW^{-1}) = WPW^{-1} + G_0(WQW^{-1})$ is Fredholm.

If S_0 is the operator of singular integration along the contour L, then

$$((WSW^{-1} - S_0)\varphi)(t) = (WSW^{-1}\varphi)(t) - (S_0\varphi)(t) =$$

$$\frac{1}{\pi i}\int_\Gamma \varphi(\beta^{-1}(\zeta))\frac{d\zeta}{\zeta-\beta(t)} - \frac{1}{\pi i}\int_L \varphi(\tau)\frac{d\tau}{\tau-t}.$$

Now we assume that the mapping β preserves the orientation. Substituting $\zeta = \beta(\tau)$, $\tau \in L$ in the first integral, we get

$$((WSW^{-1}-S_0)\varphi)(t) = \frac{1}{\pi i}\int_L \varphi(\tau)\frac{\beta'(\tau)}{\beta(\tau)-\beta(t)}d\tau - \frac{1}{\pi i}\int_L \varphi(\tau)\frac{d\tau}{\tau-t}. \quad (3.22)$$

In this way, $WSW^{-1}-S_0$ is an integral operator with the kernel $\frac{1}{\pi i}\left(\frac{\beta'(\tau)}{\beta(\tau)-\beta(t)} - \frac{1}{\tau-t}\right)$. The compactness of such an operator was established in GRUDSKIǏ [1], GRUDSKIǏ, DYBIN [1][1)] using a result of CALDERON [1]. Then the operators $WPW^{-1}-P_0$ and $WQW^{-1}-Q_0$ are also

1) Note that if β' satisfies a Hölder condition, then the compactness of $WSW^{-1}-S_0$ simply results from the weak polarity of the kernel of this integral operator (cf. LITVINCHUK [3]).

compact, where $P_0 = \frac{1}{2}(I + S_0)$, $Q_0 = \frac{1}{2}(I - S_0)$. Therefore, the operators $WPW^{-1} + G_0WQW^{-1}$ and $P_0 + G_0Q_0$ differ from each other by a compact summand, thus, they are Fredholm only simultaneously and their indices coincide (Theorem 1.5).

In case the mapping β changes the orientation, the substitution $\zeta = \beta(\tau)$ leads to the change of sign preceding the first integral into equality (3.22). Consequently, already the operator $WSW^{-1}-S_0$ will be compact. From this we deduce the compactness of the operators $WPW^{-1}-Q_0$ and $WQW^{-1}-P_0$ in $L_p^n(L)$. Thus, the operators $WPW^{-1} + G_0WQW^{-1}$ and $Q_0 + G_0P_0$ are equivalent in the sense of Fredholmness. In view of Corollary 3.6, the Fredholmness of the last operator is equivalent to the Φ-factorability of the matrix function G_0 from the right, and the index of this operator is opposite to the (right) total index of G_0. To prove that part of assertion 1) which is related to the left Φ-factorization of G, it remains to apply Theorem 3.16 to the operator $P + GQ$. The case of right Φ-factorization can be treated analogously.

Proving assertion 2), we restrict ourselves to the case of left Φ-factorization of G. In the space $L_p^n(\Gamma)$ we introduce the operators Π_j $(j = 0,\dots,m-1)$ of multiplication by a function equal to 1 on $\Gamma_{(j)}$ and to zero on the other components of the contour Γ. Evidently, Π_j is the projector of $L_p^n(\Gamma)$ on the subspace of vector functions equal to zero on all components of Γ with the exception of $\Gamma_{(j)}$. The space $L_p^n(\Gamma)$ is the direct sum of the images of operators Π_j. The projectors Π_j commute with the operator of multiplication by the matrix function G and, for $i \neq j$, $\Pi_i S \Pi_j$ is an integral operator with continuous kernel and therefore compact. Thus, Corollary 1.4 can be applied, owing to which the Fredholmness of the operator $P + GQ$ is equivalent to the Fredholmness of all operators $(\Pi_j P + G_j \Pi_j Q)|_{\operatorname{im} \Pi_j}$, and the index of the operator $P + GQ$ equals the sum of indices of these operators.

Since the contour Γ_0 is oriented anticlockwise, but the other $\Gamma_{(j)}$'s clockwise, $\Pi_j P|_{\mathrm{im}\,\Pi_j}$ and $\Pi_j Q|_{\mathrm{im}\,\Pi_j}$ are just the operators P and Q for the contour $\Gamma_{(0)}$ in case $j = 0$ and the operators Q and P, respectively, for the contours $\Gamma_{(j)}$ $(j = 1,\dots,m-1)$. In order to complete the proof of Theorem 3.21, it remains to apply Theorem 3.16 to the operator $P + G_0 Q$ and Theorem 3.16' to the operators $Q + G_j P$ $(j = 1,\dots,m-1)$. ≡

For arbitrary $G \in L_\infty$, $P + GQ$ is an operator of local type. Indeed, for an arbitrary scalar function a, we have $(P + GQ)aI - a(P + GQ) = (Pa - aP) + G(Qa - aQ) = (PaQ - QaP) + G(QaP - PaQ) = (I-G)(PaQ-QaP)$. In the function a is rational and has no poles on Γ, then from Theorem 1.35 we obtain the compactness (and even the finite rank property) of the operators PaQ and QaP and, along with them, of the difference $(P + GQ)aI - a(P + GQ)$.

Summarizing, the local principle can be applied to operators of type $P + GQ$. In order to formulate the result corresponding to Theorem 1.14 exclusively in terms of G, we introduce the following definition.

DEFINITION 3.2. The matrix functions G_1 and G_2 will be called *ε-locally equivalent* at the point $t_0 \in \Gamma$, if for all $\varepsilon > 0$, there exists an arc $\gamma(\ni t_0)$ such that $\|G_1(t) - G_2(t)\| < \varepsilon$ a.e. on γ.

THEOREM 3.22 (ε-local principle). Let G be a matrix function defined on Γ such that for all points $t \in \Gamma$ there exists a matrix function $G_t \in L_\infty$ which is Φ-factorable in L_p and ε-locally equivalent to G at the point t. Then G admits a Φ-factorization in L_p.

Proof. Obviously, under the conditions imposed on G, the relation $G \in L_\infty$ holds. We now introduce the operators $A = P + GQ$ and $A_t = P + G_t Q$. By assumption and according to Theorem 3.16, the operators A_t are Fredholm (and, thus, also locally Fredholm at the point t) for all $t \in \Gamma$. As $P + GQ$ is an operator of local type, its Fredholmness and, thus, the Φ-factorability of the matrix function G will be proved, as soon as we clarify that the operators A and A_t are equivalent at the point $t(\in \Gamma)$.

Taking a number $\varepsilon > 0$ arbitrarily, we choose an arc $\gamma(\ni t)$ in such a way that $\|G(\tau) - G_t(\tau)\| < \varepsilon\|Q\|^{-1}$ a.e. on γ (this is possible owing to the ε-local equivalence of G and G_t at t). By a we denote a function continuous on Γ which is equal to zero outside γ, takes values from the interval $[0,1]$ on γ, and equals 1 in some neighbourhood of the point t. Then we have

$$|a(A-A_t)| \leq \|a(A-A_t)\| = \|a(G-G_t)Q\| \leq \|a(G-G_t)\|\cdot\|Q\|$$

$$= \|Q\| \operatorname*{ess\,sup}_{\tau\in\gamma}\|a(\tau)(G(\tau)-G_t(\tau))\| \leq \|Q\|\cdot\operatorname*{ess\,sup}_{\tau\in\Gamma}\|G(\tau)-G_t(\tau)\| < \varepsilon .$$

Together with A and A_t, the operator $A - A_t$ is of local type. As mentioned in Section 1.1, for any function f continuous on Γ, the operator $f(A-A_t) - (A-A_t)fI$ is compact. In particular, we have $|a(A-A_t)| = |(A-A_t)aI|$ and, thus, $|(A-A_t)aI| < \varepsilon$.

Due to Definition 1.10, the operators A and A_t are equivalent at the point t, which proves Theorem 3.22. ≡

A simple example of matrix functions ε-locally equivalent at a certain point t yield those matrix functions which coincide a.e. in a neighbourhood of this point. Now we mention the ε-local principle corresponding to this special case.

<u>COROLLARY 3.9.</u> Let the contour Γ be covered by the system $\{\gamma_j\}$ of open arcs. If one can, for given G, specify a collection of matrix functions $G_j \in L_\infty$ which are Φ-factorable in L_p such that $G_j(\tau) = G(\tau)$ a.e. on γ_j, then G is also Φ-factorable in L_p.

Sufficient conditions of Φ-factorability obtained via the ε-local principle will be derived in Chapter 8.

3.7. COMMENTS

The material explained in Section 3.1 – 3.5 is, in principle, taken from SPITKOVSKIĬ [13,14]. The results of Section 3.1 generalize the classical solution scheme for the Riemann boundary value problem based on factorization of its matrix coefficient (VEKUA [4], GAKHOV [3], MUSKHELISHVILI [1], KHVEDELIDZE [4]).

Assertion 3) of Theorem 3.1, under different assumptions on G ensuring, however, its factorability, was mentioned in GOHBERG [1], SIMONENKO [1,4]. Its validity for an arbitrary function of class L_∞ defined on the circle was, for $p = 2$, proved in COBURN [1] and, for arbitrary $p \in (1,\infty)$, in GOHBERG, KRUPNIK [3]. The proof conducted here is essentially that of COBURN [1].

In connection with Theorem 3.2 we still refer to MEUNARGIYA [3], where the general solution of problem (3.1) is obtained in a form containing the values of the unknown vector function and its derivatives explicitly.

Arguments close to those used in the proof of Theorem 3.4 the reader can find in GOHBERG, FELDMAN [3] as well as in IDEMEN [2], where the case of rational matrix functions given on the circle is examined, but also in HEINIG [1], where a similar approach was used for the inversion of operators $A = a + \sum_{j=1}^{N} b_j S c_j$ generalizing (from case $N = 1$) operators of the type $AP + BG$ (see also HEINIG/ROST [1]).

Theorems 3.5, 3.15, 3.15' are explained in SPITKOVSKIĬ [6]. The original variant of Theorems 3.15 and 3.15' can be found in SPITKOVSKIĬ [2]. In the same paper the interpolation property of Φ-factorability (Theorem 3.9) was mentioned. In connection with this, we refer to the article KUCHMENT [1], in which the interpolation property of Fredholmness with the same index for an arbitrary linear operator acting and bounded in an interpolation family of Banach spaces was proved.

The notion of Φ-factorization of a bounded measurable matrix function in L_p is due to I.B. SIMONENKO (for $p = 2$, in SIMONENKO [3,4], for the general case, in SIMONENKO [6]. Lemma 3.5 and Theorems 3.16, 3.21, 3.22 are also due to him.

In the case $n = 1$, the notion of Φ-factorization, in principle can be found also in DANILYUK [1,2] and WIDOM [1].

The term "Φ-factorization" was proposed in SPITKOVSKIĬ [2,13] and motivated by Theorem 3.16, according to which the operator $P + GQ$ is Fredholm, i.e., it is a Φ-operator if and only if the matrix function G is Φ-factorable.

Notice that operators of the type $T_P(F)$ and $T_Q(F)$ introduced at the beginning of Section 3.6 are known in the literature as Toeplitz operators, and the matrix function F is usually called the symbol of the corresponding operator. In this way, Lemma 3.5 allows us to reformulate all the assertions about operators of the kind $P + GQ$ and $Q + G'P$ in terms of Toeplitz operators and vice versa. The results of some authors are formulated for Toeplitz operators. Henceforth, we shall refer to such results having in mind their analogues for the operators $P + GQ$ and $Q + G'P$.

Theorems 3.6, 3.6' and 3.11 formulated in SPITKOVSKIĬ [6] are a natural extension of Theorem 3.16 to the case of a rectifiable contour and a matrix function not necessarily bounded, whereas Theorems 3.13 and 3.14 are generalizations of corresponding results of SIMONENKO [4,6] from the case $G \in L_\infty$, $\Gamma \in \mathfrak{R}$ renouncing the restriction $G \in L_\infty$. Note that the requirement $\Gamma \in \mathfrak{R}$ can here also be omitted (see SPITKOVSKIĬ [14]). Actually, Theorem 3.13 in the scalar case with $p = 2$ was proved in WIDOM [1] and, for real-valued functions, in HARTMAN, WINTNER [1].

The boundedness criterion (3.21) for the operator S in the space L_p with weight ϱ has been derived for the circle in HUNT, MUCKENHOUPT, WHEEDEN [1] and, for a smooth contour, in KOKILASHVILI [1]. In the same paper the sufficiency of condition (3.21) in case of piecewise smooth curves was mentioned, whereas the necessity in case of piecewise Lyapunov curves was established in KRUPNIK [5].

Theorem 3.17 (on the Φ-factorability of power functions) is an auxiliary result for the theory of Φ-factorization of piecewise continuous functions, with which we shall be concerned in Chapter 5. Appropriate references will also be given there.

Lemmas 3.4, 3.6 and Theorems 3.18, 3.19 are borrowed from KRUPNIK [3]. Finally, we insert some remarks on results not mentioned in the present chapter, but being directly related to it.

First of all, we refer to NYAGA [1] and DZHVARSHEĬSHVILI [1], where the study of the scalar Riemann boundary value problem on a rectifiable contour Γ in wider classes than $L_1^{\pm}$ has been initiated (in particular, in classes of functions which can be represented as a differentiated Cauchy integral). Under sufficiently general assumptions on the coefficient G, the scalar Riemann boundary value problem in classes different from L_p was studied in ANTONTSEV [1], KATS [1-3]. The papers of DANILOV [1], ZAPUSKALOVA, KATS [1], KATS [3] and SEIFULLAEV [1] are devoted to the investigation of new effects occurring in the theory of the scalar Riemann boundary value problem under weakened conditions on the contour Γ.
Operators of the type $T_{\mathcal{P}}(F) = \mathcal{P}F|_{\mathrm{im}\,\mathcal{P}}$, where $\mathcal{P}$ is an arbitrary (bounded) projector in some Banach space $\mathfrak{B}$ and F is an arbitrary operator from $[\mathfrak{B}]$, are an abstract analogue of Toeplitz operators. In particular, Wiener-Hopf operators, i.e. integral operators on the half-line with a kernel depending on the difference of the arguments can be represented in the form $T_{\mathcal{P}}(F)$. Therefore, the operators $T_{\mathcal{P}}(F)$ are referred to as generalized Toeplitz operators or generalized Wiener-Hopf operators. The general theory of such operators was created in DEVINATZ, SHINBROT [1]. Especially, (under some additional restrictions) the validity of the abstract analogue of Theorem 3.13 is shown there, namely, the necessity of the invertibility condition of the operator F for $T_{\mathcal{P}}(F)$ to be a Φ_+- or a Φ_--operator. For more results in this direction we refer to the recent monograph of SPECK [1]. Incidentally, note that the theory of systems of Wiener-Hopf equations is equivalent to the theory of operators $P + GQ$ and $Q + G'P$. This equivalence can be established with the help of the Fourier transform; for more detail concerning this topic, see GOHBERG, KREĬN [3], GOHBERG, FELDMAN [3].

In Ch. 8 of the book GOHBERG, FELDMAN [3] also other types of systems of integral equations are enumerated, the theory of which may be erected with the aid of results related to operators studied in the present chapter. In particular, the examination of operators of the form $AP + BQ$ has been suggested by the study of systems of so-called paired integral equations. A variety of types of integral eqations and problems of mathematical physics reducible to the Riemann boundary value problem were discussed in the monograph by GAKHOV and CHERSKIǏ [1] (see also Chapters 1-7 in GOHBERG, FELDMAN [3]).

Variants of the local principle (Theorem 3.22) can be found in CLANCEY, GOHBERG [1] and CLANCEY, GOSSELIN [1]. In CLANCEY, GOHBERG [1] the notion of local factorability of matrix functions was introduced (a matrix function $G \in L_\infty$ defined on Γ is called locally factorable at the point $t_0 \in \Gamma$, if there exist an open arc $\gamma \subseteq \Gamma$ containing this point, domains $\Omega^\pm \subseteq \mathfrak{D}^\pm$ the boundary of which contains γ, and matrix functions $A_\pm \in E_p(\Omega^\pm)$ such that $A_\pm^{-1} \in E_q(\Omega^\pm)$, the operator $A_+SA_+^{-1}$ is bounded in $L_p^n(\gamma)$, and $G(t) = A_+(t)A_-(T)$ a.e. on γ). Furthermore, there was proved that the Φ-factorability of $G \in L_\infty$ is equivalent to its local factorability at every point of the contour. In CLANCEY, GOHBERG [2] it is established that in the scalar case for a Lyapunov contour Γ, the local factorability of the function G at the point t_0 implies the local Fredholmness of the operator $P + GQ$ at this point. For $p = 2$, the converse assertion is also true.

In CLANCEY, GOHBERG [2] it was mentioned that for $p \neq 2$ the question of validity of the converse assertion is still open. This problem has been solved by I.B. SIMONENKO, who in a series of papers (SIMONENKO [7-11]) successively renounced the restrictions on p, n and Γ and, finally, proved the equivalence of local Fredholmness of the operator $P + GQ$ and local factorability of the $n \times n$ matrix function G (at the same point) for arbitrary $p \in (1,\infty)$, natural n and $\Gamma \in \mathfrak{R}$.

By SHNEǏBERG [1,2] it was recognized that the Fredholmness of an operator A in one space of an analytical scale implies its Fredholmness with the same index in any sufficiently close space of this scale in which the operator A is bounded. An immediate consequence of this result is that all components of the domain of Φ-factorability of a matrix function $G \in L_\infty$ defined on a contour of class $\mathcal{R}$ are open, which was proved in SPITKOVSKIĬ [9]. For a scalar function defined on a smooth or piecewise Lyapunov contour, the corresponding result follows from Corollary 3.7 as well as the properties of condition (3.21) discussed in MIAMEE, SALEHI [1]. In this case it appears that for an unbounded function G the set $\Phi(G)$ is not necessarily open (but $\Phi_k(G)$ is open in $\mathfrak{F}_k(G)$, $k \in \mathbb{Z}$).

CHAPTER 4. Φ-FACTORIZATION OF TRIANGULAR MATRIX FUNCTIONS AND REDUCIBLE TO THEM

If it is known that the scalar function G can be factored in L_p with index $\varkappa$, then its factorization may be constructed effectively. Let us explain this in more detail, assuming for the sake of simplicity the contour Γ to be connected. From the representation (3.20) with suitable selection of $\arg G(t)$, we obtain

$$\ln t^{-\varkappa}G(t) = \ln G_+(t) + \ln G_-(t) , \tag{4.1}$$

where the functions $\ln G_\pm$ are determined up to a constant summand.

Starting from the obvious inequality $\ln x < x$ $(x > 0)$, it is easy to deduce the relations

$$-(\tfrac{r}{q})^r|G_+(z)|^{-q} < (\ln|G_+(z)|)^r < (\tfrac{r}{p})^r|G_+(z)|^p$$

for all finite r.

Since $G_+ \in E_p^+$ and $G_+^{-1} \in E_q^+$, the last inequality implies that for the function $f(z) = \ln|G_+(z)|$, for all finite r, condition 3) of Definition 1.16 is fulfilled. Under sufficiently general assumptions on the contour Γ one can conclude that this condition is also fulfilled for the analytic function $\ln G_+$, the real part of which is $\ln|G_+|$. In other words, the function $\ln G_+$ belongs to the classes E_r^+, $r < \infty$.

By analogy, it can be established that $\ln G_- \in E_r^-$, $r < \infty$. Therefore, with regard to Theorem 1.28 and equation (1.19), from equation (4.1) we conclude that

$$\begin{aligned} G_+(z) &= C^{-1} \exp\Big(\frac{1}{2\pi i}\int_\Gamma \ln \tau^{-\varkappa}G(\tau)\,\frac{d\tau}{\tau-z}\Big), \quad z \in \mathcal{D}^+, \\ G_-(z) &= C \exp\Big(-\frac{1}{2\pi i}\int_\Gamma \ln \tau^{-\varkappa}G(\tau)\,\frac{d\tau}{\tau-z}\Big), \quad z \in \mathcal{D}^-, \end{aligned} \tag{4.2}$$

where C is a constant different from zero. The existence of a multiplier C is caused by that degree of arbitrariness with which, according to Theorem 2.2, the factorization is determined. Obviously, $C = G_-(\infty)$.

Our considerations essentially use the commutativity of multiplication. Thus, they fail to be true when turning to the matrix case. Indeed, trying to use formulae (4.2) for $n > 1$ (at least in case of partial indices and $C = I$, we obtain $G_+G_- = \exp(P \ln G) \exp(Q \ln G)$. Thus, the equation $G_+G_- = G$ is valid under the following necessary and sufficient condition:

$$\exp(P \ln G) \exp(Q \ln G) = \exp(P \ln G + Q \ln G) , \quad (4.3)$$

which is satisfied only in some isolated cases.

Up to now only a comparatively small number of examples is known, in which the factorization problem for a matrix function is solved with the same degree of effectiveness as in the scalar case, i.e., with the help of a finite number of algebraic operations and using the operators P and Q. Diagonal matrix functions yield a trivial example. For their factorization it suffices to factorize every diagonal element.

In this chapter we shall examine the factorization problem for triangular matrix functions. It turns out that this task may be solved effectively in the sense pointed out above. We also study some cases reducible to this problem. The contour Γ is assumed to belong to the class $\mathfrak{R}$, and the matrix function G defined on it is supposed to be an element of class L_∞.

In Section 4.1 a criterion of Φ-factorability of block-triangular matrix functions, for which the defect numbers of the corresponding Riemann problem are finite, will be established. From the mentioned criterion we deduce that the condition of Φ-factorability of the diagonal elements of a triangular matrix function is sufficient for Φ-factorability of the matrix function itself. It is shown that this condition is, generally speaking, not necessary. However, we succeed in specifying not too hard restrictions on the diagonal elements, under which their Φ-factorability is equivalent to the Φ-factorability of the triangular matrix function. In particular, these restrictions are fulfilled for arbitrary piecewise continuous functions, a fact which will permit us in Chapter 5 to deduce the criterion of Φ-factorability of piecewise continuous functions from the corresponding criterion for the scalar case.

In Section 4.2 we shall deal with triangular matrix functions with Φ-factorable diagonal elements. Here we investigate relationships between the partial indices of their diagonal elements. Moreover, an algorithm for constructing the factorization will be presented. It appears that in the special case when the indices of the diagonal elements of a triangular matrix function are equal to zero, its partial indices are also equal to zero and, therefore, they do not depend on the elements outside the diagonal. In the general case this is not so.

For the study of the general case, a partial ordering over the set of tuples of n integers is introduced, and it is established that the set of partial indices of a triangular matrix function is majorized by the set of indices k_j of the diagonal elements in the sense of the ordering introduced. In doing so, the factorization itself is constructed effectively. Incidentally, remark that the partial ordering introduced here will also be used in Chapter 6 in connection with the study of the stability of partial indices of arbitrary matrix functions belonging to the class L_∞.

Furthermore, in Section 4.2 we shall formulate the criterion for coincidence of the partial indices with the numbers k_j, the indices of the diagonal elements. Such a coincidence holds, in particular, if the numbers k_j $(j = 1,\dots,n)$ differ from each other at most by one.

In Section 4.3 it will be shown that for second-order triangular matrix functions the problem of calculating the partial indices may be solved to the very end.

In Section 4.4 it will be established that functionally commutable matrix functions can be reduced to block-diagonal form with triangular blocks by a constant similarity transformation. In this case, Φ-factorability of the initial matrix function G is equivalent to Φ-factorability of all diagonal elements of the obtained matrix function F, and the partial indices of G are equal to the indices of the diagonal elements of F. For second-order matrix functions, a simple criterion of functional commutativity will be presented, which permits us to reduce the question of Φ-factorability of such matrix functions to the corresponding question for a pair of scalar functions which are effectively constructed by G.

As usually, the last section contains comments.

4.1. EXISTENCE PROBLEMS OF A Φ-FACTORIZATION OF TRIANGULAR MATRIX FUNCTIONS

We shall be interested in Φ-factorability both of lower and upper triangular matrix functions, where the factorization itself may also be left and right. In connection with this, mention that the multiplication from the left and from the right by the counteridentity matrix $[\delta_{i,n-j}]_1^n$ transforms upper (lower) triangular matrix functions into lower (upper) ones taking no influence on the left and right Φ-factorability and on the values of the partial indices. Moreover, the transposition operation also transfers upper (lower) triangular matrix functions into lower (upper) ones and, according to Theorem 3.7', a left (right) Φ-factorization in L_p changes into a right (left) Φ-factorization in L_q. These considerations allow us to transfer results obtained for a left Φ-factorization of lower triangular matrix functions to the remaining cases.

As an auxiliary step, examine a block-triangular matrix function G of the form

$$G = \begin{pmatrix} G_1 & 0 \\ H & G_2 \end{pmatrix}. \tag{4.4}$$

By n_j we denote the orders of the diagonal blocks G_j of this matrix function. Let $\mathfrak{B}_1$ ($\mathfrak{B}_2$, respectively) be the subspace of vector functions $f = \sum_{k=1}^{n} f_k e_k$ from the space $\mathfrak{B} = (\mathring{L}_p^-)^n$ such that $f_k = 0$ for $k = n_1+1,\dots,n$ ($k = 1,\dots,n_1$, respectively). Then of course, $\mathfrak{B} = \mathfrak{B}_1 \dot{+} \mathfrak{B}_2$, and in the representation (1.1) of the operator $A = T_Q(G)$

we have: $A_{jj} = T_Q(G_j)$ $(j = 1,2)$, $A_{12} = 0$, $A_{21} = QH$.

Consequently, Theorem 1.12 may be used, due to which we obtain the following criterion of Fredholmness of the operator $T_Q(G)$, i.e. of Φ-factorability of the matrix function G .

THEOREM 4.1. A matrix function G of the form (4.4) is Φ-factorable in L_p if and only if the following conditions are satisfied:

1) the codimension β_1 of the lineal $\operatorname{im} T_Q(G_1)$ in $(\mathring{L}_p^-)^{n_1}$ is finite,
2) the dimension α_2 of the kernel of the operator $T_Q(G_2)$ is finite,
3) the dimension γ_1 of the lineal $\{\varphi \in \ker T_Q(G_1) : QH\varphi \in \operatorname{im} T_Q(G_2)\}$ is finite,
4) the codimension γ_2 of the lineal $\operatorname{im} T_Q(G_2) + QH \ker T_Q(G_1)$ in $(\mathring{L}_p^-)^{n_2}$ is finite.

If conditions 1) - 4) are satisfied, then for the defect numbers of problem (3.1) with the matrix coefficient G the formulae

$$\alpha = \alpha_2 + \gamma_1 , \qquad \beta = \beta_1 + \gamma_2 \tag{4.5}$$

are valid. Applying Corollaries 1.3 and 1.4 to the operator $T_Q(G)$, from Theorem 1.12 we obtain the following results.

COROLLARY 4.1. If one of the diagonal blocks G_1 and G_2 of the matrix function G of the form (4.4) is Φ-factorable, then the Φ-factorability of its second diagonal block is a necessary and sufficient condition for the Φ-factorability of the matrix function G itself. The defect numbers α , β and α_j , β_j of problems (3.1) with matrix coefficients G and G_j $(j = 1,2)$ respectively, are connected by the relations

$$\alpha = \alpha_1 + \alpha_2 - \gamma , \qquad \beta = \beta_1 + \beta_2 - \gamma , \tag{4.6}$$

where γ is the dimension of the direct complement to $\operatorname{im} T_Q(G_2) \cap QH \ker T_Q(G_1)$ in $QH \ker T_Q(G_1)$.

COROLLARY 4.2. A block-diagonal matrix function G is Φ-factorable iff all its diagonal blocks are Φ-factorable.

The necessary and sufficient condition of Φ-factorability of block-diagonal matrix functions formulated in Corollary 4.2 remains sufficient also in the general case of block-triangular matrix functions. To be more exact, the following result is valid.

THEOREM 4.2. Let the matrix function G be block-triangular, and let G_{jj} $(j = 1,\dots,k)$ be its diagonal blocks. Then the Φ-factorability of all matrix functions G_{jj} implies the Φ-factorability of G. Moreover, for the defect numbers α and β of problem (3.1) with matrix coefficient G, the following estimations are true:

$$\alpha \leq \sum_{j=1}^{k} \alpha_j , \quad \beta \leq \sum_{j=1}^{k} \beta_j ,$$

where α_j and β_j are the defect numbers of problem (3.1) with the matrix coefficient G_{jj} $(j = 1,\dots,k)$.

Proof. In view of the remarks made at the beginning of the present section, it suffices to examine the case of a lower block-triangular matrix function G. We prove the assertion for this case by induction on the number k of diagonal blocks of G. For $k = 1$, the statement to be proved is obvious. Supposing the statement to be correct for $k \leq k_0$, we shall study a block-triangular matrix function G for which $k = k_0 + 1$ and the diagonal blocks of which are Φ-factorable. We represent this matrix function in the form (4.4), setting $G_1 = G_{11}$. Then the matrix function G_2 is also block-triangular and we have for it $k = k_0$. By induction hypothesis, we can assert that G_2 is Φ-factorable and the defect numbers α' and β' of the corresponding Riemann problem satisfy the inequalities $\alpha' \leq \sum_{j=2}^{k} \alpha_j$ and $\beta' \leq \sum_{j=2}^{k} \beta_j$. Since, by assumption, the block G_1 is Φ-factorable, on the strength of Corollary 4.1, one can claim that the complete matrix function G is Φ-factorable and, moreover, $\alpha = \alpha_1 + \alpha' - \gamma \leq \alpha_1 + \alpha' \leq \sum_{j=1}^{k} \alpha_j$, $\beta = \beta_1 + \beta' - \gamma \leq \beta_1 + \beta' \leq \sum_{j=1}^{k} \beta_j$.

Thus Theorem 4.2 has been proved. ≡

The next result is an immediate consequence of Theorem 4.2 in the case of one-dimensional diagonal blocks.

THEOREM 4.3. If all diagonal elements of the triangular matrix function G are Φ-factorable in L_p, then the matrix function G itself is Φ-factorable in L_p, too.

The sufficient conditions of Φ-factorability indicated in Theorem 4.3 are not necessary. To verify that, we study the question of Φ-factorability of triangular matrix functions of a special form given on the circle.

THEOREM 4.4. Let G be a matrix function of the form $\begin{pmatrix} \omega & 0 \\ 1 & \overline{\omega} \end{pmatrix}$ defined on T, where $\omega \in L_\infty$. Then for the Φ-factorability of G in L_2 it is necessary and sufficient for $T_Q(\omega)$ to be a Φ_--operator. In this case the partial indices are equal to $\pm\, \beta(T_Q(\omega))$.

Proof. Condition 1) of Theorem 4.1 shows that for the Φ-factorability of the matrix function under study, it is necessary for $T_Q(\omega)$ to be a Φ_--operator. Introducing in the space L_2^- a scalar product by the formula $(f,g) = \int_{-\pi}^{\pi} f(e^{i\theta})\overline{g(e^{i\theta})}d\theta$, we may regard the operator $T_Q(\omega)$ as the adjoint to $T_Q(\omega)$.[1] According to Theorem 1.7 and Corollary 1.2, to say that $T_Q(\omega)$ is a Φ_--operator amounts to saying that $T_Q(\overline{\omega})$ is a Φ_+-operator. Besides, the subspaces $\ker T_Q(\omega)$ and $\operatorname{im} T_Q(\overline{\omega})$ are orthogonal complements to each other. Taking into consideration that, for the considered matrix function, we have $H = I$, we find that in case of a Φ_--operator $T_Q(\omega)$ the conditions 1) - 4) of Theorem 4.1 are fulfilled, where $\gamma_1 = \gamma_2 = 0$ and $\alpha_2 = \beta_1 = \beta(T_Q(\omega))$. Consequently, the matrix function G is Φ-factorable in L_2 and, according to formulae (4.5), both defect numbers of the corresponding problem (3.1) coincide with the number $\beta(T_Q(\omega))$. Thus, the total index of G is equal to zero and, therefore, $\varkappa_1 = -\varkappa_2 = \alpha = \beta(T_Q(\omega))$. Theorem 4.4 is completely proved. ≡

If, in particular, the function ω belongs to the class H_∞, then $T_Q(\overline{\omega})$ coincides with the operator of multiplication by the function $\overline{\omega}$

---

[1] This can be established in the same way as the corresponding result for the operators QGQ and $PG'P$ from the proof of Lemma 3.5.

in the space L_2^-. Consequently, $\ker T_Q(\overline{\omega}) = \{0\}$ and $\operatorname{im} T_Q(\overline{\omega}) = \overline{\omega}L_2^-$. Hence, $T_Q(\overline{\omega})$ is a Φ_+-operator iff $\omega^{-1} \in L_\infty$. This condition, which is equivalent to the requirement concerning $T_Q(\omega)$ to be a Φ_--operator (and, thus, equivalent to the Φ-factorability of G), means only that the outer multiplier of the function ω is separated from zero, and does not impose any restrictions upon its inner multiplier. At the same time, for the factorability of the function ω (and $\overline{\omega}$) it is necessary (in view of Corollary 2.3) that its inner multiplier can be reduced to a finite Blaschke product. Choosing, for instance, the infinite Blaschke product as the function ω, we therefore obtain a triangular matrix function with non-factorable diagonal elements, which is Φ-factorable in L_2.

The example considered is based on the existence of such functions $\omega \in L_\infty$ for which $T_Q(\omega)$ is a Φ_-- or Φ_+-operator but not Fredholm. However, for a wide class of measurable bounded functions (including all piecewise continuous functions, as will be demonstrated in Chapter 5) such a situation is impossible. Let Φ denote the class of functions $\omega \in L_\infty$ for which $T_Q(\omega)$ is Fredholm. Then we have the following result

THEOREM 4.5. Let G be a triangular matrix function the diagonal elements of which (with the possible exception of one) belong to the class Φ. Then for the Φ-factorability of G in L_p it is necessary and sufficient that its diagonal elements be Φ-factorable in L_p.

Proof. The sufficiency follows from Theorem 4.3. In proving the necessity, we restrict ourselves to the case of a lower triangular matrix function G and use induction on its dimension n. In case $n = 1$, the statement is obvious. Supposing that it is true for $n = k$, we examine a lower triangular matrix function of $(k+1)$th order which is Φ-factorable in L_p. If at least k of its diagonal elements belong to the class Φ, then at least one of its left upper and right lower diagonal element is a member of this class. For the sake of definiteness, let the left upper element G_1 of the studied matrix function G be in class Φ.

Using Theorem 4.1 for $n_1 = 1$ and $n_2 = n-1$, we see that $T_Q(G_1)$ is a Φ_--operator. Because of the condition $G_1 \in \Phi$, we conclude from this that the operator is Fredholm, i.e., the function G_1 is Φ-factorable in L_p. With regard to Corollary 4.1, we thus obtain that the matrix function G_2 which is obtained from G by cancelling the first column and row can be factored in L_p. But G_2 is lower triangular, its order is equal to k and at least $k-1$ diagonal elements belong to the class Φ. Due to the induction hypothesis, the diagonal elements of the matrix function G_2 are Φ-factorable in L_p.
Therefore, the Φ-factorability of all diagonal elements of the matrix function G has been verified and, thus, Theorem 4.5 has been completely proved. Ξ

4.2. EFFECTIVE CONSTRUCTION OF A Φ-FACTORIZATION AND ESTIMATES FOR THE PARTIAL INDICES OF TRIANGULAR MATRIX FUNCTIONS WITH Φ-FACTORABLE DIAGONAL ELEMENTS

In accordance with Theorem 4.5, triangular matrix functions with Φ-factorable diagonal elements are Φ-factorable themselves. In this section we shall study the questions of calculating the partial indices of such matrix functions, their relationships with the indices of the diagonal elements and the effective construction of a factorization.

THEOREM 4.6. Let the diagonal elements of the upper (lower) triangular matrix function G be Φ-factorable in L_p, and let their indices be equal to zero. Then G is Φ-factorable in L_p, the partial indices of G are equal to zero, and the factorization factors $G_\pm$ may be chosen as upper (lower) triangular. The determination of the elements of the matrix functions $G_\pm$ leads to the solution of a finite number of scalar Riemann problems the coefficients of which are the diagonal elements of G.

Proof. Let us examine the case of a lower triangular matrix function. We try to find lower triangular matrix functions $F_\pm \in L_p^\pm$ satisfying the relation $F_+ = GF_-$, which may be elementwise rewritten in the form

$$f_{ij}^{+} - G_{ii}f_{ij}^{-} = \sum_{k=j}^{i-1} G_{ik}f_{kj}^{-} \ , \quad i \geq j \ . \tag{4.7}$$

We divide the equations of system (4.7) into groups, assigning to the l th group those of them for which $i-j = l$ $(l=0,\ldots,n-1)$. Equations of the zero group are of the form $f_{ii}^{+} = G_{ii}f_{ii}^{-}$ $(i=1,\ldots,n)$ and, under the condition $f_{ii}(\infty) = 1$, they possess a unique solution in view of the condition of factorability of the functions G_{ii} and since their indices are equal to zero.

If the solutions of all groups of equations (4.7) with numbers k less than the given l are already found, then the right-hand sides of the equations of the l th group are known (since condition $j \leq k < i$ means that $k-j < i-j = l$), moreover, they belong to L_p (because $G_{ik} \in L_\infty$ and $f_{kj}^{-} \in L_p$).

Considering, for $l > 0$, the equations of the l th group as scalar Riemann problems with respect to $f_{ij}^{+} \in L_p^{+}$ and $f_{ij}^{-} \in \overset{\circ}{L}{}_p^{-}$, we recognize, with regard to the conditions imposed upon G_{ii}, that all of them also may be solved and even in a unique manner. Consequently, with the help of mathematical induction, the solvability of system (4.7) has been proved, where on condition that $F_{-}(\infty) = I$, the solution is unique.

Analogously, one can prove the existence (and uniqueness) of upper triangular matrix functions $\Phi_{\pm} \in L_q^{\pm}$ such that $G'\Phi_{+} = \Phi_{-}$, $\Phi_{-}(\infty) = I$. We consider the matrix function $Z = \Phi_{+}'F_{+}$. On the one hand, $\Phi_{+} \in L_q^{+}$ and $F_{+} \in L_p^{+}$, so that $Z \in L_1^{+}$. On the other hand, $\Phi_{+}'F_{+} = \Phi_{+}'GF_{-} = \Phi_{-}'F_{-} \in L_1^{-}$, since $\Phi_{-} \in L_q^{-}$ and $F_{-} \in L_p^{-}$.

According to Theorem 1.28, the matrix function Z is constant. Moreover, $F_{-}(\infty) = \Phi_{-}(\infty) = I$, thus $Z = I$. Consequently, $F_{-}^{-1} = \Phi_{-}' \in L_q^{-}$, $F_{+}^{-1} = \Phi_{+}' \in L_q^{+}$. Hence it is clear that the representation $G = G_{+}G_{-}$ with $G_{+} = F_{+} = (\Phi_{+}')^{-1}$, $G_{-} = F_{-}^{-1} = \Phi_{-}'$ is a factorization of G in L_p with zero partial indices. According to Theorem 4.3, G is Φ-factorable in L_p, so that every factorization of it, including the factorization we have constructed, is a Φ-factorization (Theorem 3.8). Thus the proof is complete. $\equiv$

Therefore, if the indices of the diagonal elements of a triangular matrix function are equal to zero, then its partial indices do not depend on the elements outside the diagonal. As we shall see later on, in the general case this is not true. In order to study the question of possible values of partial indices of a triangular matrix function with fixed diagonal elements, we must introduce a partial ordering on the set $\mathbb{Z}^n$ of tuples (k_j) of n integers.

<u>DEFINITION 4.1.</u> Let us say that the tuple (k_j) majorizes the tuple (k'_j) (written $(k_j) \succ (k'_j)$), iff for all integers k, the relations

$$\sum_{j=1}^{n} \max\{k_j-k,0\} \geq \sum_{j=1}^{n} \max\{k'_j-k,0\} , \tag{4.8}$$

$$\sum_{j=1}^{n} \min\{k_j-k,0\} \leq \sum_{j=1}^{n} \min\{k'_j-k,0\} , \tag{4.8'}$$

are satisfied.

It is obvious that the relation $\succ$ is reflexive and transitive, i.e., it is indeed a partial ordering. We intend to note some more properties of the relation $\succ$, which are important in what follows.

<u>LEMMA 4.1.</u> 1) If $(k_j) \succ (k'_j)$, then the inequalities

$$\max_j k_j \geq \max_j k'_j , \quad \min_j k_j \leq \min_j k'_j , \tag{4.9}$$

$\sum_{j=1}^{n} k_j = \sum_{j=1}^{n} k'_j$ are valid.

2) If the tuples (k_j) , $(k'_j) \in \mathbb{Z}^n$ have the property that (4.8), (4.8') are satisfied for all integer k from the interval $[\min_j k_j , \max_j k_j]$, then $(k_j) \succ (k'_j)$.

3) The relations $(k_j) \succ (k'_j)$ and $(k'_j) \succ (k_j)$ hold simultaneously iff the tuple (k'_j) is some permutation of the tuple (k_j) .

<u>Proof.</u> First of all, we intend to explain what do conditions (4.8) and (4.8') mean for some concrete values of k . If $k = \max k_j$, then $\max\{k_j-k,0\} = 0$ for all $j=1,\ldots,n$. Consequently, inequality (4.8) with this choice of k is equivalent to $\sum_{j=1}^{n} \max\{k'_j-k,0\} \leq 0$. In turn, the latter inequality is satisfied if and only if $\max\{k'_j-k,0\} = 0$ $(j=1,\ldots,n)$, i.e. if $\max_j k'_j \leq \max_j k_j$. In this way, condition (4.8)

with $k = \max_j k_j$ is equivalent to $\max_j k'_j \leq \max_j k_j$.

If the latter inequality is fulfilled, then for $k \geq \max k_j$, the equalities

$$\min\{k_j-k,0\} = k_j-k \ , \quad \min\{k'_j-k,0\} = k'_j -k$$

are true and, therefore, to say that condition (4.8') is valid amounts to saying that $\sum_{j=1}^{n}(k_j-k) \leq \sum_{j=1}^{n}(k'_j-k)$. Obviously, the inequality obtained is satisfied iff

$$\sum_{j=1}^{n} k_j \leq \sum_{j=1}^{n} k'_j \ .$$

Analogously, one can readily verify that, for $k = \min k_j$, condition (4.8') is satisfied iff $\min k_j \leq \min k'_j$. When the latter inequality is fulfilled, then, for $k \leq \min k_j$, condition (4.8) is equivalent to $\sum_{j=1}^{n} k_j \geq \sum_{j=1}^{n} k'_j$.

Thus, if the conditions (4.8) and (4.8') are fulfilled for $k = \min k_j$ and $k = \max k_j$, then relations (4.9) are true, which proves statement 1) of the lemma.

Furthermore, if the relations (4.9) are valid, it follows from what was proved above that conditions (4.8) and (4.8') are satisfied for $k \leq \min k_j$ and $k \geq \max k_j$. From this we conclude the correctness of assertion 2).

Now we proceed to the last statement of the lemma. Clearly, a change of order of the elements in the tuples (k_j) , (k'_j) does not influence on the values of the sums occurring in inequalities (4.8) and (4.8'). Consequently, every tuple $(k_j) \in \mathbb{Z}^n$ majorizes any of its permutations and is majorized by it. The converse statement we shall prove by induction on n . For $n = 1$, the statement results from relations (4.9). Supposing that the statement has been proved for $n = l$, we now consider the tuples $(k_j)_{j=1}^{l+1}$ and $(k'_j)_{j=1}^{l+1}$. Without loss of generality, we may assume that $k_1 \geq k_2 \geq \dots \geq k_{l+1}$, $k'_1 \geq k'_2 \geq \dots \geq k'_{l+1}$. In this case, $k_{l+1} = \min k_j$, $k'_{l+1} = \min k'_j$, and if $(k_j)_{j=1}^{l+1} \succ (k'_j)_{j=1}^{l+1}$, $(k'_j)_{j=1}^{l+1} \succ (k_j)_{j=1}^{l+1}$, then in view of relations (4.9), $k_{l+1} = k'_{l+1}$. But then, obviously, $(k_j)_{j=1}^{l} \succ (k'_j)_{j=1}^{l}$ and $(k'_j)_{j=1}^{l} \succ (k_j)_{j=1}^{l}$.

Due to the induction hypothesis, the tuple $(k'_j)_{j=1}^{l}$ provides a permutation of the tuple $(k_j)_{j=1}^{l}$. Reordering these tuples in a non-increasing manner, we thus deduce that they coincide. Consequently, the initial tuples $(k_j)_{j=1}^{l+1}$ and $(k'_j)_{j=1}^{l+1}$ coincide, too. Lemma 4.1 is proved. ≡

THEOREM 4.7. Let the diagonal elements of the matrix function G be Φ-factorable in L_p, and assume the index of the j th diagonal element to be equal to k_j. Then the tuple $(\varkappa_j)$ of partial indices of G is majorized by the tuple (k_j). The factorization factors $G_\pm$ and Λ can be obtained with the help of a finite number of algebraic operations and solutions of scalar Riemann problems.

Proof. According to Corollary 3.1, the dimension of the kernel of problem (3.1) with matrix coefficient G is equal to $\alpha = \sum_{j=1}^{n} \max\{\varkappa_j, 0\}$, and the dimension of the kernel of the scalar problem with the coefficient G_{jj} is equal to $\alpha_j = \max\{k_j, 0\}$. On the strength of Theorem 4.2, we conclude that $\sum_{j=1}^{n} \max\{\varkappa_j, 0\} \le \sum_{j=1}^{n} \max\{k_j, 0\}$.

Analogously, $-\sum_{j=1}^{n} \min\{\varkappa_j, 0\} \le -\sum_{j=1}^{n} \min\{k_j, 0\}$, i.e. $\sum_{j=1}^{n} \min\{\varkappa_j, 0\} \ge$ $\ge \sum_{j=1}^{n} \min\{k_j, 0\}$. Multiplying G by t^{-k}, its partial indices and the indices of the diagonal elements decrease by k. Consequently, the inequalities proved continue to be valid, substituting in them $\varkappa_j$ by $\varkappa_j - k$ and k_j by $k_j - k$. Thus $(k_j) \succ (\varkappa_j)$.

Now we introduce the matrix function $D(t) = \operatorname{diag}[t^{k_1}, \ldots, t^{k_n}]$. The matrix function $G_1 = GD^{-1}$ is triangular along with G, its diagonal elements are Φ-factorable in L_p, and their indices are equal to zero. According to Theorem 4.6, the partial indices of the matrix function G_1 are equal to zero and its factorization factors $G_+^{(1)}$ and $G_-^{(1)}$ are constructed with the help of a finite number of algebraic operations and the solutions of scalar Riemann problems. Using Theorem 2.7, we see that the matrix function G can be factored with the same degree of effectiveness. The proof of Theorem 4.7 is complete. ≡

From Theorem 4.7 and statement 1) of Lemma 4.1 we obtain the following result.

COROLLARY 4.3. The partial indices of a triangular matrix function G with Φ-factorable diagonal elements are situated between the smallest and largest index of the diagonal elements, and the total index is equal to the sum of the diagonal elements indices.

We denote by μ_{ij} the order of the zero of the (i,j) th entry of the matrix function $(G_-^{(1)})^{-1}$ (introduced in the proof of Theorem 4.7) at infinity. Note that all entries outside the diagonal of $G_-^{(1)}$ vanish at ∞, so that $\mu_{ij} \geq 1$ for $i \neq j$.

We shall write $\mu_{ij} = \infty$, if the (i,j) th entry of $(G_-^{(1)})^{-1}$ is identically zero.

THEOREM 4.8. The tuple of partial indices of a lower (upper) triangular matrix function coincides (up to a permutation) with the tuple of indices of its diagonal elements if and only if

$$k_j - k_i \leq \mu_{ij} \quad \text{for} \quad i > j \quad (i < j). \tag{4.10}$$

Proof. We focus our attention on the case of a lower triangular matrix function. If condition (4.10) is fulfilled, then, setting $G_+ = G_+^{(1)}$, $\Lambda = D$, $G_- = D^{-1}G_-^{(1)}D$, we get a factorization of G. Thus we have only to verify the analyticity of the entries outside the diagonals of the matrix functions G_- and G_-^{-1} at the point $z = \infty$. It is guaranteed by the fact that the (i,j) th entry of the matrix function $(G_-^{(1)})^{-1}$ is equal to zero, if $i < j$, and has a zero of order μ_{ij} at the point $z = \infty$, if $i > j$, and the (i,j) th entry of G_-^{-1} is obtained from it via multiplication by $z^{k_j - k_i}$.

Now, let condition (4.10) fail to be fulfilled. Suppose, at first, that $k_1 - k_i \leq \mu_{i1}$ $(i=1,\dots,n-1)$, $k_1 > 0$ and $k_n + \mu_{n1} \leq 0$.

Setting $n_1 = n-1$, $n_2 = 1$, we represent the matrix function G in the form (4.4). In this case G_2 is the right lower entry G_{nn} of G, the factorization of which in L_p we write down in the following way:

$$G_{nn}(t) = G_{nn}^+(t)t^{k_n}G_{nn}^-(t).$$

By φ we denote that vector function which is obtained from the first column of the matrix function $D^{-1}(G_-^{(1)})^{-1}$ by cancelling its last element φ_n. Now we are going to show that the vector function φ lies in the kernel of the operator $T_Q(G_1)$ and that, at the same time, the relation $H\varphi \notin \mathrm{im}(P+G_{nn}Q)$ holds. In fact, the i th component of the vector function φ has a zero of order $k_i + \mu_{i1} \geq k_1 > 0$ $(i=1,\dots,n-1)$ at infinity. Consequently, $\varphi \in (\mathring{L}_p^-)^{n-1}$. Moreover, the equation $GD^{-1}(G_-^{(1)})^{-1} = G_+^{(1)}$ is satisfied, in view of which all the elements of $GD^{-1}(G_-^{(1)})^{-1}$ lie in L_p^+. But the first column of this matrix function is of the form

$$\begin{pmatrix} G_1 \varphi \\ G_{nn}\varphi_n + H\varphi \end{pmatrix},$$

so that $G_{nn}\varphi_n + H\varphi \in L_p^+$ and $G_1\varphi \in (L_p^+)^{n-1}$. The first of the relations obtained allows us to claim that $\varphi \in \ker T_Q(G_1)$. After multiplication by $(G_{nn}^+)^{-1}$, from the second relation we conclude

$$f = G_{nn}^- t^{k_n}\varphi_n + (G_{nn}^+)^{-1}H\varphi \in L_1^+ .$$

Making use of the relation (1.11), we find that, for $j \geq 0$, the equation

$$\int_\Gamma (G_{nn}^+)^{-1}(\tau)H(\tau)\varphi(\tau)\tau^j d\tau = -\int_\Gamma G_{nn}^-(\tau)\tau^{k_n+j}\varphi_n(\tau)d\tau$$

holds. Setting $j = \mu_{n1}-1$, in the right part of the last equation we get a quantity different from zero, since the order of the zero of φ_n (and, thus, also of $G_{nn} - \varphi_n$) at infinity is precisely equal to $\mu_{n1} + k_n$. But $\mu_{n1}-1 \leq -k_n-1$. Consequently, for the function $H\varphi$, the necessary condition of membership to the image of the operator $P+G_{nn}Q$ is not fulfilled.

In summary, we have found a nontrivial vector function $\varphi \in \ker T_Q(G_1)$ for which $H\varphi \notin \mathrm{im}\,(P+G_2Q)$ or, equivalently, $QH\varphi \notin \mathrm{im}\, T_Q(G_2)$. Therefore, the value γ arising in formulae (4.6) is positive, so that the number α of linearly independent solutions of the homogeneous problem (3.1) with matrix coefficient G is strictly smaller than the sum of the corresponding numbers α_1 and α_2 of problems (3.1) with matrices G_1 and G_2. However, owing to Theorem 4.2, $\alpha_1 \leq \sum_{j=1}^{n-1} \max\{k_j,0\}$ and, with regard to Corollary 3.1, $\alpha_2 = \max\{k_n,0\}$. Hence,

$$\alpha < \sum_{j=1}^{n} \max\{k_j,0\} .$$

Proceeding to the general case, we remark that element G_{kl} nearest to the main diagonal for which $k_k - k_l > \mu_{lk}$ (if such an element is not unique, we choose an arbitrary of them). Now we specify an integer k with $k_k > k$, $k_l + \mu_{lk} \leq k$ and partition the matrix function $t^{-k}G(t)$ into blocks in such a way that one of the diagonal blocks, say G_0, be placed in the rows and columns with numbers from 1 till k, and the remaining blocks be one-dimensional. Taking into account that on multiplication by t^{-k} the numbers k_j and $\varkappa_j$ decrease by k and the values μ_{ij} remain unchanged, we see that the matrix function G_0 satisfies the conditions of the special case considered above. Therefore, $\alpha_0 < \sum_{j=1}^{k} \max\{k_j - k, 0\}$. Here α_0 is the number of linearly independent solutions of the homogeneous Riemann problem with matrix coefficient G_0. Moreover, in virtue of Theorem 4.2,

$$\sum_{j=1}^{n} \max\{\varkappa_j - k, 0\} \leq \sum_{j=1}^{l-1} \max\{k_j - k, 0\} + \alpha_0 + \sum_{j=k+1}^{n} \max\{k_j - k, 0\} .$$

Comparing the last two inequalities, we obtain,

$$\sum_{j=1}^{n} \max\{\varkappa_j - k, 0\} < \sum_{j=1}^{n} \max\{k_j - k, 0\} .$$

Thus, if condition (4.10) is not fulfilled, the tuple (k_j) is not majorized by the tuple $(\varkappa_j)$ and, in view of statement 3) of Lemma 4.1, none of these tuples may be obtained from another one with the help of a permutation. In that way Theorem 4.8 is completely proved. $\equiv$

COROLLARY 4.4. If the indices of the diagonal elements of a lower (upper) triangular matrix function satisfy the condition $k_i > k_j$ for $i > j$ $(i < j)$, then the partial indices coincide with the indices of the diagonal elements.

In order to prove this proposition, we have to use Theorem 4.8 and the fact that $\mu_{ij} \geq 1$ for $i \neq j$. $\equiv$

4.3. CALCULATION OF PARTIAL INDICES OF SECOND-ORDER TRIANGULAR MATRIX FUNCTIONS

In case of second-order triangular matrix functions, the question of calculating the partial indices admits a complete solution. For the sake

of definiteness, we study lower triangular matrix functions, i.e. matrix functions of the form (4.4) with $n_1 = n_2 = 1$.
First of all, consider the case of Φ-factorable diagonal elements (in accordance with Theorem 4.3, in this case the matrix function G is Φ-factorable). Denoting the indices of the functions G_1 and G_2 by k_1 and k_2 respectively, in view of Theorem 4.7 and Corollary 4.4, we may only claim the following: the partial indices of G coincide with k_1 and k_2, if $k_1-k_2 \leq 1$, and are equal to $k_1-\lambda$ and $k_2+\lambda$, if $k_1-k_2 > 1$, where λ is an integer situated between 0 and $[\frac{1}{2}(k_1-k_2)]$.

For the exact calculation of the partial indices of a matrix function G of the form (4.4), we propose the following algorithm.

As in the proof of Theorem 4.7, we construct matrix functions

$$G_+^{(1)} = \begin{pmatrix} \xi_1^+ & 0 \\ a^+ & \xi_2^+ \end{pmatrix}, \quad G_-^{(1)} = \begin{pmatrix} \xi_1^- & 0 \\ a^- & \xi_2^- \end{pmatrix}, \quad \text{and} \quad D(t) = \begin{pmatrix} t^{k_1} & 0 \\ 0 & t^{k_2} \end{pmatrix}$$

such that $G = G_+^{(1)} G_-^{(1)} D$.

The function $\Phi(z) = a^-(z)/\xi_1^-(z)$ lies in the class $\dot{E}_1^-$. Denote the order of the zero of function Φ at ∞ by μ_0.
Then $\Phi^{-1}(z) = Q_0(z) + R_1(z)/\Phi(z)$, where Q_0 is a polynomial of degree $q_0 = \mu_0$, $R_1 \in E_1^-$ and R_1 has at ∞ a zero of order $\mu_1 > \mu_0$.
Furthermore, $\Phi R_1^{-1} = Q_1 + R_2 R_1^{-1}$, where Q_1 is a polynomial of degree $q_1 = \mu_1 - \mu_0$, $R_2 \in E_1^-$ and R_2 has at ∞ a zero of order $\mu_2 > \mu_1$.
Acting in this way, we get a sequence of polynomials $Q_0, Q_1, Q_2, \ldots$ of degrees $q_0 = \mu_0$, $q_1 = \mu_1 - \mu_0$, $Q_2 = \mu_2 - \mu_1, \ldots$. If the process stops at some stage, i.e. $R_k \equiv 0$, we put $\mu_k = \infty$.

<u>THEOREM 4.9.</u> If $k_1 - k_2 \leq \mu_0$, then the partial indices of the matrix function G are equal to k_1 and k_2. If $\mu_0, \mu_0+\mu_1, \ldots, \mu_{i-1}+\mu_i < k_1-k_2$, but $\mu_i + \mu_{i+1} \geq k_1-k_2$, then the partial indices of G are $k_1-\mu_i$ and $k_2+\mu_i$.

<u>Proof.</u> The first statement of the theorem follows from Theorem 4.8, since Φ has the same order at ∞ as the lower left element of the

matrix function $(G_-^{(1)})^{-1}$. If $\mu_0 < k_1-k_2$, but $\mu_0+\mu_1 \geq k_1-k_2$, then we obtain a factorization of G by setting [1])

$$G_-(t) = \begin{pmatrix} \xi_1^-(t)R_1(t)^{k_1-k_2-\mu_0} & \xi_2^-(t)Q_0(t)t^{-\mu_0} \\ a^-(t)t^{\mu_0} & \xi_2^-(t)t^{-k_1+k_2+\mu_0} \end{pmatrix},$$

$$G_+ = G_+^{(1)}\begin{pmatrix} 1 & Q_0 \\ 0 & 1 \end{pmatrix}, \quad \Lambda(t) = \begin{pmatrix} t^{k_2+\mu_0} & 0 \\ 0 & t^{k_1-\mu_0} \end{pmatrix}.$$

In case that $\mu_0+\mu_1 < k_1-k_2$, but $\mu_1+\mu_2 \geq k_1-k_2$, we put

$$G_+ = G_+^{(1)}\begin{pmatrix} 1 & Q_0 \\ 0 & 1 \end{pmatrix}\begin{pmatrix} 1 & 0 \\ Q_1 & 1 \end{pmatrix}, \quad \Lambda(t) = \begin{pmatrix} t^{k_1-\mu_1} & 0 \\ 0 & t^{k_2+\mu_1} \end{pmatrix},$$

$$G_-(t) = \begin{pmatrix} \xi_1^-(t)R_1(t)t^{\mu_1} & \xi_2^-(t)^{k_1+k_2+\mu_1}Q_0(t) \\ \xi_1^-(t)R_2(t)t^{k_1-k_2-\mu_1} & \xi_2^-(t)t^{-\mu_1}(1-Q_0(t)Q_1(t)) \end{pmatrix}$$

etc. ≟

COROLLARY 4.5. Let G_1, G_2 $(\in L_\infty)$ be Φ-factorable functions with indices k_1 and k_2 respectively, and suppose $k_1 > k_2$. Then for every integer $\lambda \in (0,[\frac{1}{2}(k_1-k_2)])$, one can choose a function $H \in L_\infty$ in such a manner that the partial indices of the matrix function (4.4) are equal to $k_1-\lambda$ and $k_2+\lambda$.

Proof. We intend to look for functions H of the form $H = \xi_1^-\xi_2^+t^{k_1}\Phi$, where $\Phi \in E_1^-$ and has a zero of order $\mu_0 = k_1-k_2-\lambda$ at the point $z = \infty$. The condition $H \in L_\infty$ is equivalent to $\xi_1^-\Phi/\xi_2^- \in L_\infty$ and is, obviously, fulfilled iff $\Phi = \xi_2^-\varphi/\xi_1^-$ with $\varphi \in E_\infty^-$. If this requirement is satisfied, then (in notations of Theorem 4.9) $a^+ = 0$, $a^- = \Phi\xi_1^-$. Hence, function Φ coincides with the function occurring in the proof

1) Exactly speaking, the representation obtained will be a factorization only in case if the index of the right lower element of the matrix function Λ does not exceed the index of its left upper element. Otherwise, it is still necessary to rearrange the rows of the matrix function G_- , the columns of the matrix function G_+ and the diagonal elements of Λ .

of Theorem 4.9. Since $\mu_0 \leq k_1 - k_2$ and $\mu_1 + \mu_0 > 2\mu_0 \geq k_1 - k_2$, then, owing to Theorem 4.9, the partial indices of the matrix function (4.4) are equal to $k_1 - \mu_0 = k_2 + \lambda$ and $k_2 + \mu_0 = k_1 - \lambda$. ≡

Now we proceed to the case of a Φ-factorable matrix function (4.4) the diagonal elements of which are not Φ-factorable.

THEOREM 4.10. Let the diagonal elements G_1 and G_2 of the matrix function (4.4) be not Φ-factorable in L_p. Then for the Φ-factorability in L_p of the matrix function G it is necessary and sufficient that the image of the operator $T_Q(G_1)$ coincides with $\overset{\circ}{L}{}_p^-$, the kernel of the operator $T_Q(G_2)$ is trivial, and the codimension γ_1 of the sum $\operatorname{im} T_Q(G_2) + T_Q(H)\ker T_Q(G_1)$ as well as the dimension γ_2 of the lineal $M = \{\varphi \in \ker T_Q(G_1) : T_Q(H)\varphi \in \operatorname{im} T_Q(G_2)\}$ are finite. If these conditions are fulfilled, then the partial indices of G are determined by the formulae

$$\varkappa_1 = \gamma_2 , \quad \varkappa_2 = -\gamma_1 \tag{4.11}$$

if $\gamma_1\gamma_2 \neq 0$ or $\gamma_1 + \gamma_2 \leq 1$, and by the formulae

$$\varkappa_1 = [\gamma_2/2] + r , \quad \varkappa_2 = \gamma_2 - [\gamma_2/2] - r \tag{4.12}$$

if $\gamma_1 = 0$, $\gamma_2 > 1$ [1].

Here r is the dimension of the subspace consisting of those functions $\varphi \in M$ for which

$$\begin{aligned} &\int_\Gamma [H(\tau)\varphi(\tau) - G_2(\tau)(T_Q(G_2))^{-1}GH\varphi)(\tau)]\tau^{-j}d\tau = \\ &\int_\Gamma G_1(\tau)\varphi(\tau)\tau^{-j}d\tau = 0 , \quad j=1,\dots,[\gamma_2/2] . \end{aligned} \tag{4.13}$$

Proof. The sufficiency of the conditions for Φ-factorability of the matrix function G stated in the formulation of the theorem follows immediately from Theorem 4.1.

In order to prove their necessity, according to the same Theorem 4.1, we have only to verify that the numbers $\beta_1 = \beta(T_Q(G_1))$ and $\alpha_2 = \alpha(T_Q(G_2))$ are equal to zero. Assuming $\beta_1 > 0$, owing to statement 3) of Theorem 3.1, we may conclude that $\alpha(T_Q(G_1)) = 0$. This and the

[1] Analogous formulae are valid for $\gamma_2 = 0$, $\gamma_1 > 1$.

finiteness of the value β_1 imply that the operator $T_Q(G_1)$ is Fredholm, i.e., function G_1 is Φ-factorable in L_p, which contradicts the assumption of the theorem. Thus $\beta_1 = 0$. The equality $\alpha_2 = 0$ may be proved analogously.

In this way, the conditions described in the formulation of Theorem 4.10 are actually necessary and sufficient for the Φ-factorability of G. In addition, from the equalities $\beta_1 = \alpha_2 = 0$ and the relations (4.5), the following formulae for the defect numbers of the Riemann boundary value problem with matrix coefficient G result:

$$\alpha = \gamma_2 , \qquad \beta = \gamma_1 .$$

Comparing these formulae with the expressions of the defect numbers via partial indices established in Theorem 3.16, we see that the following cases are possible:

1) $\varkappa_1 \geq 0$, $\varkappa_2 \leq 0$.

Then $\gamma_1 = -\varkappa_2$, $\gamma_2 = \varkappa_1$, i.e., formulae (4.11) are fulfilled regardless of the values $\gamma_1\gamma_2$ and $\gamma_1 + \gamma_2$.

2) $\varkappa_1 > 0$, $\varkappa_2 \leq 0$.

Then $\gamma_1 = 0$, $\gamma_2 = \varkappa_1 + \varkappa_2 \geq 1$, where $\varkappa_2 \leq \frac{1}{2}(\varkappa_1+\varkappa_2) = \gamma_2/2$. Since $\varkappa_2$ is an integer, we obtain that $\varkappa_2 \leq [\gamma_2/2]$ and, if $\gamma_2 = 1$, then $\varkappa_2 = 0$, which means $\varkappa_1 = \gamma_2$.

3) $\varkappa_1 \leq 0$, $\varkappa_2 < 0$.

Then $\gamma_2 = 0$, $\gamma_1 = -(\varkappa_1+\varkappa_2) \geq 1$. As in case 2), we can establish that, for $\gamma_1 = 1$, the equalities $\varkappa_2 = -1$ and $\varkappa_1 = 0$ hold.

In summary, if either of the conditions $\gamma_1\gamma_2 \neq 0$ and $\gamma_1+\gamma_2 \leq 1$ is fulfilled, then formulae (4.11) are true. Therefore, only one situation remains to be examined, namely, when $\gamma_1 = 0$, $\gamma_2 > 1$.

To this end, we introduce the matrix function $G_0(t) = t^{-k}G(t)$ with $k = [\gamma_2/2]$. The partial indices $\varkappa_1^{(0)} = \varkappa_1 - k$ and $\varkappa_2^{(0)} = \varkappa_2 - k$ of this matrix function meet the conditions of case 1). In fact, the inequality $\varkappa_2 \leq k$ was proved above, and $\varkappa_1 \geq \frac{1}{2}(\varkappa_1+\varkappa_2) = \gamma_2/2 \geq k$. Consequently $\varkappa_1^{(0)} = \dim M_0$, where

$$M_0 = \{\varphi \in \ker T_Q(t^{-k}G_1) : Qt^{-k}H\varphi \in \operatorname{im} T_Q(t^{-k}G_2)\} .$$

In this way, $\varkappa_1 = k + \dim M_0$, $\varkappa_2 = \gamma_2 - k - \dim M_0$, and the proof of the theorem will be perfected as soon as we establish that M_o coincides with the subspace generated in M by conditions (4.13).

The relation $\varphi \in \ker T_Q(t^{-k}G_1)$ may be rewritten in the form $t^{-k}G_1\varphi \in \in L_p^+$. If the latter condition applies, then $G_1\varphi \in L_p^+$, so that $QG_1\varphi = 0$ and $\varphi \in \ker T_Q(G_1)$. In turn, if $\varphi \in \ker T_Q(G_1)$, then $Qt^{-k}G_1\varphi = Qt^{-k}PG_1\varphi$. Consequently, $\varphi \in \ker T_Q(t^{-k}G_1)$ iff $\varphi \in \ker T_Q(G_1)$ and $Qt^{-k}PG_1\varphi = 0$. If further, $Qt^{-k}H\varphi \in \operatorname{im} T_Q(t^{-k}G_2)$, i.e. $Qt^{-k}H\varphi = Qt^{-k}G_2 f$, where $f \in \mathring{L}_p^-$, then $t^kQt^{-k}H\varphi = t^kQt^{-k}G_2f$. Hence $H\varphi - t^kPt^{-k}H\varphi = G_2f - t^kPt^{-k}G_2f$. Applying the projections Q and P to the last equation successively, we obtain

$$QH\varphi = QG_2f \,,$$
$$(P - t^kPt^{-k})H\varphi = (P - t^kPt^{-k})G_2f \,. \tag{4.14}$$

The first of the relations obtained means that

$$QH\varphi \in \operatorname{im} T_Q(G_2) \,. \tag{4.15}$$

According to what was proved above, the kernel of the operator $T_Q(G_2)$ is trivial and, thus, the pre-image f of the function $QH\varphi$ may be written in the form $f = T_Q(G_2)^{-1}QH\varphi$. With regard to this fact and the equation $P - t^kPt^{-k} = t^kQt^{-k}P$, we rewrite relation (4.14) in the following manner:

$$Qt^{-k}P(H\varphi - G_2T_Q(G_2)^{-1}QH\varphi) = 0 \,. \tag{4.16}$$

The transformations of the condition $Qt^{-k}H\varphi \in \operatorname{im} T_Q(t^{-k}G_2)$ carried out above are reversible. Consequently, this condition is equivalent to the requirements (4.15) and (4.16).

Summarizing, the inclusion $\varphi \in M_o$ holds if and only if $\varphi \in M$ and the functions $G_1\varphi$ and $H\varphi - G_2T_Q(G_2)^{-1}QH\varphi$ lie in the kernel of the operator $Qt^{-k}P$.

Since to the kernel of the operator $Qt^{-k}P$ there belong those and only those functions ψ from L_p the component $P\psi$ of which has the first k coefficients of the power series expansion equal to zero, it follows from the above considerations that conditions (4.13) are necessary and sufficient for the function $\varphi \in M$ to lie in the subspace M_o. Therefore, the equality $\dim M_o = r$ has been proved and, along with it, also Theorem 4.10. $\equiv$

Example. Let us discuss the problem of calculating the partial 2-indices of the matrix function

$$G_m = \begin{pmatrix} t^m\omega & 0 \\ 1 & \overline{\omega} \end{pmatrix}$$

given on the unit circle, where ω is an inner function not reducible to a finite Blaschke product, and m is a non-negative integer.

For $m = 0$, this matrix function meets the conditions of Theorem 4.4. According to this theorem, it is Φ-factorable in L_2. Moreover, in the proof of Theorem 4.4 it was established that, in this case we have $\gamma_1 = \gamma_2 = 0$. In virtue of Theorem 4.10, we are able to claim that the partial 2-indices of G_0 are equal to zero.
Representing the function G_m in the form

$$G_m = \begin{pmatrix} t^m & 0 \\ 0 & 1 \end{pmatrix} G_0 ,$$

from Theorem 2.11 we obtain that G_m is Φ-factorable in L_2 together with G_0, its partial 2-indices are non-negative, and the total 2-index is equal to m.

Thus, for the matrix function G_m, the relations $\gamma_1 = 0$ and $\gamma_2 = \gamma$ are valid and, consequently, for $m = 1$, we have $\varkappa_1 = 1$, $\varkappa_2 = 0$. In case $m > 1$, formulae (4.12) can be applied for calculating the partial 2-indices of G_m.

In the considered case, we have $H = 1$ and the operator $T_Q(G_2)$ coincides with the operator of multiplication by $\overline{\omega}$, therefore, $H\varphi - G_2 T_Q(G_2)^{-1} QH\varphi = P\varphi$, and the first group of conditions (4.13) is automatically fulfilled.

Furthermore, in the case under study, the subspace M coincides with $\ker T_Q(t^m\omega) \cap \overline{\omega}\, \overset{\circ}{L}{}_p^-$, i.e. with the set of functions of the form $\varphi = \overline{\omega} f$, where $f \in \overset{\circ}{L}{}_p^-$ and $t^m\omega\,\overline{\omega} f \in L_p^+$. Since $\omega\overline{\omega} = 1$, the requirements enumerated are fulfilled iff

$$f(t) = \sum_{s=-m}^{-1} c_s t^s .$$

Substituting the obtained representation of the functions $\varphi \in M$ into the second group of conditions (4.13), we get

$$0 = \int_{\mathbf{T}} \tau^m \omega(\tau)\overline{\omega(\tau)} f(\tau)\tau^{-j} d\tau =$$

$$\int_{\mathbf{T}} \sum_{s=-m}^{-1} c_s t^{m+s-j} dt = 2\pi i \, c_{j-1-m} ,$$

where $j=1,\dots,[m/2]$.

Therefore, conditions (4.13) are fulfilled for those and only those functions $\varphi \in M$ for which $c_s = 0$ for $s = -m,\dots,[m/2]-1-m$. For that reason, the subspace M_0 consists exactly of functions $\varphi \in M$ of the form

$$\varphi(t) = \overline{\omega}(t) \sum_{s=[\frac{m}{2}]-m}^{-1} c_s t^s .$$

Hence $r = \dim M_0 = m-[m/2]$.

On the strength of formulae (4.12), we thus obtain $\varkappa_1 = m$, $\varkappa_2 = 0$. In this way, for every non-negative integer m , we have constructed examples of second-order lower triangular matrix functions the diagonal elements of which are not Φ-factorable and whose partial indices are m and 0 . Multiplying these matrix functions by scalar factorable functions with suitable index (for instance, simply by t^j), one can easily obtain examples of second-order triangular matrix functions with an arbitrary tuple of partial indices and non-Φ-factorable diagonal elements.

4.4. Φ-FACTORIZATION OF FUNCTIONALLY COMMUTABLE MATRIX FUNCTIONS

All the suggestions made in Sections 4.1 - 4.3 can, with obvious modifications, be transferred to the case when the matrix function G is not triangular, but becomes a such one after multiplication from the left and the right by a constant non-degenerating matrix. In the present section a class of matrix functions possessing such a property will be investigated.

DEFINITION 4.2. A measurable matrix function G defined on a contour Γ is said to be functionally commutable if

$$G(t_1)G(t_2) = G(t_2)G(t_1) \quad (t_1,t_2 \in \Gamma) . \tag{4.17}$$

LEMMA 4.2. The matrix function G is functionally commutable if and only if

$$G(t) = \sum_{j=1}^{m} \varphi_j(t) G_j , \tag{4.18}$$

where $m \leq n^2$, G_j $(j=1,\dots,m)$ are pairwise commuting constant matrices and φ_j are measurable scalar functions.

Proof. The sufficiency of the conditions of the lemma is evident. We are going to prove their necessity. For this purpose, we form the linear hull $\mathfrak{L}$ of the set of all matrices of the form $G(t)$, $t \in \Gamma$ in the space of all constant matrices of n th order and choose in it a basis $G_1,\dots,G_m$. Since the space of $n \times n$ matrices has the dimension n^2, then $m \leq n^2$. Furthermore, condition (4.17) implies that any two matrices from $\mathfrak{L}$ commute. Consequently, all matrices $G_1,\dots,G_m$ commute pairwise. Finally, as $G(t) \in \mathfrak{L}$ for all $t \in \Gamma$ and $\{G_j\}_{j=1}^{m}$ is a basis in $\mathfrak{L}$, then there exists a unique representation $G(t)$ in the form of a linear combination of the G_j's. Supposing $\varphi_j(t)$ to be equal to the j th coefficient of this linear combination, we have $G(t) = \sum \varphi_j(t) G_j$. The measurability of functions φ_j follows from the measurability of G as well as the linear independence of $G_1,\dots,G_m$, which proves Lemma 4.2. ≡

LEMMA 4.3. If the matrices $G_1,\dots,G_m$ commute pairwise, then there exists a non-degenerating constant matrix S such that all matrices $S^{-1}G_jS$, $j=1,\dots,m$ are lower triangular.

Proof. Above all, we show by induction on m that the matrices $G_1,\dots,G_m$ have a common eigensubspace. For $m = 1$, this is obvious. Supposing that the statement is correct for $m = k-1$, we consider the pairwise commutating matrices $G_1,\dots,G_k$. By assumption, for a certain choice of scalars λ_j, $j=1,\dots,k-1$, the subspace $M = \{x\colon G_jx = \lambda_jx ,\ j=1,\dots,k-1\}$ is nontrivial. Owing to the commutativity, we have, for all $x \in M$, $G_jG_kx = G_kG_jx = G_k\lambda_jx = \lambda_jG_kx$, i.e. $G_kx \in M$. Thus, the subspace M is invariant with respect to G_k and, therefore, it contains some eigensubspace.

This will just be the common eigensubspace for $G_1,\dots,G_k$.

From what was proved above we conclude that, by all means, the matrices $G_1,\dots,G_k$, which commute pairwise, have a common eigenvector x . Let S_1 be a non-degenerate matrix the last column of which is x . As can be easily seen, in this case

$$S_1^{-1}G_jS_1 = \left(\begin{array}{ccc|c} & G_j^{(1)} & & 0 \\ \hline G_{n_1}^{(j)} & \dots & G_{n,n-1}^{(j)} & \lambda_j \end{array}\right),$$

where $G_j^{(1)}$ is a matrix of (n-1) th order. The matrices $S_1^{-1}G_jS_1$ commute pairwise and, along with them, the matrices $G_j^{(1)}$ also commute. If we succeed in finding a non-degenerate matrix $S^{(1)}$ of (n-1) th order such that all matrices $(S^{(1)})^{-1}G_j^{(1)}S^{(1)}$ are lower triangular, then, setting $S_2 = \mathrm{diag}[S^{(1)},1]$ and $S = S_1S_2$, we obtain the desired matrix S .

In this way, if the statement to be proved is true for matrices of (n-1) th order, then it is also correct for matrices of n th order. Since, for $n = 1$, the assertion is trivial, then it is valid for every natural n . Thus Lemma 4.3 is proved. ≡

On the strength of Lemmas 4.2 and 4.3, we can assert that if the matrix function G is functionally commutable, then there exists such a constant non-degenerating matrix S that the matrix function $F = S^{-1}GS$ is lower triangular, and if $G \in L_\infty$, then we have also $F \in L_\infty$. According to Theorem 4.3, the Φ-factorability of the diagonal elements of F is sufficient for the Φ-factorability of G , and its partial indices can be estimated with the aid of Theorem 4.7. However, it turns out that the matrix S may be chosen in such a way that the matrix function G is Φ-factorable only simultaneously with the diagonal elements of the matrix function F , and its partial indices coincide with the indices of the diagonal elements of F .

<u>THEOREM 4.11.</u> Let G $(\in L_\infty)$ be a functionally commutable matrix function. Assume that there exists a constant non-degenerating matrix S such that the matrix function $F = S^{-1}GS$ is block-diagonal and

every diagonal block is a lower triangular matrix function all the diagonal elements of which are equal to each other. The matrix function G is Φ-factorable if and only if the diagonal elements of F are Φ-factorable, and the partial indices of G are equal to the indices of the diagonal elements of F.

Proof. In the linear hull $\mathcal{L}$ of the set of all values of G we choose the matrix with the largest number of different eigenvalues (if such a matrix is not unique, then we take an arbitrary of them) and denote it by G_0. As well-known (GANTMAKHER [1], MAL'TSEV [1]), there exists a constant non-degenerating matrix T such that

$$T^{-1}G_0T = \operatorname{diag}[A_1^{(0)},\ldots,A_k^{(0)}] ,$$

where the block $A_i^{(0)}$ is a matrix with the unique point of the spectrum λ_i, $\lambda_{i_1} \neq \lambda_{i_2}$ for $i_1 \neq i_2$, $i_1,i_2 = 1,\ldots,k$.

In view of Lemma 4.2, $G(t) = \sum_{j=1}^{m} \varphi_j(t)G_j$ with $G_j \in \mathcal{L}$. Therefore, the matrices G_j commute with G_0. But in this case the matrices $T^{-1}G_jT$ and $T^{-1}G_0T$ commute, too. With regard to the description of all matrices permutable with a given one (see GANTMAKHER [1], p. 204 or MAL'TSEV [1], p. 192), every matrix $T^{-1}G_jT$ is also block-diagonal:

$$T^{-1}G_jT = \operatorname{diag}[A_1^{(j)},\ldots,A_k^{(j)}] ,$$

where the matrices $A_i^{(0)},\ldots,A_i^{(m)}$, $i=1,\ldots,k$, commute pairwise. Applying Lemma 4.3 to these matrices, we establish that there exists a matrix S_i (of the same order as all matrices $A_i^{(j)}$, $j=0,\ldots,m$) such that $S_i^{-1}A_i^{(j)}S_i$ are lower triangular matrices ($i=1,\ldots,k$; $j=0,\ldots,m$). Setting $S = T\operatorname{diag}[S_1,\ldots,S_k]$, we recognize that the matrices $S^{-1}G_jS$ are block-diagonal, and every diagonal block is triangular ($j=0,\ldots,m$). Consequently, on the diagonal of any of the blocks there are located its eigenvalues. Therefore, in the matrix $S^{-1}G_0S$, the diagonal elements of one and the same block agree. Now, assume that for some of the matrices $S^{-1}G_jS$, say, for $S^{-1}G_1S$, this is not true. Then, for ε small enough, the matrix $S^{-1}(G_0+\varepsilon G_1)S$ and, along with it, also the matrix $G_0+\varepsilon G_1$ has more different eigenvalues than G_0. But this contradicts the way of choosing G_o. Thus, $S^{-1}G_jS = \mathrm{diag}[G_1^{(j)},\dots,G_k^{(j)}]$, where every block $G_i^{(j)}$ is a lower triangular matrix with equal elements on the main diagonal.

Utilizing formula (4.18), we see that

$$S^{-1}G(t)S = F(t) = \mathrm{diag}[F_1(t),\dots,F_k(t)] ,$$

the blocks $F_i(t)$ are lower triangular, and the elements standing on the diagonal of one and the same block are equal to each other. Therewith, the first statement of the theorem is proved.

Furthermore, the matrix functions G and F are Φ-factorable only simultaneously and, in accordance with Corollary 4.2, the matrix function F is Φ-factorable if and only if every block $F_1,\dots,F_k$ is Φ-factorable. Let us consider one of these blocks, say F_i, denoting its diagonal element by a_i.

Applying Theorem 4.1 for $n_1 = 1$, we obtain that, if the matrix function F_i is Φ-factorable, then, in view of assertion 1), the image of the operator $P + a_iQ$ must have finite codimension. For $n_2 = 1$, due to assertion 2) of the same theorem, we conclude that the kernel of this operator must be finite dimensional. Consequently, from the Φ-factorability of F_i, the Fredholmness of the operator $P + a_iQ$ results, i.e. the Φ-factorability of its diagonal elements. The validity of the opposite statement can be seen from Theorem 4.3.

In summary, the matrix function G is Φ-factorable iff all diagonal elements of the matrix function F are Φ-factorable.

Finally, the partial indices of G and F coincide, the set of partial indices of F is obtained by uniting the tuples of partial indices of its diagonal blocks, and all partial indices of the block F_i are equal to the index of the element a_i, because for the blocks F_i the conditions of Corollary 4.4 are fulfilled. Hence the last statement of the theorem is also proved. ≡

Note that the diagonal elements of the matrix $F(t)$ are its eigenvalues and, thus, also eigenvalues of the matrix $G(t)$. Roughly speaking, we may therefore say that the partial indices of the functionally commutable matrix function G coincide with the indices of its eigenvalues.

Example. Recall that a matrix of the form

$$C = \begin{pmatrix} c_0 & c_1 & \cdots & c_{n-1} \\ c_{n-1} & c_0 & \cdots & c_{n-2} \\ \cdots & \cdots & \cdots & \cdots \\ c_1 & c_2 & \cdots & c_0 \end{pmatrix} \qquad (4.19)$$

is called a circulant (see e.g. MARCUS, MINC [1], p. 96). The product of two circulants is again a circulant as can be readily verified, where the product does not depend on the order of the factors. Consequently, a matrix function G all values of which are circulants is functionally commutable, and Theorem 4.11 is applicable. We shall get an explicit formula for the matrix S. Using this formula, F turns out to be diagonal.

Setting

$$S = n^{-1/2}\left(e^{\frac{2\pi i}{n}(k-1)(j-1)}\right)_{k,j=1}^{n},$$

we consider the product $D = S^*CS$, where C is the matrix defined by formula (4.19). Its (r,s) th entry is determined by the formula

$$d_{rs} = n^{-1}\Big(\sum_{l\geq k} c_{l-k}\, e^{\frac{2\pi i}{n}[(l-1)(s-1)-(r-1)(k-1)]} +$$

$$\sum_{l<k} c_{n+l-k}\, e^{\frac{2\pi i}{n}[(l-1)(s-1)-(r-1)(k-1)]}\Big).$$

Denoting $l-k = j$ in the first sum and $n+l-k = j$ in the second one, we obtain

$$d_{rs} = n^{-1}\Big(\sum_{j=0}^{n-1}\sum_{k=1}^{n-j} c_j\, e^{\frac{2\pi i}{n}[(j+k-1)(s-1)-(r-1)(k-1)]} +$$

$$\sum_{j=1}^{n-1}\sum_{k=n-j+1}^{n} c_j\, e^{\frac{2\pi i}{n}[(j+k-m-1)(s-1)-(r-1)(k-1)]}\Big).$$

Taking into account that $e^{\frac{2\pi i}{n}(j+k-n-1)(s-1)} = e^{\frac{2\pi i}{n}(j+k-1)(s-1)}$, we are able to unite the two sums into one:

$$d_{rs} = n^{-1} \sum_{j=o}^{n-1} \sum_{k=1}^{n} c_j \, e^{\frac{2\pi i}{n}[(j+k-1)(s-1)-(r-1)(k-1)]} =$$

$$n^{-1} \sum_{j=o}^{n-1} \Big(\sum_{k=1}^{n} e^{\frac{2\pi i}{n}(k-1)(s-r)} \Big) e^{\frac{2\pi i}{n} j(s-1)} c_j \, .$$

If $s = r$, the inner sum is equal to n and, for $s \neq r$, it is equal to zero. Thus, D is a diagonal matrix and

$$d_{rr} = \sum_{j=o}^{n-1} e^{\frac{2\pi i}{n} j(r-1)} c_j \, .$$

In particular, choosing $C = I$, we get $d_{rr} = 1$, i.e. $S^*S = I$, and the matrix S is unitary and therefore non-degenerating.

Now let G be a matrix function all values of which are circulants. Denote the elements of the first row of G by $g_o,\ldots,g_{n-1}$ and introduce the functions

$$f_r(t) = \sum_{j=o}^{n-1} e^{\frac{2\pi i}{n} jr} g_j(t) \, , \quad r=0,\ldots,n-1 \, . \tag{4.20}$$

According to the above considerations, the matrix function $F = S^*GS$ $(= S^{-1}GS)$ is of the form $F = \mathrm{diag}[f_o,\ldots,f_{n-1}]$.

From here we obtain the final conclusion: a matrix function G all values of which are circulants is Φ-factorable if and only if all functions f_j determined via formulae (4.20) are Φ-factorable, where the partial indices of G are equal to the indices of the functions f_j. A factorization of the matrix function G will be known as soon as these functions are factored.

At the end of this section we focus our attention on second-order matrix functions.

<u>LEMMA 4.4.</u> The matrix function $G = \begin{pmatrix} g_{11} & g_{12} \\ g_{21} & g_{22} \end{pmatrix}$ is functionally commutable iff $g_{11}-g_{22}$, g_{12} and g_{21} are scalar multiples of one and the same function φ, i.e.

$$g_{12}(t) = \alpha\varphi(t) \, , \; g_{21}(t) = \beta\varphi(t), \; g_{11}(t)-g_{22}(t) = \gamma\varphi(t), \tag{4.21}$$

where α, β, γ are numbers.

Proof. If the conditions (4.21) are fulfilled, then

$$G(t) = \begin{pmatrix} g_{22}(t)+\gamma\varphi(t) & \alpha g(t) \\ \beta\varphi(t) & g_{22}(t) \end{pmatrix}.$$

A direct verification shows that, for any $t_1, t_2 \in \Gamma$, $G(t_1)G(t_2) = G(t_2)G(t_1)$.

Vice versa, let the matrix function G be functionally commutable. At first, suppose that $g_{11}(t) \equiv g_{22}(t)$. If in addition, $g_{12} = g_{21} = 0$, then the conditions (4.21) are satisfied for $\alpha = \beta = \gamma = 0$. In the opposite case, a point $t_0 \in \Gamma$ can be found at which at least one of the functions g_{12} or g_{21} is different from zero. Equating the diagonal elements of the matrices $G(t)G(t_0)$ and $G(t_0)G(t)$, we get $g_{12}(t)g_{21}(t_0) = g_{12}(t_0)g_{21}(t)$. If $g_{21}(t_0) \neq 0$, we may take $\varphi = g_{21}$, $\beta = 1$, $\gamma = 0$, $\alpha = g_{12}(t_0)/g_{21}(t_0)$, and if $g_{12}(t_0) \neq 0$, then we can choose $\varphi = g_{12}$, $\alpha = 1$, $\gamma = 0$, $\beta = g_{21}(t_0)/g_{12}(t_0)$.

Now let the functions g_{11} and g_{22} be different. Then there exists a point $t_0 \in \Gamma$ such that $g_{11}(t_0) \neq g_{22}(t_0)$. Equating the elements outside the diagonals of the matrices $G(t)G(t_0)$ and $G(t_0)G(t)$, we obtain

$$(g_{11}(t)-g_{22}(t))\, g_{12}(t_0) = (g_{11}(t_0)-g_{22}(t_0))\, g_{12}(t),$$

$$(g_{11}(t)-g_{22}(t))\, g_{21}(t_0) = (g_{11}(t_0)-g_{22}(t_0))\, g_{21}(t),$$

so that we may set $\varphi = g_{11}-g_{22}$, $\gamma = 1$, $\alpha = g_{12}(t_0)/(g_{11}(t_0)-g_{22}(t_0))$, $\beta = g_{21}(t_0)/(g_{11}(t_0)-g_{22}(t_0))$, and Lemma 4.4 has been proved. ≡

THEOREM 4.12. Let the measurable bounded matrix function

$$G = \begin{pmatrix} g_{11} & g_{12} \\ g_{21} & g_{22} \end{pmatrix}$$

be functionally commutable. It is Φ-factorable in L_p iff the functions

$$\lambda_1 = \tfrac{1}{2}(g_{11}+g_{22}) + \nu\varphi \quad \text{and} \quad \lambda_2 = \tfrac{1}{2}(g_{11}+g_{22}) - \nu\varphi$$

are Φ-factorable in L_p, where ν is an arbitrary value of the square root of $\alpha\beta + \frac{1}{4}\gamma^2$, and α, β, γ and φ are defined by formulae (4.21).

In this case the partial indices of G coincide with the indices of the functions λ_1 and λ_2.

Proof. In accordance with Theorem 4.11, the matrix function G can be converted into a diagonal or a triangular form with equal elements on the diagonal by means of a constant similarity transformation. Now we intend to describe the concrete form of these transformations depending on the parameter values involved in formulae (4.21).

a) $\nu \neq 0,\ \alpha \neq 0$:

$$G(t) = -\frac{1}{2\alpha\nu}\begin{pmatrix} \alpha & \alpha \\ -\frac{\gamma}{2}+\nu & -\frac{\gamma}{2}-\nu \end{pmatrix}\begin{pmatrix} \lambda_1 & 0 \\ 0 & \lambda_2 \end{pmatrix}\begin{pmatrix} -\frac{\gamma}{2}-\nu & -\alpha \\ \frac{\gamma}{2}-\nu & \alpha \end{pmatrix};$$

b) $\nu \neq 0,\ \alpha = 0$:

$$G(t) = \begin{pmatrix} 1 & 0 \\ \frac{\beta}{2\nu} & 1 \end{pmatrix}\begin{pmatrix} \lambda_1 & 0 \\ 0 & \lambda_2 \end{pmatrix}\begin{pmatrix} 1 & 0 \\ -\frac{\beta}{2\nu} & 1 \end{pmatrix};$$

c) $\nu = 0,\ \alpha \neq 0$:

$$G(t) = \frac{1}{2\sqrt{\alpha}}\begin{pmatrix} 0 & \sqrt{\alpha} \\ 1 & i\sqrt{\beta} \end{pmatrix}\begin{pmatrix} g_{11}+g_{22} & 0 \\ 2\sqrt{\alpha}\,\varphi & g_{11}+g_{22} \end{pmatrix}\begin{pmatrix} -i\sqrt{\beta} & \sqrt{\alpha} \\ 1 & 0 \end{pmatrix};$$

d) $\nu = \alpha = 0$:

$$G(t) = \begin{pmatrix} g_{22} & 0 \\ \beta\varphi & g_{22} \end{pmatrix}.$$

Since, for $\nu = 0$, $\lambda_1 = \lambda_2 = \frac{1}{2}(g_{11}+g_{22})$ and, for $\nu = \alpha = 0$, the relation $g_{11} = g_{22}$ is valid, the assertion to be proved follows now from Theorem 4.11. Thus, Theorem 4.12 is proved. ≡

As an example we want to consider a second-order orthogonal[1] matrix with determinant equal to 1. As can be easily verified, its general form is given by the formula

$$\begin{pmatrix} \sqrt{1-\varphi^2} & \varphi \\ -\varphi & \sqrt{1-\varphi^2} \end{pmatrix},$$

[1] A matrix A is called orthogonal if $AA'=I$. For an orthogonal matrix one has $\det A = \pm 1$.

where one has to choose one and the same value of the root. Consequently, a second-order matrix function all values of which are orthogonal matrices with determinant equal to 1 satisfies the conditions (4.21) for $\gamma = 0$, $\alpha = 1$, $\beta = -1$ and is therefore functionally commutable. Using the formula of case a) from the proof of Theorem 4.12, we obtain

$$G(t) = \frac{1}{2i}\begin{pmatrix} 1 & 1 \\ i & -i \end{pmatrix}\begin{pmatrix} \sqrt{1-\varphi^2} + i\varphi & 0 \\ 0 & \sqrt{1-\varphi^2} - i\varphi \end{pmatrix}\begin{pmatrix} i & 1 \\ i & -1 \end{pmatrix}.$$

Hence, the matrix function under study is Φ-factorable only simultaneously with the functions $\sqrt{1-\varphi^2} + i\varphi$ and $\sqrt{1-\varphi^2} - i\varphi$, and its partial indices coincide with the indices of these functions.

A second-order orthogonal matrix function with determinant -1 is, generally speaking, not functionally commutable. However, after multiplication by the matrix $\begin{pmatrix} 1 & 0 \\ 0 & -1 \end{pmatrix}$ from an arbitrary side, it changes over into an orthogonal matrix function with determinant 1 and, consequently, becomes functionally commutable.

In this way, if G is a second-order orthogonal matrix function with determinant equal to -1, then

$$G = \begin{pmatrix} \sqrt{1-\varphi^2} & \varphi \\ \varphi & -\sqrt{1-\varphi^2} \end{pmatrix} = \frac{1}{2i}\begin{pmatrix} 1 & 1 \\ -i & i \end{pmatrix}\begin{pmatrix} \sqrt{1-\varphi^2} + i\varphi & 0 \\ 0 & \sqrt{1-\varphi^2} - i\varphi \end{pmatrix}\begin{pmatrix} i & 1 \\ i & -1 \end{pmatrix}.$$

Consequently, for an orthogonal matrix function with negative determinant, the result obtained above for an orthogonal matrix function with positive determinant continues to be valid.

4.5. COMMENTS

The criterion of Φ-factorability of a block-triangular matrix function in L_p (Theorem 4.1) and the Theorems 4.2, 4.3, 4.5 resulting from it are due to one of the authors (SPITKOVSKIǏ [11,12]).

In Section 1 of the next chapter it will be shown that, for matrix functions from the classes $L_\infty^{\pm} + C$, the Φ-factorability is equivalent to the invertibility in these classes. Therefore, Theorem 4.1 is obvious for such matrix functions.

Theorem 4.4 is a special case (for $p = 2$) of a result from SPITKOVSKII [8]. In connection with this theorem, note that, for $p = 2$, the decision when operators of the form $T_Q(\omega)$ are Φ_--operators, is well-known (DOUGLAS, SARASON [1]); the criterion consists in the fact that there must exist a function $\varphi \in H_\infty + C$ such that $\text{ess sup}|1-\varphi\omega^{-1}| < 1$. The proof of Theorem 4.6 is borrowed from the article of GOHBERG and KREĬN [3]. In the same article the order relation $\succ$ was introduced with the aim to study the question of stability of partial indices (concerning this topic, see Chapter 6). Furthermore, in this paper those properties of the relation $\succ$ were demonstrated which are gathered in Lemma 4.1. Theorem 4.7 based on this lemma was proved in SPITKOVSKIĬ [11,12]. The approach of "shifting" by k used in the proof of the mentioned theorem may also be applied in other cases. For instance, with the help of this approach, from results of MANDZHAVIDZE [7], it is derived in MANDZHAVIDZE [8,9] that the tuple of partial indices of the Riemann boundary value problem with shift depending on a parameter is constant. Moreover, it is majorized (for all values of the parameter with the possible exception of a finite number of points) in the sense of the relation $\succ$ by the tuple of partial indices of the corresponding problem without shift.

Theorem 4.8 and Corollary 4.3 (for $n = 2$) were established by CHEBOTAREV [2]. For $n > 2$, Corollary 4.3 was proved in GOHBERG, KREĬN [3], and Theorem 4.8 is due to NIKOLAĬCHUK [1] and PRIMACHUK [2]. Here a new proof of the necessity has been presented.

We refer also to another result from LITVINCHUK, SPITKOVSKIĬ [2] according to which the partial indices of a triangular matrix function coincide with the indices of the diagonal elements if and only if this matrix function admits a factorization with factors $G_\pm$ of the same triangularity. In the paper GAVDZINSKIĬ, SPITKOVSKIĬ [1] (see also ISKENDEROV [1]) the reduction of the factorization problem of a second-order matrix function to the corresponding problem for a triangular matrix function is discussed.

The factorization of second-order matrix functions given on a contour

$\Gamma = \Gamma_1 \cup \Gamma_2$ bounding a doubly connected domain which are upper triangular on Γ_1 and lower triangular on Γ_2 was considered in Section 4.3 of CLANCEY, GOHBERG [3].

Corollary 4.4 was presented in NIKOLAǏCHUK [1], PRIMACHUK [2] and, in a somewhat weaker variant, in GOHBERG, KREǏN [3].

Theorem 4.9 and Corollary 4.5 resulting from it have been proved in NIKOLENKO [2]; in the present book a proof of Corollary 4.5 is described that is adapted to the factorization in L_p (in NIKOLAǏCHUK [1], PRIMACHUK [2], CHEBOTAREV [2] the factorization of Hölder matrix functions was considered). Theorem 4.10 was published in SPITKOVSKIǏ [11,12]. The question of an effective factorization of triangular matrix functions with non-factorable diagonal elements has not been solved yet even for $n = 2$. This question is very important (and difficult), since the consideration of convolution type equations on a finite interval leads to the factorization of exactly such matrix functions (cf. GLAZMAN, LYUBICH [1], KARLOVICH, SPITKOVSKIǏ [1], NOVOKSHENOV [1], PAL'TSEV [5], SPITKOVSKIǏ [15]). In the papers GLAZMAN, LYUBICH [1], KOMYAK [1-4], in principle, a way of effectively factorizing matrix functions of the kind

$$\begin{pmatrix} e^{iax} & 0 \\ h(x) & e^{-iax} \end{pmatrix}$$

defined on the real line was proposed, where $a \in \mathbb{R}$, $h^{\pm 1} \in M_\infty^+$ or $h^{\pm 1} \in M_\infty^-$.

In his review [3] F.D. GAKHOV formulated the problem of finding a matrix function possessing such algebraic properties which allows us to solve the factorization problem effectively. In particular, in this paper the possibility of an effective factorization of functionally commutable matrix functions was discussed. The corresponding investigations based on the algebraic properties of functionally commutable matrix functions proved in MOROZOV [1] (Lemmas 4.2 and 4.3) were accomplished by CHEBOTAREV [1]. The Theorem 4.11 formulated here is a slight modification of results of CHEBOTAREV [1] (see also CHEBOTAREV [3]), where he considered some classes of matrix functions not functionally commutable for which condition (4.3) is valid and, consequently, a factorization may be constructed effectively. Note that these classes contain, for example, the matrix function

$$G(w) = \begin{pmatrix} 1 & -i\sqrt{k^2-w^2} \\ -\dfrac{i}{\sqrt{k^2-w^2}} & 1 \end{pmatrix}, \quad w \in \mathbb{R}$$

given on the real line, the factorization of which was built in HEINS [1] in connection with the study of the diffraction problem on the half-plane.

The factorization of triangular matrix functions was employed in TSITSKISHVILI [4] in solving some problems from filtration theory, and the factorization of functionally commutable matrix functions was used in MEISTER [1], where by its help questions of solvability of integral equations of the first kind with a generalized power kernel were examined.

The factorization of a circulant was studied in the paper GORDIENKO [1]. In GOHBERG, FELDMAN [1] the factorization of matrix functions was discussed which differ from the circulant (4.19) by the presence of the factor t^{-1} at the elements located below the main diagonal.

The paper GOHBERG, FELDMAN [2] is devoted to the generalization of these results to the case when $c_0, c_1, \ldots, c_{n-1}$ are quadratic blocks. Note that the matrix functions considered in GOHBERG, FELDMAN [1,2], generally speaking, fail to be functionally commutable. A detailed explanation of the results from GOHBERG, FELDMAN [1,2] can be found in Section 6.1 of the monograph CLANCEY, GOHBERG [3].

Lemma 4.4 is due to LAPPO-DANILEVSKIĬ [1], p. 82, and Theorem 4.12 to CHEBOTAREV [1].

The factorization of complex orthogonal matrix functions was considered by DIDENKO and CHERNETSKIĬ [1].

CHEBOTAREV's results [1,2] were applied in LE-DINZON [1] in order to obtain sufficient conditions for the effective solution of the Hilbert problem and in VOROB'EV [1] for the factorization of 3×3 matrix functions of a special kind occurring in problems of elasticity theory.

In VOROB'EV [1] the method of separating zeros was also employed for solving the factorization problem. Furthermore, this method was used in TROITSKIǏ [1], where the problem of finding the Jost matrix for the mixture of partial waves in case of a rational dependence of the mixture angle tangent on the squared impulse module was reduced to the factorization problem for a special matrix function.

There is a close connection between the factorization problem for functionally commutable matrix functions and matrix functions of permutation type, i.e. such matrix functions G for which, for every $t \in \Gamma$, exactly one element from any row and column of $G(t)$ is different from zero. If this nontrivial element is equal to one, then $G(t)$ provides a matrix of some permutation, where G is functionally commutable iff the permutations associated with the values of the matrix function G generate an Abelian group. Basing on this fact and making use of well-known results on the structure of finite Abelian groups, KRUGLOV [2,3] succeeded in constructing a factorization of such functionally commutable matrix functions of permutation type and calculated their partial indices. He also studied in [3-6] the factorization of some classes of permutation type matrix functions which are not functionally commutable. In his investigations he used profound connections between the Riemann problem for a pair of functions on a Riemann surface and the corresponding problem for n pairs of functions on the plane with matrix coefficient of permutation type established in È.I. ZVEROVICH [1] (see also ZVEROVICH, POMERANTSEVA [1]). In turn, the results concerning the solution of the factorization problem are applied in KRUGLOV [2-6] to the constructive determination of a field of algebraic functions defined by a monodromy group.

The factorization problem for permutation type matrix functions that are not functionally commutable was considered by PRIMACHUK [3,4] either. Let us point out that the factorization problem for permutation type matrix functions appeared first in CHEREPANOV's papers [1-3] related to certain problems of elasticity theory.

In connection with the question of an effective solution of the factorization problem for matrix functions we would still like to mention the article of KRAVCHENKO, NIKOLAǏCHUK [1], where it was shown that, if some (n-1) th order minor of the $n \times n$ matrix function G is factored, then the factorization problem of the matrix function G itself can be reduced to the solution of a special type integral equation with respect to one unknown function.

In PAL'TSEV [2,6], POPOV [1], KHRAPKOV [1], HURD [1], RAWLINS, WILLIAMS [1] one can find several approaches different from the ones pursued here, which allow for effectively factorizing some new classes of matrix functions occurring in applied problems. The search for such approaches is still an actual task because of the non-existence of general methods for constructing factorizations.

Finally, let us focus our attention on another unsolved problem. We have in mind the problem of describing all possible tuples $(\varkappa_j)$ of partial indices of a triangular matrix function with fixed diagonal elements. According to Theorem 4.7, in case of factorable diagonal elements the relation $(k_j) \succ (\varkappa_j)$ necessarily holds, where k_j are the diagonal element indices. For $n = 2$, it follows from Corollary 4.5 that, thanks to the choice of an element outside the diagonal, we succeed in obtaining an arbitrary tuple $(\varkappa_j)$ satisfying this condition. However, it is not clear whether this property preserves for $n > 2$. In case of given non-factorable diagonal elements, the question of the description of all possible tuples $(\varkappa_j)$ remains open even for $n = 2$.

CHAPTER 5. SOME CLASSES OF FACTORABLE MATRIX FUNCTIONS

In this chapter we consider the factorization problem for some special classes of matrix functions with specific analytic properties. In Section 5.1 the criterion of Φ-factorability in L_p of matrix functions of the classes $L_\infty^{\pm} + C$ will be derived. Section 5.2 is dedicated to the factorization in some Banach algebras of functions, and in Section 5.3 we shall study the Φ-factorization of piecewise continuous matrix functions. The consideration of classes of special matrix functions is useful by making general factorability and Φ-factorability criteria proved in the previous more precise. Moreover, via this approach we are able to clarify some principal facts of factorization theory.

In Section 3.6 we have established that the Φ-factorability of a bounded measurable matrix function G on a contour of the class $\mathfrak{R}$ is equivalent to the Fredholmness of the singu ar integral operator $P + GQ$. Naturally, there arises the question of existence of closed subclasses of L_∞ such that the Φ-factorability criterion of matrix functions from these subclasses can be immediately expressed in terms of the matrix function G itself. It appeared (see Section 3.1) that $L_\infty^{\pm} + C$ are just subclasses in L_∞ of such a type: for any p, the Φ-factorability of a matrix function $G \in L_\infty^{\pm} + C$ is equivalent to the membership of the inverse matrix G^{-1} to the same class. For continuous matrix functions, this criterion looks quite simple: the Φ-factorability in L_p of the matrix function $G \in C$ is, for any p, equivalent to the non-singularity of G.

In Chapter 3 it was also shown that the condition $G^{-1} \in L_\infty$ is a necessary condition for Φ-factorability of the matrix function $G \in L_\infty$ in L_p and the gap between this condition and the exact condition of Φ-factorability is rather large. In Section 5.1 below we shall prove that, for a matrix function $G \in L_\infty^{\pm} + C$, the condition $G^{-1} \in L_\infty$ is equivalent to the demand that $P + GQ$ is a Φ_+- or Φ_--operator. Thus, the consideration of classes $L_\infty^{\pm} + C$ allowed for more precisely characterizing the gap between necessary and exact condition of Φ-factorability mentioned above: in case $G \in L_\infty^{+} + C$ $(L_\infty^{-} + C)$, the operator $P + GQ$

is a Φ_-- (Φ_+-) operator, but not Fredholm if and only if $G^{-1} \in L_\infty$ and $G^{-1} \notin L_\infty^+ + C$ $(L_\infty^- + C)$.

The following principal fact is connected with the statement proved below (in 5.1): for matrix functions $G \in L_\infty^\pm + C$ Φ-factorable in L_p , the factorization factors are summable with arbitrary power p . The weakness of this result strikes the eye, however, it turns out that it cannot be improved.

In Section 5.2 it will be established that the factorization factors really go beyond the initial class $L_\infty^\pm + C$. From here, in particular, it follows that it is inevitable for the factorization factors of continuous matrix functions to go beyond the "natural" classes $C^\pm$ of matrix functions analytic in $\mathfrak{D}^\pm$ and continuous in $\mathfrak{D}^\pm \cup \Gamma$. In connection with this, it is necessary to find such classes (narrower than L_∞ or C), which, along with every matrix function, contain also its factorization factors.

In Section 5.2 we shall clarify the conditions under which the role of such classes may be played by the classes of matrix functions with entries from Banach algebras of functions containing R and belonging to L_∞ or C (in what follows they are called (L_∞,R)- and (C,R)-algebras, respectively). Furthermore, the connection between factorability in such algebras and the Fredholmness of the operators $P + GQ$ and $Q + G'P$ in the corresponding spaces will be investigated, where we shall essentially use the analogy with results of Chapter 3.

A special case of a (C,R)-algebra are R-algebras, which are, by definition, Banach algebras of functions containing R as a dense subset. It appears that all non-singular matrix functions with entries from some R-algebra $\mathfrak{C}$ admit a factorization with factors of the same class iff the algebra is decomposing, that means, it consists only of elements analytically continuable in $\mathfrak{D}^+$ and $\mathfrak{D}^-$ as well as sums of such elements. For a more general case of a (C,R)-algebra $\mathfrak{A}$, decomposition continues to be a necessary condition of factorability in $\mathfrak{A}$ for all non-singular matrix functions with entries from $\mathfrak{A}$. Unfortunately, a necessary and sufficient condition can be formulated only in a more

complicated way. This condition will be discussed in the comments 5.4. In the main text of the chapter only one sufficient condition will be proved, and various corollaries will be obtained from it.
A concrete example of a decomposing R-algebra important for application is the algebra W of functions defined on the unit circle and representable in the form of absolutely convergent Fourier series. The algebra H^{α} of functions satisfying a Hölder condition with exponent $0<\alpha \leq 1$ provides an example of a (C,R)-algebra which is not an R-algebra.
In Section 5.2 we present also a local principle which allows for constructing new classes of matrix functions, which admit a factorization with continuous factors, by sticking together matrix functions from the classes indicated.

The results of Section 5.3 on Φ-factorability of piecewise continuous matrix functions rest essentially upon the factorization of the power function (considered in Chapter 2) and enlarge our knowledge on the structure of the domain of Φ-factorability.
Unlike the case of matrix functions from $L_{\infty}^{\pm} + C$, the domain of Φ-factorability of piecewise continuous matrix functions may be not only an empty set or the ray $(1,\infty)$, but may consist of a finite number of connected components. Correspondingly, the complement to a non-empty domain of Φ-factorability of piecewise continuous matrix functions with respect to the ray $(1,\infty)$ consists of a finite number of isolated values of the parameter p.
The characterization of the results of the present chapter would be incomplete, if we did not indicate that for Φ-factorable continuous and piecewise continuous matrix functions effective formulae will be presented with the aid of which the total index can be evaluated.
In addition to comments on references, in Section 5.4 we shall discuss the question of relationships between Φ-factorization in L_p of piecewise Hölder matrix functions and the classical factorization problem for these matrix functions in special classes introduced by N. I. MUSKHELISHVILI and named after him as well as the Φ-factorization

problem in weighted L_p spaces.

5.1. Φ-FACTORIZATION OF MATRIX FUNCTIONS FROM CLASSES $L_\infty^{\pm} + C$

In this section as well as in the whole previous chapter, the contour Γ is assumed to belong to the class $\mathfrak{R}$. Under this condition, for any matrix function G belonging together with its inverse to L_∞, the following relations are satisfied:

$$T_Q(G^{-1})T_Q(G) = (I - QG^{-1}PGQ)|_{\operatorname{im} Q}, \tag{5.1}$$

$$T_Q(G)T_Q(G^{-1}) = (I - QGPG^{-1}Q)|_{\operatorname{im} Q}. \tag{5.2}$$

Employing formulae (3.22), these equations can be immediately verified.

The key to the investigation of Φ-factorization of matrix functions from $L_\infty^{\pm} + C$ is, in addition to relations(5.1) and (5.2), the following simple statement.

<u>LEMMA 5.1.</u> If $G \in L_\infty^+ + C$ $(L_\infty^- + C)$, then the operator QGP (PGQ) is compact in all spaces L_p, $1 < p < \infty$.

<u>Proof.</u> According to the definition of $L_\infty^+ + C$, a matrix function G from this class can be represented in the form $G = G_0 + G_+$, where $G_0 \in C$, $G_+ \in L_\infty^+$. Now we introduce a sequence of rational matrix functions F_k uniformly converging to G_0. Its poles are assumed to lie off Γ. Moreover, we set $G_k = F_k + G_+$. Obviously $G_k \in M_\infty^+$, thus, according to Theorem 1.35, the operators QG_kP defined on the whole space L_p^n are bounded and finite dimensional. Furthermore, the sequence $\{QG_kP\}$ converges to QGP in the operator topology. Consequently, the operator QGP is the limit of a uniformly convergent sequence of finite-dimensional bounded operators and therefore compact. The case $G \in L_\infty^- + C$ can be treated analogously. ≡

<u>THEOREM 5.1.</u> Let $G \in L_\infty^+ + C$ $(L_\infty^- + C)$. Then $P + GQ$ is a Φ_-- $(\Phi_+$-$)$ operator iff $G^{-1} \in L_\infty$.

<u>Proof.</u> The necessity of the condition $G^{-1} \in L_\infty$ for $P + GQ$ to be a Φ_-- $(\Phi_+$-$)$ operator results from Theorem 3.18 (for $A = I$, $B = G$) and holds regardless of the validity of the requirement $G \in L_\infty^{\pm} + C$.

If $G \in L_\infty^+ + C$, then, owing to Lemma 5.1, the operator QGP is compact. If in this case $G^{-1} \in L_\infty$, then the operator $QGPG^{-1}Q$ is also compact. Therefore, in view of relation (5.2), the operator $T_Q(G^{-1})$ is a right regularizer for $T_Q(G)$. With regard to Theorem 1.8, we conclude from this that $T_Q(G)$ is a Φ_--operator. Due to assertion 1) of Lemma 3.5, $P + GQ$ is a Φ_--operator, too.

Analogously, making use of (5.1), we establish that, for $G \in L_\infty^- + C$, $P + GQ$ is a Φ_+-operator if $G^{-1} \in L_\infty$. This proves Theorem 5.1. ≡

COROLLARY 5.1. If $G \in L_\infty^+ + C$ $(L_\infty^- + C)$, then the property of the operator $P + GQ$ to be a Φ_+- (Φ_--) operator is equivalent to its Fredholmness.

In fact, if $P + GQ$ is a Φ_+- (Φ_--) operator, then it follows that $G^{-1} \in L_\infty$. If, in addition, $G \in L_\infty^+ + C$ $(L_\infty^- + C)$, then, in view of Theorem 5.1, the operator $P + GQ$ is a Φ_-- (Φ_+-) operator, and, consequently, Fredholm. ≡

However, from the fact that $P + GQ$ is a Φ_-- (Φ_+-) operator and $G \in L_\infty^+ + C$ $(L_\infty^- + C)$ does not follow the Fredholmness. A class of examples of this kind was presented in the previous chapter after Theorem 4.4.

In order to get a criterion of Fredholmness of the operator $P + GQ$ in case $G \in L_\infty^\pm + C$, we need, in addition to Lemma 5.1, some more properties of the classes $L_\infty^\pm + C$ formulated below as Lemmas 5.2 and 5.3. As in Theorem 3.21, we denote by G_j the restriction of the matrix function G on the contour $\Gamma(j)$, $j = 0,\dots,m-1$.

LEMMA 5.2. The inclusion $G \in L_\infty^+ + C$ holds if and only if $G_0 \in (L_\infty^+ + C)(\Gamma(0))$ and $G_j \in (L_\infty^- + C)(\Gamma(j))$, $j=1,\dots,m-1$. The inclusion $G \in L_\infty^- + C$ is valid iff $G_0 \in (L_\infty^- + C)(\Gamma(0))$ and $G_j \in (L_\infty^+ + C)(\Gamma(j))$, $j=1,\dots,m-1$.

Proof. Let $G \in L_\infty^+ + C$, i.e. $G = \Psi + F$, where $\Psi \in L_\infty^+$ and $F \in C$. Set $\Psi_j(z) = \frac{1}{2\pi i} \int_{\Gamma(j)} \Psi(\tau) \frac{d\tau}{\tau - z}$, $j=0,\dots,m-1$.

Then $\Psi_j \in E(\mathcal{D}^-(j))$, $j=1,\dots,m-1$, $\Psi_0 \in E(\mathcal{D}^+(0))$, and $\Psi = \sum_{j=0}^{m-1} \Psi_j$.

Each of the matrix functions Ψ_j is continuous on Γ with the exception of the curve $\Gamma(j)$. From here and $\Psi \in L_\infty^+$ it follows that $\Psi_j \in L_\infty$. With regard to Theorems 1.25 and 1.33, we have $\Psi_j \in E_\infty(\mathcal{D}^-(j))$, $j=1,\dots,m-1$, $\Psi_0 \in E_\infty(\mathcal{D}^+(0))$.

Now we define a continuous matrix function H on the contour Γ by setting $H_j = F_j + \sum_{s\neq j} \Psi_s$. Then $G_j = \Psi_j + H_j$ and, therefore

$G_0 \in (L_\infty^+ + C)(\Gamma(0))$, $G_j \in (L_\infty^- + C)(\Gamma(j))$, $j=1,\dots,m-1$.

Conversely, if $G_j = \Psi_j + H_j$, where H_j is a matrix function continuous on $\Gamma(j)$, $\Psi_j \in E_\infty(\mathcal{D}^-(j))$ for $j=1,\dots,m-1$, and $\Psi_0 \in E_\infty(\mathcal{D}^+(0))$, then $G = \Psi + F$ for $\Psi = \sum_{j=0}^{m-1} \Psi_j$ and $F_j = H_j - \sum_{s\neq j} \Psi_s$. Hence $G \in L_\infty^+ + C$.

The assertion related to a matrix function $G \in L_\infty^- + C$ may be proved in an analogous manner. ≡

<u>LEMMA 5.3.</u> The classes $L_\infty^\pm + C$ are closed in L_∞.

<u>Proof.</u> Lemma 5.2 enables us to restrict ourselves to the case of a connected contour Γ. Since a conformal mapping of a simply connected domain with Jordan boundary onto the circle may be extended to a continuous mapping of the closure of this domain onto the closed disk (VLADIMIROV, VOLOVICH [1], p.46), it is sufficient to verify the closedness of the classes $H_\infty^\pm + C$. Denote by C^+ the subspace of C consisting of boundary values of functions analytic in the unit disk, and consider the mapping i of the factor space $C|C^+$ into the factor space $L_\infty|H_\infty$ defined by the equation

$$i(f + C^+) = f + H_\infty, \quad f \in C.$$

For any function $f \in L_\infty$, we have $\inf_{h\in H_\infty}\|f-h\| \le \inf_{h\in C^+}\|f-h\|$, thus, the mapping i is a contraction. Now we intend to show that, of course, it is isometric, i.e., $\inf_{h\in H_\infty}\|f-h\| = \inf_{h\in C^+}\|f-h\|$ holds for every $f \in C$.

For this purpose, we choose $h \in H_\infty$ arbitrarily and consider the k th Cesaro means f_k and h_k of the functions f and h, respectively. Clearly, $\|f-h_k\| \le \|f-f_k\| + \|f_k-h_k\| \le \|f-f_k\| + \|f-h\|$.

Since $\lim_{k\to\infty} \|f-f_k\| = 0$ $h_k \in C^+$, from the latter inequality we obtain the assertion to be proved. Therefore, the image $\mathfrak{C}$ of the whole subspace $C|C^+$ under the mapping i must be closed. But then the pre-image of the subspace $\mathfrak{C}$ under the natural projection of L_∞ onto $L_\infty|H_\infty$ is also closed. In order to prove the closedness of $H_\infty + C$ in L_∞ , it remains to note that $H_\infty + C$ is just this pre-image. The class $H_\infty^- + C$ is obtained from $H_\infty + C$ by applying conjugation. Consequently, it is closed along with the class $H_\infty + C$, and the proof of Lemma 5.3 is complete. ≡

COROLLARY 5.2. The classes $L_\infty^+ + C$ and $L_\infty^- + C$ are subalgebras of L_∞ .

Proof. It is easy to check that the product of a rational function by a function from L_∞^+ may be represented as the sum of a rational function and a function from L_∞^+ . Hence, it follows that the multiplication does not go beyond the class M_∞^+ . Therefore, M_∞^+ is a subalgebra of L_∞ . Since M_∞^+ is dense in $L_\infty^+ + C$, and $L_\infty^+ + C$ is a closed subset of L_∞ (in view of the previous lemma), $L_\infty^+ + C$ is just the closure of M_∞^+ . Thus, it is a subalgebra simultaneously with M_∞^+ . The class $L_\infty^- + C$ can be considered in the same way. ≡

THEOREM 5.2. Let $G \in L_\infty^+ + C$ $(L_\infty^- + C)$. Then the condition $G^{-1} \in L_\infty^+ + C$ $(L_\infty^- + C)$ is necessary for Φ-factorability of the matrix function G for at least one $p \in (1,\infty)$ and sufficient for its Φ-factorability for all $p \in (1,\infty)$. If this condition is satisfied, then each factorization of G has the properties

$$G_+^{\pm 1} \in \bigcap_{p<\infty} L_p^+ , \quad G_-^{\pm 1} \in \bigcap_{p<\infty} L_p^- \tag{5.3}$$

and is a Φ-factorization in all spaces L_p , $1 < p < \infty$.

Proof. Assume $G \in L_\infty^+ + C$. We choose a sequence $G_k \in M_\infty^+$ uniformly converging to G . If, for some $p \in (1,\infty)$, the matrix function G is Φ-factorable in L_p , then, according to Corollary 3.4, $G^{-1} \in L_\infty$. Therefore, for k large enough, the matrix functions G_k are invertible and $G_k^{-1} \to G^{-1}$. Moreover, since the property of Φ-factorability is stable (Theorem 3.20), we conclude the Φ-factorability of G_k

(again for sufficiently large k). Owing to Theorem 3.15, $G_k^{-1} \in M_\infty^+$. From here and the closedness of the class $L_\infty^+ + C$, it follows that $G^{-1} \in L_\infty^+ + C$.

Conversely, if $G^{\pm 1} \in L_\infty^+ + C$, then formulae (5.1) and (5.2) show that the operator $T_Q(G^{-1})$ is a two-sided regularizer of the operator $T_Q(G)$ in all subspaces $(\overset{\circ}{L}{}_p^-)^n$, $1 < p < \infty$. Consequently, the operator $T_Q(G)$ (and, thus, also $P + GQ$) is Fredholm. By Theorem 3.16, the matrix function G is Φ-factorable in all spaces L_p, $1 < p < \infty$. Now we are going to show that its factorization does not depend upon p. Since the regularizer of the operator $T_Q(G)$ constructed above is one and the same in all spaces $(\overset{\circ}{L}{}_p^-)^n$, $1 < p < \infty$, then its index also does not depend upon p and, therefore, the total index of G does not depend upon p, too. According to Corollary 2.2 and Theorem 3.8, any factorization of G is its Φ-factorization in all L_p, $1 < p < \infty$. From this fact we get, in particular, the property (5.3). Thus Theorem 5.2 has been proved. ≡

<u>COROLLARY 5.3.</u> Suppose $G \in M_\infty^+$ (M_∞^-). Then the conditions $G^{-1} \in M_\infty^+$ (M_∞^-) and $G^{-1} \in L_\infty^+ + C$ $(L_\infty^- + C)$ are equivalent.

In fact, with regard to Theorem 3.15 (3.15'), the condition $G^{-1} \in M_\infty^+$ (M_∞^-) is necessary and sufficient for the Φ-factorability of the matrix function $G \in M_\infty^+$ (M_∞^-). At the same time, Φ-factorability of G is, thanks to Theorem 5.2, equivalent to the requirement $G^{-1} \in L_\infty^+ + C$ $(L_\infty^- + C)$. ≡

<u>COROLLARY 5.4.</u> Suppose $G \in L_\infty^+ + C$ $(L_\infty^- + C)$. Then $P + GQ$ is a Φ_-- $(\Phi_+$-$)$ operator but not Fredholm if and only if $G^{-1} \in L_\infty$ but $G^{-1} \notin L_\infty^+ + C$ $(L_\infty^- + C)$.

<u>COROLLARY 5.5.</u> The domain of Φ-factorability of the matrix function $G \in L_\infty^\pm + C$ is either empty (in case $G^{-1} \notin L_\infty^\pm + C$) or consists of a single component occupying the whole ray $(1,\infty)$ (in case $G^{-1} \in L_\infty^\pm + C$).

<u>COROLLARY 5.6.</u> The matrix function $G \in L_\infty^\pm + C$ admits or does not admit a right Φ-factorization only simultaneously with a left one,

and its left total index coincides with its right total index.

Proof. In virtue of Theorem 3.7', the matrix function $G \in L_\infty$ is right Φ-factorable in L_p iff the matrix function G' is left Φ-factorable in L_q. But if $G \in L_\infty^\pm + C$, then also $G' \in L_\infty^\pm + C$, and by Theorem 5.2, the left Φ-factorability of G' in L_q is equivalent to the requirement $(G')^{-1} \in L_\infty^\pm + C$ or, what is the same, $G^{-1} \in L_\infty^\pm + C$. The latter demand is, again by Theorem 5.2, equivalent to the left Φ-factorability of G in L_p. With regard to Theorem 3.20, it suffices to prove the coincidence of the left total index with the right one for matrix functions from the class $M_\infty^\pm$, which is dense in $L_\infty^\pm + C$. For matrix functions of this class, the assertion to be proved is valid by Theorems 2.6 and 2.7. ≡

Let us consider in more detail that important case when the matrix function under study is continuous.

THEOREM 5.3. Let the matrix function G be continuous. Its non-singularity on Γ is necessary for the Φ-factorability of G in at least one L_p and sufficient for its Φ-factorability in all L_p, $1 < p < \infty$. Furthermore, the total index of a non-singular continuous matrix function G can be calculated by the formula

$$\varkappa = \frac{1}{2\pi}\{\arg \det G(t)\}_\Gamma . \tag{5.4}$$

Proof. If the matrix function G is Φ-factorable for at least one value $p \in (1,\infty)$, then, by Corollary 3.5, $G^{-1} \in L_\infty$. Hence the function $\det G^{-1}$ is bounded and, consequently, the continuous function $\det G$ is separated from zero. Therefore, G is non-singular on Γ. Vice versa, if the continuous matrix function G is non-singular on Γ, then the inverse matrix function G^{-1} is also continuous. By Theorem 5.2, G is Φ-factorable in all L_p, $1 < p < \infty$.

Now we intend to prove formula (5.4). Both parts of this formula are continuous functions, if we consider them on the set $\hat{C}$ of continuous non-degenerating matrix functions with the topology of uniform convergence. According to Theorem 2.6 and statement 4) of Theorem 1.29, formula (5.4) holds on the set of rational matrix functions not having poles and being non-singular on Γ. Since this set is dense in $\hat{C}$, formula (5.4) is true for all continuous non-singular matrix functions. Thus Theorem 5.3 is proved. ≡

COROLLARY 5.7. Assume G to be a continuous non-degenerating matrix function. Its partial indices are equal to zero (non-positive, non-negative) if and only if the operator $P + GQ$ is invertible (right invertible, left invertible) in the space L_p^n for at least one $p \in (1,\infty)$.

Corollary 5.7 is a reformulation of Corollary 3.8, taking into account that the Fredholmness of the operator $P + GQ$ is guaranteed by the continuity and non-singularity of G.

COROLLARY 5.8. A continuous non-singular matrix function admits a right Φ-factorization in any space L_p, $1 < p < \infty$, where the total right index coincides with the total left one.

Proof. In order to prove the existence of a right Φ-factorization, it is necessary to apply Theorem 5.3 to the matrix function G' and, after this, to use Theorem 3.7. The sum of right indices of the matrix function G coincides with the sum of left indices of G'. Consequently, it is equal to $\frac{1}{2\pi}\{\arg \det G'(t)\}_\Gamma = \frac{1}{2\pi}\{\arg \det G(t)\}_\Gamma$, i.e., it coincides with the sum of left indices of G. ≡

In connection with Theorem 5.3, note that the sufficiency of the non-singularity condition of a continuous matrix function G for its Φ-factorability can be easily proved by means of the local principle, i.e., with the aid of Theorem 3.2. Indeed, the continuous matrix function G is, at each point $t_0 \in \Gamma$, ε-locally equivalent to its value at this point. The Φ-factorability of a constant non-singular matrix $G(t_0)$ is evident.

The factorization factors $G_\pm$ of G belonging together with its inverse to the class $L_\infty^+ + C$ $(L_\infty^- + C)$ possess the property (5.3). Therefore, they themselves and their inverses are summable with any finite power. With the help of Theorem 2.4 we are able to prove a local variant of this assertion.

THEOREM 5.4. Let the matrix function G be factorable in L_p and G_1 be such that $G_1^{\pm 1} \in L_\infty^+ + C$ or $G_1^{\pm 1} \in L_\infty^- + C$, and G coincides with G_1 on some open arc γ of the contour Γ. Then the factorization factors $G_\pm$ of the matrix function G as well as their inverses are summable with any finite power on each arc $\tilde{\gamma}$ whose closure lies in γ.

COROLLARY 5.9. If the matrix function G is factorable in L_p, continuous and non-degenerating on the open arc $\gamma \subset \Gamma$, then the factorization factors $G_\pm$ of G and their inverses are summable with any finite power on every arc whose closure lies in γ.

THEOREM 5.5. Let $G_1^{\pm 1} \in L_\infty^+ + C$, $G_2^{\pm 1} \in L_\infty^- + C$. If the matrix function G is Φ-factorable in L_p, then G_1GG_2 is also Φ-factorable in L_p, where the total indices of the matrix functions G, G_1, G_2 and G_1GG_2 are connected by the relation $\varkappa(G_1GG_2) = \varkappa(G_1)+\varkappa(G)+\varkappa(G_2)$.

Proof. For arbitrary matrix functions $G_1, G, G_2 \in L_\infty$, the equation

$$T_Q(G_1GG_2) = QG_1GG_2|_{\operatorname{im} Q} = QG_1QGQG_2|_{\operatorname{im} Q} + QG_1PGQG_2|_{\operatorname{im} Q} + QG_1GPG_2Q|_{\operatorname{im} Q}$$

holds. By Lemma 5.1, the operators QG_1P and PG_2Q are compact if $G_1 \in L_\infty^+ + C$, $G_2 \in L_\infty^- + C$. Hence, the operators $T_Q(G_1GG_2)$ and $QG_1QGQG_2Q|_{\operatorname{im} Q} = T_Q(G_1)T_Q(G)T_Q(G_2)$ are Fredholm only simultaneously and have the same index. It remains to note that, in view of Theorem 5.2, the condition $G_1^{\pm 1} \in L_\infty^+ + C$ implies the Fredholmness of the operator $T_Q(G_1)$ and the condition $G_2^{\pm 1} \in L_\infty^- + C$ the Fredholmness of $T_Q(G_2)$. Due to Theorems 1.6 and 1.9, the product $T_Q(G_1)T_Q(G)T_Q(G_2)$ is Fredholm only simultaneously with the operator $T_Q(G)$ and has an index equal to the sum of the indices of the factors. Thus Theorem 5.5 is completely proved. ≡

COROLLARY 5.10. If the matrix functions G_1 and G_2 are continuous and non-singular on Γ, then the matrix functions G and G_1GG_2 are Φ-factorable only simultaneously, and their total indices are connected via the relation

$$\varkappa(G_1GG_2) = \varkappa(G) + \frac{1}{2\pi}\{\arg \det G_1(t)G_2(t)\}_\Gamma .$$

Owing to the inclusion $M_\infty^\pm \in L_\infty^\pm + C$, Theorem 5.5 is a generalization of Theorem 3.10. Theorem 5.5 and Corollary 5.10 are often useful, even if one of the matrix functions G_1 and G_2 is chosen equal to I.

5.2. FACTORIZATION IN BANACH ALGEBRAS OF MATRIX FUNCTIONS

Theorem 5.2 answers the question of Φ-factorability of continuous matrix functions completely. However, the boundary properties of the factorization factors established in this theorem (inclusion (5.3)) seem to be too weak. Consequently, we wish to strengthen them. The following example shows that it is impossible to do this without additional restrictions, even for continuous G and $n = 1$.

<u>EXAMPLE.</u> Let Γ be the unit circle $\mathbf{T}$ and

$$G(t) = \exp[\sum_{k=2}^{\infty} \frac{1}{k \ln k}(t^k - t^{-k})] .$$

G is a continuous and non-vanishing function. Setting

$$G_+(z) = \exp(\sum_{k=2}^{\infty} \frac{1}{k \ln k} z^k) , \quad G_-(z) = \exp(- \sum_{k=2}^{\infty} \frac{1}{k \ln k} z^{-k}) ,$$

$\Lambda(t) = 1$, we obtain its factorization (and, thus, also its Φ-factorization). Since the series $\sum_{k=2}^{\infty} \frac{1}{k \ln k}$ diverges, the function $\sum_{k=2}^{\infty} \frac{1}{k \ln k} z^k$ is not bounded from above on the interval $(0,1)$ and, consequently, its absolute value is not bounded from above in the open unit disk. Therefore $G_+ \notin H_\infty^+$. Hence, the relation $|G_+G_-| = |G| = 1$ implies $G_-^{-1} \notin H_\infty^-$.

In this way, it is actually necessary to go beyond the classes $L_\infty^\pm$ (and, moreover, the classes $C^\pm$ of functions analytic in $\mathcal{D}^\pm$ and continuous in $\mathcal{D}^\pm \cup \Gamma$) if we look for the factorization of all continuous matrix functions. Nevertheless, there exist continuous matrix functions the factorization factors $G_\pm$ of which satisfy the condition

$$G_+^{\pm 1} \in L_\infty^+ , \quad G_-^{\pm 1} \in L_\infty^- . \tag{5.5}$$

According to Theorem 2.9, all rational matrix functions not possessing poles and non-degenerating on Γ are of this type. The following problem arises in a natural way: find classes of matrix functions (as wide as possible) which admit a factorization with property (5.5). Going towards the solution of this problem, we come to the notion of factorization in a Banach algebra of functions.

In what follows, we shall consider algebras $\mathcal{A}$ intermediate between R and L_∞ :

$$R \subseteq \mathcal{A} \subseteq L_\infty .$$

In doing so, it is clear that 1 (the function identically equal to 1) is the unit element in $\mathcal{A}$. The algebra $\mathcal{A}$ is assumed to be supplied with a norm $\|\cdot\|_{\mathcal{A}}$ with respect to which it is complete. A Banach algebra of functions defined on Γ having the properties enumerated is called a (L_∞, R)-algebra.

A trivial example of such an algebra provides $\mathcal{A} = L_\infty$ with the natural norm

$$\|a\|_\infty = \text{ess sup}\{|a(t)| : t \in \Gamma\} .$$

The role of $\mathcal{A}$ can also be played by any subalgebra of L_∞, e.g. by $L_\infty^{\pm} + C$ or C. Other examples, in which $\|\cdot\|_{\mathcal{A}}$ is not equivalent to $\|\cdot\|_\infty$, will be given below.

<u>LEMMA 5.4.</u> Let $\mathcal{A}$ be a (L_∞, R)-algebra. Then we have:

1) for the invertibility of $a \in \mathcal{A}$ it is necessary that $a^{-1} \in L_\infty$,

2) for all $a \in \mathcal{A}$, the relation

$$\|a\|_{\mathcal{A}} \geq \|a\|_\infty \tag{5.6}$$

holds.

<u>Proof.</u> Assertion 1) is an obvious consequence of the inclusion $\mathcal{A} \subseteq L_\infty$. To prove 2), let us consider an arbitrary element $a \in \mathcal{A}$. One can specify a number ζ_0 such that $|\zeta_0| = \|a\|_\infty$ and $a - \zeta_0$ is not invertible in L_∞. In fact, in the opposite case we could, for every point ζ of the circle $C_0 = \{\zeta : |\zeta| = \|a\|_\infty\}$, indicate a neighbourhood the values of which are attained by the function a only on a set of measure zero [1]. Choosing from the obtained covering of C_0 a finite subcovering, we would get that a.e. the function a attains values which have a positive distance from C_0. But this contradicts the

[1] We assume $\zeta_0 \neq 0$, since otherwise $a \equiv 0$ and $\|a\|_{\mathcal{A}} = \|a\|_{L_\infty} = 0$.

definition of $\|a\|_\infty$. Thus, the existence of the desired ζ_0 has been proved. According to assertion 1) just proved, the non-invertibility of $a - \zeta_0$ in L_∞ implies its non-invertibility in $\mathcal{A}$. Along with $a - \zeta_0$, the element $a\zeta_0^{-1} - 1$ is also non-invertible in $\mathcal{A}$. But then $\|a\zeta_0^{-1}\|_{\mathcal{A}} \geq 1$, because otherwise from the contracting mapping principle we would obtain the invertibility of $a\zeta_0^{-1} - 1$.
Thus, $\|a\|_{\mathcal{A}} = \|a\zeta_0^{-1}\|_{\mathcal{A}} |\zeta_0| \geq |\zeta_0| = \|a\|_\infty$, which proves Lemma 5.4. ≡

The necessary invertibility condition stated in assertion 1) of Lemma 5.4 is, generally speaking, not sufficient. A suitable example yield the algebras $L_\infty^\pm + C$. The invertibility criteria in these algebras will be presented in Section 5.4. Nevertheless, in a number of important cases (for instance, for the algebra C) an assertion converse to 1) is true. With this in mind, the following definition is of interest.

DEFINITION 5.1. The (L_∞,R)-algebra $\mathcal{A}$ is said to have the invertibility property if the conditions $a \in \mathcal{A}$ and $a^{-1} \in L_\infty$ imply $a^{-1} \in \mathcal{A}$.

From assertion 2) of Lemma 5.4 there results now the following proposition.

COROLLARY 5.11. The convergence in a (L_∞,R)-algebra $\mathcal{A}$ implies the uniform convergence.

Setting $\mathcal{A}^\pm = \mathcal{A} \cap L_\infty^\pm$ and $\mathring{\mathcal{A}}^- = \mathcal{A} \cap \mathring{L}_\infty^-$, we observe that $\mathcal{A}^\pm$ and $\mathring{\mathcal{A}}^-$ are subalgebras of $\mathcal{A}$ closed in virtue of Corollary 5.11.

DEFINITION 5.2. The representation (2.1) with

$$G_+^{\pm 1} \in \mathcal{A}^+, \quad G_-^{\pm 1} \in \mathcal{A}^-, \tag{5.7}$$

where the matrix function Λ is defined by formula (2.2), is called a factorization of the matrix function G in the (L_∞,R)-algebra $\mathcal{A}$.

It is clear from this definition that only matrix functions belonging together with their inverses to the algebra $\mathcal{A}$ may admit a factorization in $\mathcal{A}$. With this in mind, remark that for a matrix function $G \in \mathcal{A}$ the condition $G^{-1} \in \mathcal{A}$ is fulfilled if and only if $(\det G)^{-1} \in \mathcal{A}$. If the algebra $\mathcal{A}$ has the invertibility property, then, for a matrix function $G \in \mathfrak{A}$, the conditions $G^{-1} \in \mathfrak{A}$ and $G^{-1} \in L_\infty$ are equivalent to each other.

Clearly, a factorization in an arbitrary (L_∞,R)-algebra $\mathfrak{A}$ is simultaneously a factorization of G in all L_p , $1 < p < \infty$. A simple calculation of the argument increment of the determinants on the left- and right-hand sides in equation (2.1) shows that, for a factorization in $\mathfrak{A}$, the rule (5.4) for calculating the total index is valid. According to formulae (2.6) related to the passage from one factorization to another (with the same total index), from the relation $R \subseteq \mathfrak{A}$ we conclude that either every factorization of a matrix function in L_p is its factorization in $\mathfrak{A}$ or there does not exist any factorization in $\mathfrak{A}$.

<u>DEFINITION 5.3.</u> The (L_∞,R)-algebra $\mathfrak{A}$ is said to be <u>decomposing</u> if every element $a \in \mathfrak{A}$ admits a representation in the form $a = a_+ + a_-$, where $a_+ \in \mathfrak{A}^+$ and $a_- \in \overset{\circ}{\mathfrak{A}}{}^-$.

<u>LEMMA 5.5.</u> The following assertions are equivalent:

1) the (L_∞,R)-algebra $\mathfrak{A}$ is decomposing,

2) $\mathfrak{A} = \mathfrak{A}^+ \dotplus \overset{\circ}{\mathfrak{A}}{}^-$,

3) $P\mathfrak{A} = \mathfrak{A}^+$, $Q\mathfrak{A} = \overset{\circ}{\mathfrak{A}}{}^-$,

4) the operator S maps $\mathfrak{A}$ into itself,

5) the operator S is bounded in $\mathfrak{A}$.

<u>Proof.</u> The equivalence of assertions 1) and 2) follows from the relation $\mathfrak{A}^+ \cap \overset{\circ}{\mathfrak{A}}{}^- \subseteq L_\infty^+ \cap \overset{\circ}{L}{}_\infty^- = \{0\}$. If 2) is fulfilled, then $P\mathfrak{A} = P(\mathfrak{A}^+ \dotplus \overset{\circ}{\mathfrak{A}}{}^-) = P\mathfrak{A}^+ = \mathfrak{A}^+$. Analogously, $Q\mathfrak{A} = \overset{\circ}{\mathfrak{A}}{}^-$. Therefore, 2) implies 3).
If 3) holds, then, for every $\varphi \in \mathfrak{A}$, the inclusion $S\varphi = P\varphi - Q\varphi \in \mathfrak{A}^+ \dotplus \overset{\circ}{\mathfrak{A}}{}^- = \mathfrak{A}$ is valid. Thus, 3) ==> 4).
Vice versa, if $S\mathfrak{A} \subseteq \mathfrak{A}$, then also $P\mathfrak{A} \subseteq \mathfrak{A}$. Since $\mathfrak{A} \subseteq L_\infty$, we have in this case $P\mathfrak{A} \subseteq L_p^+$ for any $p < \infty$. Hence $P\mathfrak{A} \subseteq \mathfrak{A} \cap L_p^+$. Taking into consideration that Γ is a Smirnov contour, we obtain from Theorem 1.25 that $\mathfrak{A} \cap L_p^+ = \mathfrak{A} \cap L_\infty^+ = \mathfrak{A}^+$. In other words, $P\mathfrak{A} \subseteq \mathfrak{A}^+$. Analogously, $Q\mathfrak{A} \subseteq \overset{\circ}{\mathfrak{A}}{}^-$. Since the converse inclusions are obvious, the implication 4) ==> 3) is valid. If 3) is satisfied, then, for every $a \in \mathfrak{A}$, we have $a = Pa + Qa \in \mathfrak{A}^+ \dotplus \overset{\circ}{\mathfrak{A}}{}^-$, i.e. 3) ==> 1). Implication 5) ==> 4) is

evident, thus, it remains to prove the implication 4) ==> 5). In view of 4), the operator S is defined everywhere on $\mathfrak{A}$. Now we want to show that it is closed: if $a_k \to a$, $Sa_k \to b$ $(a_k, a, b \in \mathfrak{A})$, then $Pa_k = \frac{1}{2}(I + S)a_k \to \frac{1}{2}(a+b)$, $Qa_k = \frac{1}{2}(I - S)a_k \to \frac{1}{2}(a-b)$. Since, for all $k=1,2,\ldots$, the inclusions $Pa_k \in \mathfrak{A}^+$, $Qa_k \in \overset{\circ}{\mathfrak{A}}{}^-$ hold (here we use the implication 4) ==> 3) proved above) and $\mathfrak{A}^+$, $\overset{\circ}{\mathfrak{A}}{}^-$ are closed in $\mathfrak{A}$, then $a+b \in \mathfrak{A}^+$, $a-b \in \overset{\circ}{\mathfrak{A}}{}^-$.

Hence, $S(a+b) = a+b$, $S(a-b) = b-a$ and $Sa = S(\frac{1}{2}(a+b) + \frac{1}{2}(a-b))$ $= \frac{1}{2}(a+b) + \frac{1}{2}(b-a) = b$. Due to the closed graph theorem, the operator S is bounded in $\mathfrak{A}$. Thus Lemma 5.5 is completely proved. ≡

In the space $\mathfrak{A}^n$ of n-dimensional vector functions with components from $\mathfrak{A}$ we introduce a norm, setting $\|f\|_{\mathfrak{A}^n} = \sum_{j=1}^{n} \|f_j\|_{\mathfrak{A}}$, where $f = \sum_{j=1}^{n} f_j e_j$. To any matrix function $A \in \mathfrak{A}$ we assign the operator of multiplication by this matrix function in $\mathfrak{A}^n$.

As it can be easily verified, the norm of this operator is determined by the formula

$$\|A\| = \max_j \sum_{i=1}^{n} \|a_{ij}\|_{\mathfrak{A}} \tag{5.8}$$

with $A = (a_{ij})_{i,j=1}^n$.

In the sequel we suppose that the (L_∞, R)-algebra $\mathfrak{A}$ is decomposing. According to Lemma 5.5, this supposition is equivalent to the boundedness of S viewed as an operator in $\mathfrak{A}$ and, thus, also in $\mathfrak{A}^n$.

Consequently, it plays here the same role as the assumption $\Gamma \in \mathfrak{R}$ in Section 3.5 and 3.6. Up to the end of the present section (with the exception of Theorems 5.12 and 5.13), the latter condition (i.e. $\Gamma \in \mathfrak{R}$) can be replaced by the weaker requirement $\Gamma \in \mathfrak{C}$ used in the proof of the previous lemma.

Every $n \times n$ matrix function $G \in \mathfrak{A}$ can be associated with the linear bounded operators $P + GQ$ and $Q + G'P$ acting in $\mathfrak{A}^n$. The defect numbers of these operators are connected by the relations

$$\alpha(P+GQ) \le \beta(Q+G'P), \quad \beta(P+GQ) \ge \alpha(Q+G'P) \tag{5.9}$$

as in case when the operators $P + GQ$ and $Q + G'P$ are studied in the spaces L_p^n and L_q^n, respectively.

The relations (5.9) are a simple consequence of the fact that the kernel of the operator $P + GQ$ $(Q + G'P)$ exactly coincides with the set of vector functions on $\mathfrak{A}^n$ orthogonal (in the sense of the bilinear form (1.13)) to the image of the operator $Q + G'P$ $(P + GQ)$. In its turn, this assertion can be proved in the same way as the corresponding result from Theorem 3.1.

As in Chapter 3, in addition to the operators $P + GQ$ and $Q + G'P$, it is convenient to introduce the operators $T_Q(G)$ and $T_P(G')$ defined via formulae (3.22). Since in the case in question $\operatorname{im} P = (\mathfrak{A}^+)^n$ and $\operatorname{im} Q = (\mathring{\mathfrak{A}}^-)^n$, these operators act in $(\mathfrak{A}^+)^n$ and $(\mathring{\mathfrak{A}}^-)^n$, respectively. Assertion 1) of Lemma 3.5 on the relationships between defect numbers of the operators $P + GQ$ and $T_Q(G)$ as well as $Q + G'P$ and $T_P(G')$ can be transferred (together with its proof) without any changes to the situation under study. However, assertion 2) of this lemma is no longer valid, since the operators QGQ and $PG'P$ do not act in spaces dual to each other (as it was the case in Chapter 3), rather in one and the same space $\mathfrak{A}^n$. Therefore, nothwithstanding that formulae (5.9) are available, we cannot a priori assert that the operators $P + GQ$ and $Q + G'P$ are Fredholm only simultaneously.

<u>THEOREM 5.6.</u> Let $\mathfrak{A}$ be a decomposing (L_∞, R)-algebra, and let G be a matrix function of n th order with entries from $\mathfrak{A}$. The matrix function G is factorable in $\mathfrak{A}$ if and only if the operators $P + GQ$ and $Q + G'P$ regarded in $\mathfrak{A}^n$ are Fredholm and their indices are opposite. In addition, the value $\alpha(P + GQ)$ coincides with the sum of positive partial indices and $\beta(P + GQ)$ is opposite to the sum of negative partial indices of G.

<u>Proof.</u> Let G be factorable in $\mathfrak{A}$. Due to conditions (5.7), the operators $G_+PG_+^{-1}$, $G'_-Q(G'_-)^{-1}$, $G_-^{-1}\Lambda^{-1}QG_+^{-1}$ and $(G'_+)^{-1}\Lambda^{-1}P(G'_-)^{-1}$ are bounded in $\mathfrak{A}^n$. The same calculations which where carried out in the proof of Theorem 3.16 show that the operator $G_-^{-1}\Lambda^{-1}QG_+^{-1}+G_+PG_+^{-1}$ is a two-sided regularizer of the operator $P + GQ$. In the same way one can deduce that $(G'_+)^{-1}\Lambda^{-1}P(G'_-)^{-1} + G'_-Q(G'_-)^{-1}$ is a two-sided regularizer for $Q + G'P$. Therefore, according to Theorem 1.8, the

Fredholmness of the operators $P + GQ$ and $Q + G'P$ acting in $\mathfrak{A}^n$ will be proved on condition that G can be factored in $\mathfrak{A}$.

Furthermore, in case of a matrix function G factorable in $\mathfrak{A}$, all solutions of the homogeneous equation $(P+GQ)\varphi = 0$ determined by formulae (3.6)(for $\varphi_0^{\pm} = 0$), clearly belong to the space $\mathfrak{A}^n$. Consequently, the dimension of the kernel of the operator $P + GQ$ remains the same as in class L_p^n, i.e., it is equal to the sum of positive partial indices of G. As in Chapter 3, the conditions (3.9) continue to be necessary for the solvability of the equation $(P+GQ)\varphi=g$. If they are fulfilled, then the vector functions φ_0^+ and φ_0^- defined by formulae (3.5) belong to the classes $\mathfrak{A}^+$ and $\mathring{\mathfrak{A}}^-$, respectively. Hence, conditions (3.9) are necessary and sufficient solvability conditions as in case when G is Φ-factorable. Therefore, $\beta(P+GQ)$ is opposite to the sum of negative partial indices of G.

Similar considerations related to the equation $(Q+G'P)\psi = h$ show that in relations (5.9) equality holds. Hence, the indices of the operators $P + GQ$ and $Q + G'P$ are opposite, which finishes the proof of the direct assertion of the theorem.

Vice versa, let the operators $P + GQ$ and $Q + G'P$ be Fredholm in the space $\mathfrak{A}^n$, and assume their indices to be opposite. Then there exist [1] matrix functions F and H belonging to class R together with their inverses such that $\operatorname{im} T_Q(G) = \operatorname{im} T_Q(GF)$, $\operatorname{im} T_P((GF)') = \operatorname{im} T_P((HGF)')$, $\ker T_Q(HGF) = \ker T_Q(GF) = \{0\}$, $\ker T_P((GF)') = \ker T_P(G')$, $\ker T_P((HGF)') = \{0\}$. According to Theorem 2.9, the matrix functions F and H admit factorizations with factors from R being, consequently, factorizations in $\mathfrak{A}$. Owing to what was proved above, the operators $T_Q(F)$, $T_Q(H)$, $T_P(F')$ and $T_P(H')$ are Fredholm, where $\operatorname{ind} T_Q(F) = -\operatorname{ind} T_P(F')$, $\operatorname{ind} T_Q(H) = -\operatorname{ind} T_P(H')$. But the operators $T_Q(G_1)$ and $T_P(G_1')$ with $G_1 = HGF$ differ only by a finite-dimensional summand from the products $T_Q(H)T_Q(G)T_Q(F)$ and $T_P(F')T_P(G')T_P(H')$, respectively. In view of Theorems 1.5 and 1.6, from this we obtain that the

[1] This can be proved in precisely the same way as the existence of matrix functions F and H in Lemma 3.2.

operators $T_Q(G_1)$ and $T_P(G_1')$ are Fredholm either, moreover,

$$\operatorname{ind} T_Q(G_1) = \operatorname{ind} T_Q(H) + \operatorname{ind} T_Q(G) + \operatorname{ind} T_Q(F)$$

$$= -\operatorname{ind} T_P(H') - \operatorname{ind} T_P(G') - \operatorname{ind} T_P(F') .$$

At the same time, the kernels of the operators $T_Q(G)$ and $T_P(G')$ are trivial, so that their indices are non-positive. Consequently, these indices are equal to zero. But then the codimensions of the images of operators $T_Q(G_1)$ and $T_P(G_1')$ are also equal to zero. In other words, $T_Q(G_1)$ and $T_P(G_1')$ are invertible (in $(\dot{A}^-)^n$ and $(A^+)^n$, respectively). From here it follows that the operators $P + G_1Q$ and $G + G_1'P$ are invertible in A^n. Acting now as in the proof of Theorem 3.4, we recognize that the matrix function G_1 is factorable in A and thus also in all L_p (moreover, its partial indices are equal to zero; however, in the sequel we shall not make use of this fact). According to Theorem 2.11, this implies the factorability of G in L_p. In addition, the factorization factors of G are obtained from the factorization factors of G_1 by multiplication by a matrix function from R as it is clear from the proof of Theorem 2.11. The multiplication by such matrix functions does not go beyond the algebra A. Hence, the matrix function G is factorable in A simultaneously with G_1, and the proof of Theorem 5.6 is complete. ≡

COROLLARY 5.12. If the operators $T_Q(G)$ and $T_P(G')$ are Fredholm and their indices are opposite, then the matrix function G is invertible in A.

The (L_∞, R)-algebra A is said to be a (C,R)-algebra if all its elements are continuous functions, i.e. if $A \subseteq C$. For (C,R)-algebras, Theorem 5.6 allows the following strengthening.

THEOREM 5.7. Let the (C,R)-algebra A possess the invertibility property, and assume, for every non-degenerating matrix function G, the operators $T_Q(G)$ and $T_P(G)$ to be Fredholm. Then any non-degenerating matrix function from A is factorable in A.

Proof. Consider an arbitrary non-degenerating matrix function $G \in A$. Since $A \subseteq C$, G can be approximated in the metric of L_∞ with

arbitrary accuracy by rational matrix functions. Now we choose $F \in R$ close to G in such a way that, for all $\lambda \in [0,1]$, the matrix function $G_\lambda = \lambda F + (1-\lambda)G$ does not degenerate on Γ.
Owing to the conditions of the theorem, the operators $A_\lambda = T_Q(G_\lambda)$ and $B_\lambda = T_P(G'_\lambda)$ are then Fredholm for any $\lambda \in [0,1]$. Since A_0 and A_1 as well as B_0 and B_1 are homotopic, we get $\text{ind } A_0 = \text{ind } A_1$ and $\text{ind } B_0 = \text{ind } B_1$. At the same time, the matrix function G_1 $(=F)$ can be factored in $\mathfrak{A}$. Thus, by Theorem 5.6, $\text{ind } A_1 = -\text{ind } B_1$. Hence $\text{ind } A_0 = -\text{ind } B_0$, i.e., the indices of the operators $T_Q(G)$ and $T_P(G')$ are opposite. Now the factorability of G in $\mathfrak{A}$ results from Theorem 5.6. ≡

In some cases the Fredholmness condition for the operators $T_Q(G)$ and $T_P(G')$ can be replaced by another one more suitable for checking.

THEOREM 5.7'. Let the decomposing (C,R)-algebra $\mathfrak{A}$ possess the invertibility property, and let, for all $a \in \mathfrak{A}$, the operator QaP (PaQ) be compact in $\mathfrak{A}$. Then every invertible matrix function from $\mathfrak{A}$ admits a factorization in $\mathfrak{A}$.

Proof. Consider, e.g., the case of compactness of QaP for all $a \in \mathfrak{A}$. From the hypothesis of the theorem it is not hard to conclude that the operator QGP is compact in $\mathfrak{A}^n$ for arbitrary $G \in \mathfrak{A}$. Formulae (5.1) and (5.2), which continue to be valid in the situation under study, show that the operator $T_Q(G^{-1})$ is a regularizer for $T_Q(G)$. Hence, the operator $T_Q(G)$ is Fredholm. The Fredholmness of the operator $T_P(G')$ may be verified in a similar way. According to Theorem 5.7, G can be factored in $\mathfrak{A}$, which proves Theorem 5.7'. ≡

COROLLARY 5.13. If the decomposing (C,R)-algebra $\mathfrak{A}$ has the property that the closure of R in $\mathfrak{A}$ contains $\mathfrak{A}^+$ or $\mathfrak{A}^-$, then any invertible matrix function from $\mathfrak{A}$ admits a factorization in $\mathfrak{A}$.

Proof. Let, for instance, $\text{clos}_{\mathfrak{A}} R \supseteq \mathfrak{A}^+$. Taking an arbitrary $a \in \mathfrak{A}$, we can write (since $\mathfrak{A}$ is decomposing): $a = a_+ + a_-$, where $a_+ \in \mathfrak{A}^+$, $a_- \in \overset{\circ}{\mathfrak{A}}{}^-$. Now we take a sequence $r_n \in R$ such that $\|r_n - a_+\|_{\mathfrak{A}} \to 0$. Then the sequence of operators $P(r_n + a_-)Q$ converges in the norm to PaQ.

At the same time, $P(r_n+a_-)Q = Pr_nQ$ is a finite-dimensional operator. Consequently, the operator PaQ is compact in $\mathfrak{R}$. It remains to use Theorem 5.7'. ≡

Now we intend to illustrate Theorem 5.7' by the following examples, which are of some significance for applications.

1°. To H^α, the space of Hölder (or H-continuous) functions with exponent α $(0 < \alpha \leq 1)$, there belong such functions defined on Γ for which the quantity

$$\sup_{\tau, t\in\Gamma} \frac{|a(\tau)-a(t)|}{|\tau-t|^\alpha}$$

is finite (Hölder condition). Setting

$$\|a\|_{H^\alpha} = \|a\|_C + \sup_{\tau, t\in\Gamma} \frac{|a(\tau)-a(t)|}{|\tau-t|^\alpha},$$

H^α becomes a Banach algebra. Obviously $R \subset H^\alpha \subset C$, thus, H^α is a (C,R)-algebra. If the contour Γ is piecewise smooth and $\alpha < 1$, then the operator S is bounded in H^α (GOHBERG, KRUPNIK [4], MUSKHELISHVILI [1]) and the operator $aS-SaI$ is compact for all $a \in H^\alpha$ (see GOHBERG, KRUPNIK [4]). Since $aS-Sa = 2(QaP-PaQ)$, we deduce from this the compactness of operators PaQ and QaP. Moreover, H^α has the invertibility property (see e.g. MUSKHELISHVILI [1]). Owing to Theorem 5.7', we thus obtain the following result.

<u>THEOREM 5.8.</u> If the contour Γ is piecewise smooth and G is a non-degenerating matrix function all entries of which fulfil a Hölder condition of order α $(0 < \alpha < 1)$, then the representation (2.1) is valid, where $G_\pm$ are matrix functions analytic in $\mathfrak{D}^\pm$, continuous and non-degenerating in $\mathfrak{D}^\pm \cup \Gamma$ and fulfilling on Γ a Hölder condition of the same order α.

2°. Let $H^{\alpha,k}$ denote the set of k times continuously differentiable functions defined on Γ the k th derivative of which belongs to the class H^α. Furthermore, we introduce the norm

$$\|a\|_{H^{\alpha,k}} = \sum_{j=0}^{k} \frac{1}{j!} \max_{\tau\in\Gamma}|a^{(j)}(\tau)| + \sup_{\tau, t\in\Gamma} \frac{|a^{(k)}(\tau)-a^{(k)}(t)|}{|\tau - t|^\alpha}$$

in $H^{\alpha,k}$.

In GOHBERG, KRUPNIK [4] it is shown that, if the contour Γ is piecewise smooth, then $H^{\alpha,k}$ is a decomposing (C,R)-algebra which meets the assumptions of Corollary 5.13. As in the case of the algebra H^{α}, all functions from $H^{\alpha,k}$ non-vanishing on Γ are invertible elements in $H^{\alpha,k}$. Therefore, the following statement holds.

THEOREM 5.9. Let Γ be a piecewise smooth contour. A non-degenerating matrix function all entries of which are functions of class $H^{\alpha,k}$ $(0 < \alpha < 1)$ admits a factorization (2.1) in which $G_{\pm}$ are matrix functions analytic in $\mathfrak{D}^{\pm}$, continuous and non-degenerating in $\mathfrak{D}^{\pm} \cup \Gamma$ and belonging to class $H^{\alpha,k}$ on Γ.

Obviously, $H^{\alpha,k}$ is, for $k > 0$, a proper subclass of H^{α}, thus, the H-continuity of $G_{\pm}$ in Theorem 5.9 is not a new result (in comparison with Theorem 5.8). The crucial fact of Theorem 5.9 is that differential properties of G are acquired by its factorization factors to a certain extent.

3°. In 1° and 2° two concrete examples of (C,R)-algebras have been described which possess the invertibility property and fulfil the conditions of Corollary 5.13. Here we focus our attention on one subclass of the class of (C,R)-algebras for which the invertibility property is automatically fulfilled and which, under the additional supposition of decomposition, satisfies also the condition of compactness of operators of the type $aS-SaI$.

DEFINITION 5.4. A (C,R)-algebra $\mathfrak{C}$ is called a R-algebra, if the lineal R is dense in $\mathfrak{C}$ with respect to the norm of $\mathfrak{C}$.

LEMMA 5.6. 1) A R-algebra possesses the invertibility property.
2) If the R-algebra $\mathfrak{C}$ is decomposing, then $\mathfrak{C}^{+}$, $\mathfrak{C}^{-}$ and $\mathring{\mathfrak{C}}^{-}$ coincide with the closure in $\mathfrak{C}$ of the lineals R^{+}, R^{-} and $\mathring{R}^{-}$, respectively.

Proof. 1) We use the well-known fact from the theory of commutative Banach algebras according to which an element $a \in \mathfrak{C}$ is invertible in $\mathfrak{C}$ if and only if $\varphi(a) \neq 0$ for any multiplicative linear functional φ. After choosing such a functional φ arbitrarily, we denote its value on the function $u(t) = t$ by t_o. Then

$$\varphi(u-t_o e) = \varphi(u) - t_o\varphi(e) = t_o - t_o = 0 .$$

Therefore, the function $t-t_o$ is not invertible in $\mathfrak{C}$. If $t_o \notin \Gamma$, the function $(t-t_o)^{-1}$ belongs to $\mathfrak{C}$. Consequently, the non-invertibility of $t-t_o$ in $\mathfrak{C}$ means that $t_o \in \Gamma$. Now, let r be a rational function not having poles on Γ, and assume $r(t) = \sum_{k=o}^{N_1} c_k t^k / \sum_{k=o}^{N_2} b_k t^k$. Then

$$\varphi(r) = \frac{\varphi(\sum_{k=o}^{N_1} c_k u^k)}{\varphi(\sum_{k=o}^{N_2} b_k u^k)} = \frac{\sum_{k=o}^{N_1} c_k \varphi(u)^k}{\sum_{k=o}^{N_2} b_k \varphi(u)^k} = \frac{\sum_{k=o}^{N_1} c_k t_o^k}{\sum_{k=o}^{N_2} b_k t_o^k} = r(t_o) .$$

Using the density of R in $\mathfrak{C}$, we choose, for arbitrary $a \in \mathfrak{C}$, a sequence $\{r_k\}$ in R converging to a in $\mathfrak{C}$. Then $\varphi(a) = \lim_{k\to\infty} \varphi(r_k) = \lim_{k\to\infty} r_k(t_o) = a(t_o)$ (the last equation holds in view of Corollary 5.11). Thus, every multiplicative linear functional is of the form $\varphi(a) = a(t_o)$, where $t_o \in \Gamma$.

Hence, $a(t) \neq 0$ for any $t \in \Gamma$ is a necessary and sufficient condition for the invertibility of the element a in $\mathfrak{C}$, which proves statement 1).

2) The inclusions $\mathfrak{C}^{\pm} \supseteq \text{clos}_{\mathfrak{C}} R^{\pm}$, $\overset{\circ}{\mathfrak{C}}{}^{-} \supseteq \text{clos}_{\mathfrak{C}} \overset{\circ}{R}{}^{-}$ [1] result from the closedness of the classes $\mathfrak{C}^{\pm}$ and $\overset{\circ}{\mathfrak{C}}{}^{-}$ in $\mathfrak{C}$ and hold irrespective of whether the algebra $\mathfrak{C}$ is decomposing or not. In order to prove the inclusion $\mathfrak{C}^{+} \subseteq \text{clos}_{\mathfrak{C}} R^{+}$ in case of a decomposing R-algebra $\mathfrak{C}$, we choose an arbitrary element $a \in \mathfrak{C}^{+}$ and consider the sequence $\{r_k\} \subset R$ converging in $\mathfrak{C}$ to a. According to Lemma 5.5, the operators S and P are bounded in $\mathfrak{C}$ and, therefore, $\{Pr_k\}$ converges in $\mathfrak{C}$ to Pa. But $Pa = a$ and $Pr_k \in R^{+}$. Hence, $a \in \text{clos}_{\mathfrak{C}} R^{+}$, i.e. $\mathfrak{C}^{+} \subseteq \text{clos}_{\mathfrak{C}} R^{+}$. The inclusions $\mathfrak{C}^{-} \subseteq \text{clos}_{\mathfrak{C}} R^{-}$ and $\overset{\circ}{\mathfrak{C}}{}^{-} \subseteq \text{clos}_{\mathfrak{C}} \overset{\circ}{R}{}^{-}$ can be verified analogously. Lemma 5.6 is proved. ≡

Notice that the (C,R)-algebras H^{α} and $H^{\alpha,k}$ are not R-algebras (cf. BOJARSKI [2]), although statements 1) and 2) of Lemma 5.6 are correct

---

1) By clos we denote the closure of a set: the index indicates, in which space the closure is taken.

for them. Every (C,R)-algebra $\mathfrak{A}$ contains some R-algebra $\mathfrak{C}$ as a subalgebra. This algebra is the closure of the lineal R in the norm of $\mathfrak{C}$. Thus, it is defined uniquely. The R-algebra $\mathfrak{C}$ coincides with $\mathfrak{A}$ if and only if the initial (C,R)-algebra $\mathfrak{A}$ is itself a R-algebra.

From Corollary 5.13 and statements 1) and 2) of Lemma 5.6 we obtain the following result.

THEOREM 5.10. Assume $\mathfrak{C}$ to be a decomposing R-algebra. The matrix function $G \in \mathfrak{C}$ admits a factorization in $\mathfrak{C}$ if and only if it is non-singular on Γ.

An obvious example of a R-algebra provides C. Theorem 5.10, along with the example given at the beginning of this section, show that the R-algebra of functions continuous on the circle fails to be decomposing. As an example of a decomposing R-algebra we quote the class $C^+ \dot{+} \overset{\circ}{C}{}^-$ supplied with the direct sum norm

$$\|a_+ + a_-\|_{C^+\dot{+}\overset{\circ}{C}{}^-} = \|a_+\|_C + \|a_-\|_C .$$

This class coincides with the closure of R with respect to the norm $\|\cdot\|_{C^+\dot{+}\overset{\circ}{C}{}^-}$ and is therefore a R-algebra. Clearly, every other decomposing R-algebra is contained in $C^+\dot{+}\overset{\circ}{C}{}^-$.

A less trivial example of a decomposing R-algebra provides the Wiener algebra W, whose elements are functions defined on the unit circle representable in the form of absolutely convergent series

$$a(t) = \sum_{k=-\infty}^{+\infty} a_k t^k \quad (|t|=1), \quad \sum_{k=-\infty}^{+\infty} |a_k| < \infty .$$

The norm in W is introduced by the formula $\|a\|_W = \sum_{k=-\infty}^{+\infty} |a_k|$. Apparently, functions of the kind $\sum_{k=-N_1}^{N_2} a_k t^k$ form a dense lineal in W.

Furthermore, W contains all twice continuously differentiable functions, because for them an estimate $|a_k| \leq \mathrm{const}/k^2$ holds. Especially $R \subset W$. Hence W is a R-algebra. If $a \in W$ and

$$a(t) = \sum_{k=-\infty}^{+\infty} a_k t^k ,$$

then

$$a_+(t) = \sum_{k=0}^{\infty} a_k t^k \in W^+ , \quad a_-(t) = \sum_{k=-\infty}^{-1} a_k t^k \in W^- \text{ and } a = a_+ + a_- .$$

Thus, the R-algebra W is decomposing [1] and, therefore, the following proposition is true.

<u>THEOREM 5.11.</u> A non-degenerating matrix function defined on the unit circle $\mathbb{T}$ with elements from W admits a factorization in W.

In summary, we have constructed some classes of matrix functions admitting a factorization in C. The following local principle enables us to "stick together" new classes from already known ones.

<u>THEOREM 5.12.</u> Let the contour Γ belong to class $\mathcal{R}$, and assume the matrix function G to have the property that, for all points $t \in \Gamma$, there exists a matrix function G_t coinciding with G on some open arc γ_t $(\ni t)$ and admitting a factorization in C. Then G also admits a factorization in C.

<u>Proof.</u> Every matrix function G_t is continuous and non-degenerating on Γ. Hence, for all $t \in \Gamma$, the matrix function G is continuous on γ_t and, thus, also on the whole contour Γ. Therefore (Theorem 5.3), G admits a factorization in the spaces L_p, $1 < p < \infty$. Applying Theorem 2.4 to matrix functions G and G_t, we recognize that the factorization factors $G_{\pm}$ are continuous and non-degenerating in $\mathcal{D}^{\pm} \cup \gamma_t$ for all $t \in \Gamma$. Hence, the matrix functions $G_{\pm}$ are continuous and non-degenerating in $\mathcal{D}^{\pm} \cup \Gamma$, which proves Theorem 5.12. $\equiv$

<u>COROLLARY 5.14.</u> Suppose $\{\gamma_j\}_{j=1}^{N}$ to be a system of open arcs covering the unit circle $\mathbb{T}$. If on every arc γ_j the elements of the matrix function G possess one of the properties

1) they fulfil a Hölder condition of order α $(0 < \alpha < 1)$,

2) they are restrictions of certain functions possessing a absolutely convergent Fourier series,

then the matrix function G admits a factorization in C.

In order to apply Theorem 5.12 to some matrix function G, it is necessary to cover the contour Γ with a system of arcs $\{\gamma_j\}$ and to find

---

[1] It is obvious that $\|P\|_W = \|Q\|_W = 1$.

the extension of G from every of these arcs γ_j to such a matrix function G_j which is factorable in C. In doing so, it is clear from the proof of Theorem 5.12 that the continuity of the factorization factors for G_j is essential not on the entire contour Γ, but only on the arc γ_j. This way of reasoning leads to a modification of Theorem 5.12 more convenient for applications.

DEFINITION 5.5. The matrix function G is said to admit a local factorization in C with respect to the arc γ of the contour Γ, if there exists a neighbourhood (on the plane) U_γ of this arc as well as a pair of matrix functions $G_\gamma^{\pm}$ analytic in $\mathfrak{D}^{\pm} \cap U_\gamma$, continuous and non-degenerate in $(\mathfrak{D}^{\pm} \cup \Gamma) \cap U_\gamma$ such that $G(t) = G_\gamma^{+}(t)G_\gamma^{-}(t)$ for $t \in \Gamma \cap U_\gamma$.

LEMMA 5.7. If the matrix function G admits a factorization in L_p with respect to the contour Γ and a local factorization in C with respect to some arc γ, then the factorization factors $G_{\pm}$ are continuously extendable from $\mathfrak{D}^{\pm}$ onto γ. Moreover, these extensions do not degenerate on γ.

The proof differs only in detail from the proof of Theorem 2.4.

In Lemma 5.7 and the definition preceding it, the class C may be replaced by H^{α} or $H^{\alpha,k}$. In this case, in Definition 5.5 we have to demand that the matrix functions $G_\gamma^{\pm}$ fulfil the condition in H^{α} ($H^{\alpha,k}$, respectively) on the arc γ, and in Lemma 5.7 we may claim that the extensions of the matrix functions $G_{\pm}$ onto γ satisfy the condition in H^{α} ($H^{\alpha,k}$, respectively).

COROLLARY 5.15. Suppose the matrix function G to be factorable in L_p with respect to the contour Γ, and assume that G fulfils a Hölder condition of order α on the piecewise smooth arc γ. Then the factorization factors $G_{\pm}$ are continuously extendable from $\mathfrak{D}^{\pm}$ onto the arc γ, and these extensions are non-degenerating and fulfil a Hölder condition of the same order α on every arc $\tilde{\gamma}$ whose closure lies in γ.

THEOREM 5.13. If the matrix function G admits a local factorization in C with respect to the system of open arcs $\{\gamma_j\}$ covering the contour Γ which belongs to class $\mathcal{R}$, then it admits a factorization in C with respect to the contour Γ.

Proof. From the local factorability in C of a certain matrix function with respect to the arc γ there results its continuity and non-degeneracy on this arc. Therefore, the studied matrix function G is continuous and non-degenerating on the whole contour Γ and, according to Theorem 5.3, it is factorable with respect to this contour. Hence Lemma 5.7 is applicable, in virtue of which the factorization factors $G_\pm$ are continuously extendable on every arc γ_j and therefore also on the whole contour Γ. Moreover, these extensions are non-degenerating matrix functions. In this way, the factorization of the matrix function G, existing in view of Theorem 5.3, is, actually, its factorization in C. Thus Theorem 5.13 is proved. $\equiv$

Finally, note that Theorems 5.7 - 5.13 may be transferred to the case of right-factorization, since the matrix function G admits a right-factorization in $\mathcal{A}$ (or C) if and only if G' admits a left-factorization in $\mathcal{A}$ (or C).

5.3. Φ-FACTORIZATION OF PIECEWISE CONTINUOUS MATRIX FUNCTIONS

We are going to discuss the case of a piecewise continuous matrix function G, which is important for applications. First of all, recall that a matrix function G is said to be piecewise continuous if it is continuous on the contour Γ with the exception of a finite number of points t_j, $j=1,\dots,N$ at every of which there exists the limit from the left $G(t_j-0)$ and from the right $G(t_j+0)$ [1)]

From Corollaries 5.9 and 5.15 we get the following result.

THEOREM 5.14. Assume G to be a piecewise continuous matrix function non-degenerating at continuity points and factorable in L_p.

1) "From the left" and "from the right" one has to understand in the sense of the orientation of the contour Γ as it was explained in Chapter 1.

Furthermore, γ is assumed to be an arc of the contour Γ not containing discontinuity points of G. Then

1) if the operator of singular integration along the arc γ is bounded, then the factorization factors $G_{\pm}$ and the inverse to them are summable with any finite power on every arc $\tilde{\gamma}$ the closure of which lies in γ,

2) if γ is a piecewise smooth arc and the matrix function G fulfils on γ a Hölder condition with exponent α, then the matrix functions $G_{\pm}$ are continuously extendable from $\mathcal{D}^{\pm}$ onto Γ, and these extensions are non-degenerating and fulfil a Hölder condition with the same exponent α on every arc $\tilde{\gamma}$ the closure of which lies in γ.

Up to the very end of the present chapter we shall assume (by tacit agreement) that the contour Γ belongs to class $\tilde{\mathcal{R}}$ and that for the discontinuity points t_j of G the boundedness requirement for branches of the functions $\operatorname{Arg}(t-t_j)$, which are continuous on $\Gamma \setminus \{t_j\}$, is fulfilled. In any case, this requirement is satisfied, if there are tangents to Γ at the points t_j.

Above all, we intend to prove a result which is related to the factorization problem for scalar piecewise continuous functions.

THEOREM 5.15. Let G be a piecewise continuous function and $\{t_j\}_{j=1}^{N}$ be the set of its discontinuity points. The function G is Φ-factorable in L_p if and only if

1) G does not vanish at continuity points and $G(t_j \pm 0) \neq 0$ and

2) $\arg G(t_j+0)/G(t_j-0) \neq 2\pi/p$, $j=1,\dots,N$ [1].

If conditions 1) and 2) are fulfilled, the p-index of G is calculable with the aid of the formula

$$\varkappa = \frac{1}{2\pi}\Big(\sum_{j=1}^{N}\{\operatorname{Arg} G(t)\}_{\gamma_j} + \sum_{j=1}^{N} \arg\frac{G(t_j+0)}{G(t_j-0)}\Big) - l\,, \tag{5.10}$$

where l is the number of discontinuities at which $\arg G(t_j+0)/G(t_j-0) > 2\pi/p$, and $\{\gamma_j\}$ is the set of arcs into which the contour Γ

[1] Here and below in this section the value arg is taken from the interval $[0,2\pi)$.

is divided by the discontinuity points of the function G.

If at least one of the conditions 1) and 2) is violated, then $P + GQ$ is neither a Φ_+- nor a Φ_--operator in L_p.

Proof. The necessity of condition 1) for $P + GQ$ to be a semi-Fredholm operator results from Theorem 3.10. Therefore, it will be assumed that the first condition is fulfilled. In view of statement 2) of Theorem 3.21, it is possible to restrict oneself to the case of a connected contour Γ. Setting

$$\alpha_j = \frac{1}{2\pi}(\arg[G(t_j-0)/G(t_j+0)] - i\log|G(t_j-0)/G(t_j+0)|), \quad j = 1,\ldots,N,$$

we define functions $\psi_j(t) = t^{\alpha_j}$ with a discontinuity at the point t_j as it was done in Theorem 2.5. We further introduce the family of functions $\{G_\tau\}_{\tau\in\Gamma}$ as follows:

$$G_\tau(t) = \begin{cases} G(\tau) & \text{if} \quad \tau \neq t_j, \\ \dfrac{G(t_j-0)}{\psi_j(t_j-0)}\psi_j(t) & \text{if} \quad \tau = t_j. \end{cases} \qquad j=1,\ldots,N$$

The function G_τ is ε-locally equivalent to the function G at the point τ. For $\tau \neq t_j$ $(j=1,\ldots,N)$, this is evident and, for $\tau = t_j$, it follows from the relation $G_{t_j}(t_j \pm 0) = G(t_j \pm 0)$.

According to Theorem 3.17 and the remark after it, condition 2) guarantees Φ-factorability in L_p of the functions ψ_j and, thus, also of G_{t_j}. The functions G_τ are constant if $\tau \neq t_j$ and, therefore, Φ-factorable in L_p either. Taking this into account, we are able to apply Theorem 3.22, due to which function G is Φ-factorable in L_p. In this way, the sufficiency of conditions 1) and 2) for the Φ-factorability of G in L_p has been proved.

Now we are going to verify formula (5.10). First of all, the equation $G(t) = f(t)\prod_{j=1}^{N}\psi_j(t)$ holds, where f is continuous and non-degenerating on Γ. Utilizing the factorization $f(t) = f^+(t)t^k f^-(t)$ of the function f existing by Theorem 5.3 as well as the factorization (2.10) of functions of the form t^α, we obtain

$$G(t) = \left(f^+(t)\prod_{j=1}^{N}(t-t_j)^{\alpha_j-k_j}\right) t^{k+\sum_{j=1}^{N} k_j}\left(f^-(t)\prod_{j=1}^{N}\left(1-\frac{t_j}{t}\right)^{k_j-\alpha_j}\right), \quad (5.11)$$

where $k = \frac{1}{2\pi}\{\mathrm{Arg}\, f(t)\}_\Gamma$ and $k_j = \begin{cases} 0 & \text{if } \mathrm{Re}\,\alpha_j < 1 - \frac{1}{p} \\ 1 & \text{if } \mathrm{Re}\,\alpha_j > 1 - \frac{1}{p}. \end{cases}$

With regard to the boundary properties of the functions $f^{\pm}$ stated in Theorem 5.3 and of the functions $(t-t_j)^{\alpha_j-k_j}$ and $(1-\frac{t_j}{t})^{k_j-\alpha_j}$ described in Theorem 2.5, we find that the representation (5.11) is a factorization and, therefore, also a Φ-factorization of the function G. Hence the p-index of G is equal to $k + \sum_{j=1}^{N} k_j$. It remains to note that

$$\frac{1}{2\pi}\sum_{j=1}^{N}\{\mathrm{Arg}\, G(t)\}_{\gamma_j} = \frac{1}{2\pi}\sum_{j=1}^{N}\{\mathrm{Arg}\, f(t)\}_{\gamma_j} + \sum_{j=1}^{N}\mathrm{Re}\,\alpha_j = k + \sum_{j=1}^{N}\mathrm{Re}\,\alpha_j,$$

$$\frac{1}{2\pi}\sum_{j=1}^{N}\arg\frac{G(t_j+0)}{G(t_j-0)} = \sum_{j=1}^{N}(1-\mathrm{Re}\,\alpha_j) = N - \sum_{j=1}^{N}\mathrm{Re}\,\alpha_j \quad \text{and} \quad \sum_{j=1}^{N} k_j = N-1.$$

Finally, we want to prove the last statement of the theorem. If condition 1) is fulfilled, but condition 2) is violated, then there exist piecewise continuous functions arbitrarily close (in the sense of the norm in L_∞) to the given function G such that for them conditions 1) and 2) are satisfied, but the quantity (5.10) takes different values. In other words, in any neighbourhood of the operator $P + GQ$ there exist Fredholm operators with different indices, which means that $P + GQ$ can be neither a Φ_+- nor a Φ_--operator. Thus, the proof of Theorem 5.15 is complete. ≡

Applying the notation introduced in Section 4.1, one may claim that piecewise continuous functions are contained in the class Φ.

<u>THEOREM 5.16.</u> Let G be a piecewise continuous matrix function, let, further, $\{t_j\}_{j=1}^{N}$ be the set of its discontinuities and $\{\gamma_j\}$ the set of arcs into which the contour Γ is divided by these discontinuity points. The matrix function G is Φ-factorable in L_p if and only if

1) for all $t \in \Gamma$, the relation

$$\det G(t \pm 0) \neq 0 \qquad (5.12)$$

holds and

2) all eigenvalues of the matrix $G^{-1}(t_j-0)G(t_j+0)$, $j=1,\dots,N$ are outside the ray l_p drawn from the origin at an angle of $2\pi/p$ with the positive x-axis.

If conditions 1) and 2) apply, then the total index of G is defined by the formula

$$\varkappa = \frac{1}{2\pi}\Big(\sum_{j=1}^{N} \{\operatorname{Arg} \det G(t)\}_{\gamma_j} + \sum_{j=1}^{N}\sum_{k=1}^{n} \arg \lambda_{jk}\Big) - l \ , \qquad (5.13)$$

where $\{\lambda_{jk}\}_{k=1}^{n}$ is the set of eigenvalues of the matrix $G^{-1}(t_j-0)G(t_j+0)$, and l is the number of such scalars λ_{jk} $(j=1,\dots,N;\ k=1,\dots,n)$ for which $\arg \lambda_{jk} > 2\pi/p$.

Proof. The necessity of condition 1) for the Φ-factorability of the matrix function G results from Theorem 3.13. Assuming condition 1) to be fulfilled, we denote by A_j some constant non-singular matrix transforming $G^{-1}(t_j-0)G(t_j+0)$ into upper triangular form, i.e. such that the matrix $A_jG^{-1}(t_j-0)G(t_j+0)A_j^{-1}$ is upper triangular, $j=1,\dots,N$ [1].

Let A be a continuous matrix function non-singular on Γ such that $A(t_j) = A_j$ $(j=1,\dots,N)$, and assume X to be an upper triangular non-singular matrix function continuous on Γ with the possible exception of those points t_j at which $X(t_j-0) = I$, $X(t_j+0) = A_jG^{-1}(t_j-0)G(t_j+0)A_j^{-1}$.

Set $B(t) = G(t)A^{-1}(t)X^{-1}(t)$. Clearly, the matrix function B is continuous on Γ with the possible exception of the points t_j .

Since $B(t_j-0) = G(t_j-0)A_j^{-1}$,

$$B(t_j+0) = G(t_j+0)A_j^{-1}A_jG^{-1}(t_j+0)G(t_j-0)A_j^{-1} = B(t_j-0) \ ,$$

the matrix function B is continuous (and non-singular) on the contour Γ . Thus $G = BXA$, where the matrix functions A and B are continuous and non-singular, and X is upper triangular and piecewise continuous.

1) The existence of matrices A_j having this property follows for example, from the reduction theorem to Jordan's normal form.

Owing to Corollary 5.10, the matrix function G is Φ-factorable only simultaneously with X. Since the diagonal elements of the matrix function X are piecewise continuous and therefore belong to the class Φ, Theorem 4.5 can be applied, according to which the Φ-factorability of the matrix function X is equivalent to the Φ-factorability of its diagonal elements. Denoting the k th diagonal element of the matrix $X(t)$ by $x_k(t)$ and using Theorem 5.15, we see that the matrix function X (and, thus, also G) is Φ-factorable in L_p if and only if

$$\arg x_k(t_j+0) \neq 2\pi/p \quad (j=1,\dots,N;\ k=1,\dots,n) . \tag{5.14}$$

Furthermore, $\{x_k(t)\}_{k=1}^n$ is the set of eigenvalues of the matrix $X(t)$, and the matrices $X(t_j+0)$ and $G^{-1}(t_j-0)G(t_j+0)$ are similar. Therefore, their eigenvalues coincide. Consequently, condition (5.14) is equivalent to condition 2) of the theorem. In this way, the necessity of condition 2) as well as the sufficiency of both conditions 1) and 2) for the Φ-factorability of G in L_p have been proved. According to Corollary 4.4, the total p-index $\varkappa(X)$ of the matrix function X is equal to the sum of p-indices of the functions x_k, $k=1,\dots,n$. Applying formula (5.10) for calculating them, we get

$$\varkappa(X) = \sum_{k=1}^{n} \Big(\frac{1}{2\pi}\Big(\sum_{j=1}^{N} \{\operatorname{Arg} x_k(t)\}_{\gamma_j} + \sum_{j=1}^{N} \arg \lambda_{jk}\Big) - l_k\Big) ,$$

where l_k is the number of such scalars λ_{jk} for which $\arg \lambda_{jk} > 2\pi/p$. Since

$$\begin{aligned}\sum_{k=1}^{n} \sum_{j=1}^{N} \{\operatorname{Arg} x_k(t)\}_{\gamma_j} &= \sum_{j=1}^{N} \{\operatorname{Arg} \prod_{k=1}^{n} x_k(t)\}_{\gamma_j} \\ &= \sum_{j=1}^{N} \{\operatorname{Arg} \det X(t)\}_{\gamma_j}\end{aligned}$$

and $\sum_{k=1}^{n} l_k = l$, we may write:

$$\varkappa(X) = \frac{1}{2\pi}\Big(\sum_{j=1}^{N} \{\operatorname{Arg} \det X(t)\}_{\gamma_j} + \sum_{j=1}^{N} \sum_{k=1}^{n} \arg \lambda_{jk}\Big) - l .$$

To prove formula (5.13), it remains to remark that

$$\varkappa(G) = \varkappa(X) + \frac{1}{2\pi}\{\operatorname{Arg} \det A(t)B(t)\}_{\Gamma}$$

(in view of Corollary 5.10) and

$$\sum_{j=1}^{N}\{\mathrm{Arg}\det X(t)\}_{\gamma_j} + \{\mathrm{Arg}\det A(t)B(t)\}_{\Gamma} = \sum_{j=1}^{N}\{\mathrm{Arg}\det G(t)\}_{\gamma_j}.$$

Theorem 5.16 is proved. ≡

Let us note that formula (5.13) may be rewritten in the form

$$\varkappa = \frac{1}{2\pi}\Big(\sum_{j=1}^{N}\{\mathrm{Arg}\det G(t)\}_{\gamma_j} + \sum_{j=1}^{N}\sum_{k=1}^{n}\arg_p \lambda_{jk}\Big), \qquad (5.13')$$

where $\arg_p$ is the branch of Arg with values from the interval $(\frac{2\pi}{p} - 2\pi, \frac{2\pi}{p})$.

COROLLARY 5.16. The domain of Φ-factorability of a piecewise continuous matrix function is either empty (if condition (5.12) is violated at least at one point $t \in \Gamma$) or consists of a finite number of components $(1,p_1)$, $(p_1,p_2),\dots,(p_{r-1},p_r)$, (p_r,∞), where $\{p_s\}_{s=1}^{r}$ is the sequence of all different numbers of the form $2\pi/\arg \lambda_{jk}$ $(j=1,\dots,N;\ k=1,\dots,n;\ \arg\lambda_{jk} \neq 0)$ arranged in increasing order. Passing from the s th component to the $(s+1)$ th one, the total index decreases by the number of scalars λ_{jk} the argument of which is equal to $2\pi/p_s$.
In particular, if condition (5.12) is fulfilled and all numbers λ_{jk} are positive, then the domain of Φ-factorability of the matrix function G consists of one component occupying the whole ray $(1,\infty)$.

The criterion of Φ-factorability stated in Theorem 5.16 may be expressed in a somewhat different form which is sometimes more convenient. Denote by Δ_p the locus from which the interval $[0,1]$ of the x-axis is visible at the angle of $2\pi/p$ [1]. It is well-known that Δ_p is an arc of a certain circle connecting the points 0 and 1. Suppose that $0,1 \in \Delta_p$.

THEOREM 5.16'. The piecewise continuous matrix function G is Φ-factorable in L_p if and only if, for all $t \in \Gamma$ and $\mu \in \Delta_p$, the relation

$$\det((1-\mu)G(t-0) + \mu G(t+0)) \neq 0 \qquad (5.15)$$

is valid.

1) Assume that the interval $[0,1]$ is visible from the upper semi-plane at an angle smaller than π, and from the lower at an angle greater than π.

Proof. For $\mu = 0$ and $\mu = 1$, condition (5.15) changes over into condition (5.12); for $\mu \neq 0$, it is equivalent to the fact that $1-\mu^{-1}$ does not provide an eigenvalue of the matrix $G^{-1}(t-0)G(t+0)$.
It remains to point out that, when μ ranges over $\Delta_p \setminus \{0\}$, the point $\lambda = 1-\mu^{-1}$ ranges over the ray l_p. ≡

COROLLARY 5.17. The piecewise continuous matrix function G is right Φ-factorable in L_p iff the matrix $\mu G(t-0) + (1-\mu)G(t+0)$ is non-degenerating for all $t \in \Gamma$ and $\mu \in \Delta_p$.

Proof. Due to Theorem 3.7', the matrix function G is right Φ-factorable in L_p if and only if the matrix function G' is left Φ-factorable in L_q. It remains to remark that the matrix function G' is piecewise continuous along with G and to apply Theorem 5.16' taking into account that the transformation $\mu \to 1-\mu$ maps Δ_p onto Δ_q.
In the scalar case, for fixed t and μ ranging over Δ_p, the set of points of the form $(1-\mu)G(t-0) + \mu G(t+0)$ forms an arc such that from its points the interval with the ends $G(t-0)$ and $G(t+0)$ is visible at the angle $2\pi/p$. In the general case one can also imagine the set $\{(1-\mu)G(t-0) + \mu G(t+0) : \mu \in \Delta_p\}$ as an arc connecting $G(t-0)$ and $G(t+0)$. For $p = 2$, we have $\Delta_2 = [0,1]$, and this arc reduces to an interval. ≡

5.4. COMMENTS

The sufficient condition of compactness of the operators QGP and PGQ in case of a Lyapunov contour given in Lemma 5.1 is also necessary. For $p = 2$ and the unit circle, this has been proved in ADAMYAN, AROV, KREĬN [1], JONCKHEERE, DELSARTE [1].

Theorem 5.1 in the scalar case on the circle was proved by LEE and SARASON [1], as well as by AMBARTSUMYAN [2]; concerning the general case, see SPITKOVSKIĬ [10]. The closedness of the class $H_\infty + C$ was observed by SARASON [1]; the proof of Lemma 5.3 presented here is based on results of ZALCMAN [1] and is borrowed from SARASON [2].
The Φ-factorization of matrix functions of the class $H_\infty + C$ was con-

sidered by DOUGLAS [1] and AMBARTSUMYAN [1]. In the latter paper the sufficiency of the condition $G^{-1} \in H_\infty + C$ for the Φ-factorability of a matrix function G $(\in H_\infty + C)$ was verified for all $p \in (1,\infty)$, and in the former its necessity and sufficiency was proved for $p = 2$.
In addition, in DOUGLAS [1] it is proved that the matrix function $G \in H_\infty + C$ is invertible in $H_\infty + C$ if and only if the harmonic extension of the function $\det G$ into the unit disk Δ is separated from zero in a certain annulus $\{z: r \leq |z| < 1\}$, and its total index can be calculated by the formula $\varkappa = \frac{1}{2\pi}\{\arg \det G(\varrho e^{i\theta})\}_{\theta=-\pi}^{\pi}$, where ϱ is an arbitrary number of the interval $(r,1)$.

After some modification this result can be transferred to the general case considered in Theorem 5.2. This theorem is contained in SPITKOVSKIǏ [10]; for the case of a Lyapunov contour, see also SPITKOVSKIǏ [2].
In Ch. 7 of DOUGLAS [2] problems related to Φ-factorization in L_2 of functions of the class $H_\infty + C$ are explained in full detail. One of the authors has obtained the criterion of Φ-factorability of unbounded matrix functions of classes $L^{\pm} + C$ (SPITKOVSKIǏ [6,14]).

Corollary 5.6 becomes false without the assumption $G \in L_\infty^{\pm} + C$: there exist matrix functions $G \in L_\infty$ which are left Φ-factorable, but not right factorable (and vice versa) (SPITKOVSKIǏ [4]).

The Φ-factorization problem of a continuous function in terms of singular integral equations was studied by MIKHLIN [1]. His results were later completed by GOHBERG [1]. Theorem 5.3 in the scalar case has been established by IVANOV [1], SIMONENKO [1] and KHVEDELIDZE [2], and in the matrix case by MANDZHAVIDZE, KHVEDELIDZE [1,2] and SIMONENKO [2].
Formula (5.4) for calculating the total index in case of a Hölder matrix function G was presented by MUSKHELISHVILI [1] (see also MUSKHELISHVILI, VEKUA [1]).
Corollary 5.10, whose generalization is Theorem 5.5 from SPITKOVSKIǏ [10], is, in principle, contained in SIMONENKO [4]. The example on page 180 is taken from GOHBERG, FELDMAN [3].

Theorem 5.10 on the factorization in a decomposing R-algebra (and the notion of the R-algebra itself) is due to GOHBERG [3,4]. The factorization of functions from the Wiener algebra W was studied by KREĬN [1] and of matrix functions with elements from W (Theorem 5.11) by GOHBERG and KREĬN [3].

The factorization of certain singular matrix functions with elements from the Wiener algebra on the real line has been studied in ENGIBARYAN [1]. The factorization in a (C,R)-algebra with the invertibility property was considered by BUDYANU and GOHBERG [3,4], who have proved Theorems 5.7 and 5.7'. The way of proving Theorems 5.8 - 5.10 described here is also borrowed from BUDYANU, GOHBERG [3,4]. The passage to (L_∞, R)-algebras was accomplished in HEINIG, SILBERMANN [1], where Theorem 5.6 was established. Remark that in the latter paper the condition $R \subseteq A$ was replaced by a weaker one: $1, t, t^{-1} \in A$. This condition, of course, means that A contains all rational functions with poles located outside the annulus $\{\zeta : \min_{\tau\in\Gamma}|\tau| < |\zeta| < \max_{\tau\in\Gamma}|\tau|\}$. We leave to the reader to verify that, in virtue of the mentioned circumstance, the argument of Theorem 5.6, in principle, remains unchanged under such a weakening of the condition $R \subseteq A$.

Theorem 5.8 is a classical result of the factorization theory of matrix functions (see VEKUA [4], MUSKHELISHVILI [1]). Theorem 5.9 was independently obtained by ISAKHANOV [1] (see also VEKUA [4], p. 336). In the paper of LAGVILAVA [1] the proof of Theorems 5.8 and 5.11 based on the theory of Banach algebras is outlined but somewhat differing from the one given here. Direct proofs of Theorem 5.10, which are not founded on the factorization theory in (C,R)-algebras, the reader can find in Ch. 8 of GOHBERG, FELDMAN [3] and in LEITERER [1].

Notice that the decomposability condition of the (C,R)-algebra A stipulated here is necessary for any non-singular matrix (and even scalar) function $G \in A$ to admit a factorization in A (BUDYANU, GOHBERG [3,4]). As it is clear from Theorem 5.10, for R-algebras, this condition is also sufficient. It turns out that in the scalar case it remains sufficient for arbitrary (C,R)-algebras with the invertibility

property (concerning this topic, see GOHBERG, KRUPNIK [4]). In case that $n > 1$, necessary and sufficient conditions of factorability of all non-singular $n \times n$ matrix functions from a (C,R)-algebra possessing the invertibility property were obtained in BUDYANU, GOHBERG [3,4]. Besides the demand of decomposability of $\mathfrak{A}$, these conditions include the requirement of boundedness of the quantities $\operatorname{spr} U_G/\|G\|_C$ and $\operatorname{spr} V_G/\|G\|_C$ by a certain constant independent of G.
Here U_G and V_G are operators defined in the space $\mathfrak{A}^{n\times n}$ of $n\times n$ matrix functions with elements from $\mathfrak{A}$ via the formulae $U_G X = PGX$, $V_G X = QXG$; the symbol spr denotes the spectral radius, and G ranges over the class $\mathfrak{A}^{n\times n}$.
An abstract analogue of the established result as well as other aspects of the factorization problem in abstract algebras was studied in BUDYANU [1] and BUDYANU, GOHBERG [2]. Factorization in Banach algebras and its applications to the factorization problem of operators along a chain of projectors are subjects of Ch. 4 of the book GOHBERG, KREĬN [5]. This chapter also containes a detailed bibliography concerning this topic.

Classes of matrix functions different from those described in the main text of the present book and admitting a factorization in C were described in BABAEV, SALAEV [1], BELARMINO [1], GERUS [1], MAGNARADZE [1], PAL'TSEV [1], YANDAROV [1]. In particular, as stated in MAGNARADZE [1] (see also PAL'TSEV [1]), the condition that the modulus of continuity $\omega(\tau)$ for the elements of the considered matrix function be of order $\log^{-p}(|\tau|^{-1})$ for some $p > 1$ is sufficient.

In HEINIG, SILBERMANN [1], from where the study of factorization in (L_∞, R)-algebras originates, one can find concrete examples of algebras in which each matrix function invertible in them is factorable. Algebras with elements from the following classes are of this type:

1) $H^\alpha + E_\infty^\pm$ and $h^\alpha + E_\alpha^\pm$ (h^α denotes the closure of R with respect to the norm in H^α),

2) $B_1^1 + H_\infty$ and $K_{p,q}^{\alpha,\beta} = \{f \in L_\infty\colon Pf \in B_q^\beta,\ Qf \in B_p^\alpha\}$ for $\alpha p > 1$, $p \geq 1$,

$\beta > 0$, $q < \infty$ (here B_r^ν denotes the Besov classes defined as follows:

$$B_r^\nu = \{f \in L_r(\mathbb{T}) : \int_{-\pi}^{\pi} \|\Delta_t^n f\|_{L_r} |t|^{-1-r\nu} dt < \infty\} ,$$

where $n > \nu$ is integer, $\Delta_t^n = \Delta_t \Delta_t^{n-1}$, $(\Delta_t f)(e^{ix}) = f(e^{i(x+t)}) - f(e^{ix})$).

Remark that results of VERBITSKIǏ [1] were essentially employed in the proof of factorability in case 1), whereas in case 2) ideas of PELLER [1] were used.

The search for new classes of matrix functions admitting a factorization in C or in L_∞ is of great interest.

THEOREM 5.13 was proved by SHUBIN [2] with the aid of a local principle of RÖHRL [1], which rests upon the theory of holomorphic fibre bundles.

In SHUBIN [2] Theorem 5.12 was obtained as a consequence of the more general Theorem 5.13. The proof of the latter one given here is, in principle, involved in GOHBERG, KRUPNIK [4]. Note that the Definition 5.5 of local factorization in C borrowed here from SHUBIN [2] has chronologically preceded the definition of factorization mentioned in the final section of Chapter 3.

The factorization of piecewise Hölder functions was studied by GAKHOV [4] and MUSKHELISHVILI [1] and of matrix functions by VEKUA [4]. The factorization of piecewise continuous functions in L_p , under different assumptions on the contour Γ , was considered in KHVEDELIDZE [3,5], SHAMIR [1], GORDADZE [1,2,4–6], GORDADZE and KHVEDELIDZE [1], KHVEDELIDZE and ISHCHENKO [1,2], ISHCHENKO [1], GOHBERG and KRUPNIK [4], and of matrix functions by GOHBERG, KRUPNIK [1], MANDZHAVIDZE [6], ISHCHENKO [2,3], SOLDATOV [2]. Let us remark that the two latter authors studied problems of the asymptotic behaviour of the factorization factors $G_\pm$ in a neighbourhood of discontinuities of a piecewise continuous matrix function G , which have not been considered here (concerning this topic, see VEKUA [4] for results under additional restrictions on G). Theorems 5.15 and 5.16 are explained here supporting mainly on GOHBERG, KRUPNIK [1,4]. For the formula of the total index we chose a form found by KARAPETYANTS and SAMKO [1]. In our view, it is more convenient for calculations than the formula contained in GOHBERG, KRUPNIK [1,4].

The study of convolution type equations on an interval with a kernel being of a quite general type of a generalized function can also be reduced to the factorization problem of piecewise continuous matrix functions (see KARLOVICH, SPITKOVSKIǏ [1], NOVOKSHENOV [1], PAL'TSEV [3,4], SPITKOVSKIǏ [15]). Studying systems of generalized Abel equations, the factorization of piecewise continuous matrix functions also appears in a natural way (see VASIL'EV [1,2], LOWENGRUB, WALTON [1], WALTON [1]).

The Riemann problem with a piecewise Hölder matrix function in special classes of piecewise Hölder functions $h(c_1,\dots,c_q)$ (Muskhelishvili classes) was considered in VEKUA [4]. The classes $h(c_1,\dots,c_q)$ include functions φ satisfying a Hölder condition on any closed arc which does not contain a certain fixed set of points $c_1,\dots,c_m$ $(m \geq q)$ and allowing the estimation $|\varphi(t)| \leq \text{const}|t-c_j|^{-\lambda}$ in a neighbourhood of these points, where λ (> 0) is arbitrarily small for $j=q+1,\dots,m$, and $\lambda < 1$, if $j=1,\dots,q$. The factorization of a piecewise Hölder matrix function G corresponding to the study of a Riemann problem in the class $h(c_1,\dots,c_q)$, where the set $\{c_j\}_{j=1}^q$ covers the discontinuities of G, coincides with its factorization in L_p corresponding to the most left component of the domain of Φ-factorability. If $\{c_j\}_{j=1}^q$ does not intersect the set of discontinuities of the matrix funtion G, then we get a factorization corresponding to the most right component. Intermediate cases of location of $\{c_j\}_{j=1}^q$ and of the set of discontinuities of G are not comprehended by factorization in L_p. Incidentally, the factorization of piecewise Hölder matrix functions associated with the study of a Riemann problem in the Muskhelishvili classes, conversely, does not comprehend the Φ-factorization, when p lies in intermediate components of the Φ-factorability domain.

As a rule, singular integral operators with discontinuous (and, in particular, with piecewise continuous) coefficients are regarded in the spaces $L_p(\varrho)$ of functions summable in the p th power with weight $\varrho(t) = \prod |t-t_k|^{\beta_k}$, $-1 < \beta_k < p-1$ (see e.g. GORDADZE [2,5], ZVEROVICH [1], MISHCHISHIN [1], NIKOLAǏCHUK [1]). Analogously, piecewise continuous matrix functions can also be factored in $L_p(\varrho)$. The results thus obtained generalize the statements of Theorems 5.15 and 5.16, however, qualitative differences do not occur. As a matter of fact, for given weight ϱ, one can always choose a piecewise continuous function h such that Φ-factorability of the matrix function G in $L_p(\varrho)$ is equivalent to Φ-factorability of the matrix function hG in L_p (concerning this topic, see e.g. NYAGA [3]). Therefore, in particular, the structure of the domain of Φ-factorability of piecewise continuous matrix functions described in Corollary 5.16 is preserved: a change of weight causes only a shift (with a possible splitting or confluence) of exceptional points $p_1,\ldots,p_r$. Any factorization of a piecewise Hölder matrix function in the classes $h(c_1,\ldots,c_q)$ can be viewed as its factorization in $L_p(\varrho)$ with specially chosen p and β_k. In NYAGA [2] it is shown that, if Γ is an unbounded closed contour (i.e. a closed curve on a Riemann sphere passing through ∞) and

$$\varrho(t) = |t|^{\beta_0} \prod_{k=1}^{N} \left|\frac{t-t_k}{t}\right|^{\beta_k},$$

then the Fredholmness of $AP_\Gamma + BQ_\Gamma$ in $L_p(\Gamma,\varrho)$ is equivalent to the Fredholmness of $\tilde{A}P_{\tilde{\Gamma}} + \tilde{B}Q_{\tilde{\Gamma}}$ in $L_p(\tilde{\Gamma},\tilde{\varrho})$. Here $\tilde{f}(z) = f(1/(z-z_0))$ (z_0 is an arbitrarily chosen fixed point lying beyond the domain bounded by Γ, $\tilde{\Gamma}$ is the image of Γ under the mapping $z \to \frac{1}{z-z_0}$ and

$$\tilde{\varrho}(t) = |t|^{p-\beta_0-2} \prod_{k=1}^{N} \left|t - \frac{1}{t_k-z_0}\right|^{\beta_k}).$$

This result allows us to transfer all statements concerning Φ-factorization of matrix functions defined on bounded contours to the case of unbounded contours (in particular, to the real line).

The papers MISHCHISHIN [1,2], MOSSAKOVSKIǏ, MISHCHISHIN [1], KHVOSHCHINSKAYA [1], TSITSKISHVILI [1-4], CHAKRABARTI [1] are dedicated to the construction of a factorization of piecewise continuous matrix functions and its applications. In these papers methods of analytical theory of differential equations are employed.

A natural generalization of piecewise continuous matrix functions is the case of matrix functions $G(t)$ for which, at each point $t \in \Gamma$, there exists the (finite) limit from the left $G(t-0)$ and from the right $G(t+0)$. The set of discontinuities of such a matrix function is at most countable, and the class of these matrix functions coincides with the closure $\overline{PC}$ of the class of piecewise continuous matrix functions PC in L_∞ (BOURBAKI [1]). Theorem 5.16' and Corollary 5.17 can be transferred to $\overline{PC}$-matrix functions without any changes.

The Φ-factorization of $\overline{PC}$-matrix functions was considered in GOHBERG, KRUPNIK [2], concerning the scalar case, see also GOHBERG, KRUPNIK [4], DANILYUK [1,3,6], FROLOV [1], SHELEPOV [1].

Note (cf. SPITKOVSKIĬ [2]) that the domain of Φ-factorability of a $\overline{PC}$-matrix function is either empty or the ray $(1,\infty)$ with exceptional points $\{p_s\}$ condensing at most at 1 and ∞. Vice versa, any set of this kind is the domain of Φ-factorability of a certain $\overline{PC}$-matrix function (and even of a scalar $\overline{PC}$-function). An appropriate example (for the case of the single condensation point ∞) yields the function G equal to $\exp \frac{2\pi i\lambda_s(t)}{p_s}$ on the arc (t_s, t_{s+1}), $s=0,1,\ldots,$ where $\{t_s\}_{s=1}^\infty$ is an increasing (in the sense of the orientation of the contour) sequence of points of the contour Γ converging to a fixed point t_0, and λ_s is a function continuous on (t_s, t_{s+1}) and monotonically decreasing from 1 to 0.

In the case of a Lyapunov contour and $p = 2$, Theorem 5.16 has been transferred to all matrix functions that have no more than two limits $G_1(t)$ and $G_2(t)$ at each point $t \in \Gamma$. In this case in (5.15) one has to substitute $G(t \pm 0)$ by $G_{1,2}(t)$ (CLANCEY [1]). The corresponding result for $p \neq 2$ is no longer a literal analogue of Theorem 5.16, but runs as follows:

For Φ-factorability of the matrix function G in L_p it is necessary and sufficient that

1) $\det G_1(t)G_2(t) \neq 0$ for all $t \in \Gamma$,

2) $\omega_j(t) \neq 2\pi/p$ at those points $t \in \Gamma$ where $G_1(t)$ is the right-side limit for G and $G_2(t)$ is the left-side limit,

3) $\omega_j(t)$ is not located on the interval with ends $2\pi/p$ and $2\pi/q$ at those points $t \in \Gamma$ for which the matrix function G has two essential limits at least from one side.

Here $\omega_1(t),\ldots,\omega_n(t)$ is the collection of arguments of all eigenvalues of the matrix $G_2^{-1}(t)G_1(t)$.

This result has been derived by one of the authors (SPITKOVSKIǏ [11]); its complete proof can be found in SPITKOVSKIǏ [16].

CHAPTER 6. ON THE STABILITY OF FACTORIZATION FACTORS

The present chapter is devoted to the study of the behaviour of the factorization factors under small perturbations of the matrix function G. Here we encounter another qualitative difference between the matrix case $(n > 1)$ and the scalar one $(n = 1)$. It appears that the partial indices of a matrix function and the factorization factors $G_{\pm}$ are unstable, in general. This fact gives rise to principle difficulties in the solution of the two most important problems in the factorization theory of a matrix function G and the solvability theory of the vector-valued Riemann boundary value problem with this matrix:

1) the problems of calculating the partial indices (of calculating the defect numbers of the Riemann problem) and

2) the problems of constructing an approximate factorization of a matrix function (of finding an approximate solution of the Riemann boundary value problem).

Till now, an effective solution has not been found yet for both problems in the general case. Therefore, it is still actual to look for effectively verifiable estimates for the partial indices.

In the framework of the latter problem, those cases when the partial indices preserve one and the same sign and, especially, are stable are of particular interest. The equality of sign of all partial indices is necessary and sufficient for the stability of the defect numbers α and β of the Riemann boundary value problem. In this and only this case they can be determined by the total index $\varkappa$ of the matrix function G via the formulae $\alpha = \max\{0,\varkappa\}$, $\beta = \max\{0,-\varkappa\}$. As we shall see later on, the stability of partial indices is necessary and sufficient for the stability of the solutions of the Riemann problem, which is important for an effective application of approximate solution methods.

Now as ever, the theory of Fredholm operators will be an important means of research in the present chapter. Thus, we shall be concerned with one of the following two situations:

1) Φ-factorization of a matrix function $G \in L_\infty$ given on a contour of class $\mathfrak{R}$;

2) factorization of a matrix function $G \in \mathfrak{A}$ in a decomposing (C,R)-algebra $\mathfrak{A}$.

In considerations of Sections 6.1, 6.2 which apply to both situations just mentioned we shall simply write "factorization" instaed of "Φ-factorization" or "factorization in $\mathfrak{A}$". Perturbations of the matrix function G we shall estimate in the operator norm, i.e., in the norm (3.19) in situation 1) and by means of (5.8) in situation 2).

Let us clarify that a certain characteristic $f(G)$ of the matrix function G is called stable, if it is a continuous function of G, i.e., small (in the sense of the indicated norm) perturbations of G correspond to small variations of $f(G)$. In particular, to say that integer characteristics (partial indices) are stable amounts to saying that they remain unchanged under sufficiently small variations of G.

The crucial result of Section 6.1 is the stability criterion for the partial indices of a matrix function G consisting in the fact that the largest of the partial indices of G differs from the smallest one by at most one. The proof of this criterion is preceded by the proof of the stability of the partial indices $\varkappa_1, \dots, \varkappa_n$ written in a non-increasing order to remain in the interval $[\varkappa_n, \varkappa_1]$ and, in particular, the stability of the property of partial indices to preserve one and the same sign. As mentioned above, the latter statements are very important and without doubt of independent interest, thus, we emphasized them in separate theorems, departing from the traditional method of presenting the stability criterion for partial indices.

The statements of Section 6.1 are based on the stability property of the defect numbers of one-sided invertible Fredholm operators. These assertions show, especially, that the change of partial indices of the matrix function G obeys certain laws in case they are unstable. Applying the more sophisticated theorem on the semi-stability of the defect numbers of Fredholm operators, one can obtain an exhaustive description of the character of change of partial indices under perturbations of G. This description is contained in Section 6.2. It is formulated in terms of the ordering $\succ$ introduced in Chapter 4, which is defined over tuples of n integers ordered in a special way.

The main disadvantage of the stability criterion for partial indices and the other theorems from Section 6.1 consists in their non-effectivity: for checking the assumptions of these theorems, first of all, one has to evaluate the largest and the smallest index $\varkappa_1$ and $\varkappa_n$ or at least to determine their signs. Therefore, we are interested in sufficient conditions of stability, non-negativity or non-positivity of partial indices which enable a more effective method of checking to be applied than the direct calculation of $\varkappa_1$ and $\varkappa_n$ or the determination of their signs. This topic is pursued in Section 6.3 and 6.4. The method, with the help of which sufficient conditions will be derived in 6.1, is based on the application of the contracting mapping principle. As will be clarified in the next chapter, if the contour is a circle and the factorization is considered in L_2, then the conditions of Section 6.3 are exact. Another way of obtaining conditions of sign definiteness and stability of partial indices is presented in 6.4. It rests upon the utilization of certain algebraic properties and notions. For instance, such a property is the normality of the matrix function in question a.e. on the contour Γ. The main result of Section 6.4 related to normal matrices consists in the factorability with equal partial indices of a matrix function whose eigenvalues are located in some disk. A natural generalization of this fact is provided by an analogous result (factorability with equal partial indices) for matrix functions $G \in L_\infty$ not necessarily normal, but such that its numerical domain is separated from zero for all points of Γ.

For matrix functions of classes $L_\infty^{\pm} + C$ as well as for continuous matrix functions, the results of 6.3 and 6.4 will be made more precise in Section 6.5.

Section 6.6 is devoted to the stability of the factorization factors $G_\pm$.

It will be shown that for matrix functions close to the original one which have different tuples of partial indices, the factorization factors $G_{\pm}$ cannot be too close. The opposite assertion is particularly important: the closeness of matrix functions whose tuples of partial indices coincide implies the closeness of the factorization factors $G_{\pm}$. This means the stability of factorization with a stable tuple of partial indices.

Section 6.7 contains comments on published papers.

6.1. STABILITY CRITERION FOR PARTIAL INDICES

In Chapter 3 the stability of the factorability property and the stability of the total index (Theorem 3.20) have been proved. Thus, in the scalar case the diagonal factorization factor Λ is stable. Passing to the matrix case, the situation essentially changes. This will be demonstrated by the following example.

<u>Example.</u> Suppose $\Lambda_\varepsilon(t) = \begin{pmatrix} t^\nu & 0 \\ \varepsilon t^{\mu+1} & t^\mu \end{pmatrix}$, where ε is a constant, ν and μ are integers and $\nu-\mu \geq 2$. If $\varepsilon \neq 0$, then the matrix function Λ_ε admits the factorization

$$\Lambda_\varepsilon(t) = \begin{pmatrix} 1 & \varepsilon^{-1}t^{\nu-\mu-1} \\ 0 & 1 \end{pmatrix} \begin{pmatrix} t^{\nu-1} & 0 \\ 0 & t^{\mu+1} \end{pmatrix} \begin{pmatrix} 0 & -\varepsilon^{-1} \\ \varepsilon & t^{-1} \end{pmatrix}$$

and, thus, its partial indices are $\nu-1$ and $\mu+1$. Clearly, the matrix function Λ_ε depends continuously on the parameter ε. At the same time, the matrix function Λ_0 is diagonal and its partial indices are equal to λ and μ. Therefore, in any neighbourhood of Λ_0 there exists a matrix function the tuple of partial indices of which does not coincide with the tuple of partial indices of Λ_0. In other words, the system of partial indices of the matrix function is unstable.

However, in some cases the tuple of partial indices can be stable. For instance, the equality of all partial indices to zero is, in view of Corollary 3.8 (or Theorem 5.6), equivalent to the invertibility of the operator $P + GQ$ (and $Q + \underline{G}'P$). Therefore, it is a stable characteristic of factorization. This assertion is a special case of the following proposition,

THEOREM 6.1. Let the matrix function G admit a factorization with non-negative (non-positive) partial indices. Then there exists a $\delta > 0$ such that every matrix function G_1 from the δ-neighbourhood of G is factorable with the same total index and its partial indices are also non-negative (non-positive).

Proof. We study the case of non-negative partial indices. According to Corollary 3.8, the operator $T_Q(G)$ is invertible from the right and its kernel is of dimension $\varkappa$. In view of the stability of one-sided invertibility, all operators from a certain ε-neighbourhood of $T_Q(G)$ are also invertible from the right. Since the kernel dimension of a one-sided invertible operator does not change under small perturbations, it is constant on connected sets consisting of operators invertible from the right and, particularly, on an ε-neighbourhood of $T_Q(G)$. Setting $\delta = \varepsilon\|Q\|^{-1}$, we find that $\|T_Q(G_1) - T_Q(G)\| < \varepsilon$ as soon as $\|G - G_1\| < \delta$. Thus, the operator $T_Q(G_1)$ is invertible from the right and its kernel dimension is $\varkappa$. In other words, the matrix function G_1 can be factored, its partial indices are non-negative and the total index is equal to $\varkappa$. ≡

Let us note how one can specify an estimate for δ.
Since ε can be taken as $\varepsilon = \|T_Q(G)^{-1}\|^{-1}$, where $T_Q(G)^{-1}$ denotes an arbitrary one-sided inverse of $T_Q(G)$, one may choose

$$\delta = \|Q\|^{-1}\|G_-^{-1}Q\Lambda^{-1}G_+^{-1}|_{\mathrm{im}\,Q}\|^{-1} \tag{6.1}$$

in the case of non-negative and

$$\delta = \|Q\|^{-1}\|G_-^{-1}\,\Lambda^{-1}QG_+^{-1}|_{\mathrm{im}\,Q}\|^{-1} \tag{6.1'}$$

in the case of non-positive partial indices. For $G_{\pm}^{-1} \in L_\infty^{\pm}$, one can take

$$\delta = \|Q\|^{-2}\|G_+^{-1}\|\|G_-^{-1}\|\|\Lambda^{-1}\| . \tag{6.1''}$$

THEOREM 6.2. Let the matrix function G admit a factorization with partial indices $\varkappa_1 \geq \dots \geq \varkappa_n$. Then there exists a $\delta > 0$ such that every matrix function G_1 from the δ-neighbourhood of G can be factored with the same total index and partial indices lying in $[\varkappa_n, \varkappa_1]$.

Proof. Consider the matrix function $t^{-\varkappa_n}G(t)$. Its partial indices are non-negative. Due to the previous theorem, there exists a $\delta_n > 0$ such that all matrix functions from the δ_n-neighbourhood of $t^{-\varkappa_n}G(t)$ are factorable with the same total index and have non-negative partial indices. Analogously, all matrix functions from a certain δ-neighbourhood of $t^{-\varkappa_1}G(t)$ have non-positive partial indices. Set $\delta = \min\{\delta_n\|t^{-\varkappa_n}\|, \delta_1\|t^{-\varkappa_1}\|\}$. Then, provided that $\|G - G_1\| < \delta$, we have $\|t^{-\varkappa_1}G_1 - t^{-\varkappa_1}G\| < \delta_1$ and $\|t^{-\varkappa_n}G_1 - t^{-\varkappa_n}G\| < \delta_n$, which means that the matrix function $t^{-\varkappa_n}G_1$ has non-negative and $t^{-\varkappa_1}G_1$ has non-positive partial indices. Consequently, the partial indices of G_1 are included between $\varkappa_n$ and $\varkappa_1$. The equality of total indices of the matrix functions $t^{-\varkappa_1}G$ and $t^{-\varkappa_1}G_1$ implies the equality of total indices of G and G_1. Theorem 6.2 is proved. ≡

Now we proceed to the main result of the present section.

THEOREM 6.3 (Stability criterion for partial indices). The system $\varkappa_1 \geq \ldots \geq \varkappa_n$ of partial indices is stable if and only if

$$\varkappa_1 - \varkappa_n \leq 1. \tag{6.2}$$

Proof. Note that the partial indices are defined uniquely by condition (6.2) and the value of the total index $\varkappa$. Namely,

$$\begin{aligned} \varkappa_n &= \varkappa_{n-1} = \ldots = \varkappa_{r+1} = [\varkappa/n], \\ \varkappa_r &= \ldots = \varkappa_1 = [\varkappa/n] + 1, \end{aligned} \tag{6.3}$$

where r $(0 \leq r < n)$ is the remainder from the division of $\varkappa$ by n. If condition (6.2) is fulfilled, then, in accordance with Theorem 6.2, it remains valid for all matrix functions from some neighbourhood of G. The value of the total index is also preserved and, consequently, formulae (6.3) either. Thus, the sufficiency of the condition (6.2) for the stability of the system of partial indices is proved.

Now, let (2.1) be a factorization of the matrix function G, and assume the condition (6.2) to be not fulfilled. Denote by A that matrix of n th order the only non-zero element of which is equal to one and situated in the left lower corner. By φ we denote some function from E_∞^- which has a zero of order $\varkappa_1 - \varkappa_n - 2$ at infinity and satisfies the condition $|\varphi(t)g_{in}^+(t)g_{1j}^-(t)| \leq 1$ $(i,j=1,\dots,n)$ a.e. on Γ. Here g_{in}^+ and g_{1j}^- are entries of the last column and the first row of the matrix functions G_+ and G_- respectively.

Such a function φ can always be constructed. To this end, we assume $\varphi \equiv 0$ in the domains $\mathcal{D}^+(j)$, $j=1,\dots,m-1$, and reduce the construction of φ in $\mathcal{D}^+(0)$ with the help of a conformal mapping to the problem of the reconstruction of a function analytic in a disk by the values of its absolute value on its boundary. If g_{in}^+, $g_{1j}^- \in L_\infty$ $(i,j=1,\dots,n)$, but also in case of factorization in a (C,R)-algebra $\mathfrak{A}$, it suffices to set $\varphi(z) = cz^{-\varkappa_1+\varkappa_n+2}$, where c is a constant sufficiently small. The matrix function $F_\varepsilon(t) = \Lambda(t) + \varepsilon\varphi(t)At^{\varkappa_1-1}$ is lower triangular. By a method similar to the one used in Theorem 4.9, for $\varepsilon \neq 0$, we may get its factorization

$$F_\varepsilon(t) = F_{+,\varepsilon}(t)\ \Lambda^{(1)}(t)F_{-,\varepsilon}(t) ,$$

where

$$F_{+,\varepsilon}(t) = I + \varepsilon^{-1}Q_0(t)A' ,$$

$$\Lambda^{(1)}(t) = \operatorname{diag}[t^{\varkappa_1-1} , t^{\varkappa_2},\dots,t^{\varkappa_{n-1}} , t^{\varkappa_n+1}] ,$$

$$F_{-,\varepsilon}(t) = \begin{pmatrix} t-Q_0(t)\varphi(t) & & & & & -\varepsilon^{-1}Q_0(t)t^{1-\varkappa_1+\varkappa_n} \\ & 1 & & & & \\ & & \ddots & & & \\ & & & \ddots & & \\ & & & & 1 & \\ \varepsilon\varphi(t)t^{\varkappa_1-\varkappa_n-2} & & & & & t^{-1} \end{pmatrix}$$

$$Q_0(z) = c_1 z^{\varkappa_1-\varkappa_n-1} + c_2 z^{\varkappa_1-\varkappa_n-2} ,$$

and the numbers c_1 and c_2 are chosen in such a way that the function $z - Q_0(z)\varphi(z)$ vanishes at infinity.

Apparently, the representation

$$G_+F_\varepsilon G_- = (G_+F_{+,\varepsilon})\,\Lambda^{(1)}(F_{-,\varepsilon}G_-)$$

is a factorization of the matrix function $G_+F_\varepsilon G_-$. At the same time,

$$G_+F_\varepsilon G_- - G = \varepsilon\varphi t^{\varkappa_1-1}\, G_+AG_- = \varepsilon\varphi t^{\varkappa_1-1}(g_{in}^+ g_{1j}^-)_{i,j=1}^n .$$

In this way, arbitrarily close to the matrix function G we may indicate a factorable matrix function the tuple of partial indices of which $-\varkappa_1-1, \varkappa_2,\dots,\varkappa_{n-1}, \varkappa_n+1$ does not coincide with the tuple of partial indices of G (the equation $\varkappa_1-1=\varkappa_n$ is impossible, because $\varkappa_1-\varkappa_n \geq 2$). This implies the necessity of condition (6.2) for the stability of the tuple of partial indices. The proof of Theorem 6.3 is complete. ≡

6.2. ON THE BEHAVIOUR OF PARTIAL INDICES UNDER SMALL PERTURBATIONS. THE STRUCTURE OF THE FAMILY OF Φ-FACTORABLE MATRIX FUNCTIONS

In the previous section it was shown that, if the system of partial indices is of the form (6.3), then it remains unchanged under small perturbations of the matrix function. From Theorem 6.2 it is clear that in the general case the changes of partial indices are also not of arbitrary nature. In the proof of Theorem 6.2 the stability of the defect numbers of one-sided invertible operators was used. Taking advantage of the semistability of the defect numbers of Fredholm operators (Theorem 1.4), one can obtain a more precise result. It will be formulated in terms of the ordering $\succ$ introduced in Chapter 4 (Definition 4.1).

<u>THEOREM 6.4.</u> Assume the matrix function G to admit a factorization with the tuple of partial indices $(\varkappa_j)_{j=1}^n$. Then there exists a neighbourhood of this matrix function consisting only of such factorable matrix functions whose tuple of partial indices is majorized (in the sense of the ordering $\succ$) by the tuple $(\varkappa_j)_{j=1}^n$.

Note that, by assertion 1) of Lemma 4.1, the partial indices of all matrix functions from the indicated neighbourhood are concentrated in the interval $[\varkappa_n, \varkappa_1]$ and the total index coincides with the total index of G. Thus, Theorem 6.4 makes Theorem 6.2 more precise.

<u>Proof.</u> Choosing an integer k from the interval $[\varkappa_n, \varkappa_1]$, we consider

the operator $P + t^{-k}G(t)Q$. It is Fredholm, since the matrix function $t^{-k}G(t)$ is factorable along with $G(t)$. Consequently, all operators from some neighbourhood of it are Fredholm and, due to Theorem 1.2, their defect numbers do not exceed the corresponding defect numbers of the operator $P + t^{-k}G(t)$.

We can find a number $\delta_k > 0$ such that the operator $P + t^{-k}F(t)Q$ belongs to the mentioned neighbourhood of the operator $P + t^{-k}G(t)Q$ as soon as $\|F-G\| < \delta_k$. Utilizing the formulae stated in Theorem 3.16 for the defect numbers of such operators, we obtain

$$\begin{aligned} \sum_{j=1}^{n} \max\{\varkappa'_j-k,0\} &\le \sum_{j=1}^{n} \max\{\varkappa_j-k,0\}, \\ \sum_{j=1}^{n} \min\{\varkappa'_j-k,0\} &\ge \sum_{j=1}^{n} \min\{\varkappa_j-k,0\}, \end{aligned} \tag{6.4}$$

where $(\varkappa'_j)_{j=1}^{n}$ is the tuple of partial indices of the matrix function F. Setting $\delta = \min\{\delta_k\}_{k=\varkappa_n}^{\varkappa_1}$, we observe that, for all matrix functions F from the δ-neighbourhood of G, the inequalities (6.4) are fulfilled for $k = \varkappa_n, \ldots, \varkappa_1$. According to assertion 2) of Lemma 4.1, this implies that the tuple $\{\varkappa'_j)$ is majorized by the tuple $(\varkappa_j)$, which proves Theorem 6.4. ≡

The result of Theorem 6.4 is of final nature. In order to support this, we must describe the relation $\succ$ in a somewhat different way. To this end, we first introduce some definitions and notations.

Let σ_n be the collection of tuples $(k_j)_{j=1}^{n}$ of integers satisfying $k_1 \geq \ldots \geq k_n$.

<u>DEFINITION 6.1.</u> Let us say that (k'_j) $(\in \sigma_n)$ is obtained from the tuple (k_j) $(\in \sigma_n)$ with the aid of an elementary operation, if for some integers r and s $(1 \le r < s \le n)$,

$$k'_r = k_r-1, \quad k'_s = k_s+1, \quad k'_j = k_j \quad (j \neq r,s). \tag{6.5}$$

We shall write $(k_j) \to (k'_j)$, if the tuple $\{k'_j\}$ $(\in\sigma_n)$ either coincides with (k_j) $(\in\sigma_n)$ or is obtained from it as a result of a finite number of elementary operations.

Obviously, $\to$ is a partial order relation on the set σ_n. For comparability of the tuples (k'_j) and (k_j), it is necessary (and, for $n=2$, also sufficient) that $\sum_{j=1}^{n} k'_j = \sum_{j=1}^{n} k_j$. Minimal elements in the sense of the relation $\to$ are those and only those tuples $(\varkappa_j)_{j=1}^{n}$ which satisfy the condition (6.2).

<u>LEMMA 6.1.</u> Let $(k_j)_{j=1}^n$, $(k'_j)_{j=1}^n \in \sigma_n$. Then the relation $(k_j) \ddot{\to} (k'_j)$ holds if and only if $(k_j) \succ (k'_j)$.

<u>Proof.</u> Suppose $(k_j) \ddot{\to} (k'_j)$. By the very definition, either $(k_j) = (k'_j)$ (and then, clearly $(k_j) \succ (k'_j)$) or (k'_j) is obtained from (k_j) as the result of a certain number of elementary operations. If we succeed in proving that as the result of an arbitrary elementary operation there will be obtained a tuple which is majorized (in the sense of the relation $\succ$) by the initial one, then, in view of the transitivity of the ordering $\succ$, $(k_j) \succ (k'_j)$.

Thus, it is sufficient to examine tuples connected via relations (6.5). It is easily checked that in this case the inequalities (4.8) and (4.8') are true.

Now we are going to show that, if (k_j), $(k'_j) \in \sigma_n$ and $(k_j) \succ (k'_j)$, then $(k_j) \ddot{\to} (k'_j)$. For this purpose, we use induction on n. For $n = 2$, the condition $(k_j) \succ (k'_j)$ means that $k'_1 = k_1 - k$, $k'_2 = k_2 + k$, where $0 \leq k \leq (k_1 - k_2)/2$. Apparently, the tuple (k'_j) is obtained from (k_j) by means of k elementary operations. Therefore, $(k_j) \ddot{\to} (k'_j)$.

Supposing that the assertion to be proved is true for $n \neq 1$, we consider two tuples of length n. The condition $(k_j)_{j=1}^n \succ (k'_j)_{j=1}^n$ means, in particular, that $k_1 \geq k'_1$, $k_n \leq k'_n$. If, for instance, $k_n = k'_n$, then $(k_j)_{j=1}^n \succ (k'_j)_{j=1}^n$ implies $(k_j)_{j=1}^{n-1} \succ (k'_j)_{j=1}^{n-1}$, By induction hypothesis, $(k_j)_{j=1}^{n-1} \ddot{\to} (k'_j)_{j=1}^{n-1}$. But in this case we have also $(k_j)_{j=1}^n \ddot{\to} (k'_j)_{j=1}^n$. Analogous reasonings are applicable for $k_1 = k'_1$. Consequently, we shall assume that $k_1 > k'_1$ and $k_n < k'_n$. Then $k_1 - k_n \geq (k'_1 + 1) - (k'_n - 1) \geq 2$. Therefore, there exist numbers l and $m, l < m$, such that $k_1 = \ldots = k_l > k_{l+1}$, $k_1 - k_{m-1} \leq 1$, $k_1 - k_m \geq 2$,

Consider the tuple $(k_j^{(1)})_{j=1}^n$ obtained from $(k_j)_{j=1}^n$ with the aid of the elementary operation for $r = 1$, $s = m$ and show that $(k_j^{(1)})_{j=1}^n \succ (k'_j)_{j=1}^n$. We restrict ourselves to the verification of inequality (4.8) in which k_j is replaced by $k_j^{(1)}$.

For $k \geq k'_1$, we have $\sum_{j=1}^n \max\{k'_j-k,0\} = 0$, for $k \leq k_m$, $\sum_{j=1}^n \max\{k_j^{(1)}-k,0\} \geq \sum_{j=1}^n \max\{k'_j-k,0\}$. Finally, if $k_m < k < k'_1$, $k_j > k$, then $k_j \geq k_{m-1} \geq k_l - 1 \geq k'_j$, where $k_j > k'_j$ for $j=1,\ldots,l,$ so that $\sum_{j=1}^n \max\{k'_j-k,0\} \leq \sum_{j=1}^n \max\{k_j-k,0\} -1 \leq \sum_{j=1}^n \max\{k_j-k,0\} -1 = \sum_{j=1}^n \max\{k_j^{(1)}-k,0\}$.

Repeating these considerations, after $d\ (\leq l)$ steps we come to the tuple $(k_j^{(d)})_{j=1}^n \succ (k'_j)_{j=1}^n$, where either $k_n^{(d)} > k_n$ (if $d < l$) or $k_1^{(d)} < k_1$ (if $d = l$) . From this we conclude that after the application of a finite number N of elementary operations one can obtain a tuple $(k_j^{(N)})_{j=1}^n \succ (k'_j)_{j=1}^n$ for which either $k_n^{(N)} = k'_n$ or $k_1^{(N)} = k'_1$. In any case, as shown above, $(k_j^{(N)})_{j=1}^n \to (k'_j)_{j=1}^n$. But then $(k_j)_{j=1}^n \to (k'_j)_{j=1}^n$, because $(k_j)_{j=1}^n \to (k_j^{(N)})_{j=1}^n$.

Lemma 6.1 is proved. ≡

THEOREM 6.5. Let the matrix function G be factorable, and assume the tuple of its partial indices to be $(\varkappa_j)_{j=1}^n$. If $(\varkappa_j)_{j=1}^n \succ (\varkappa'_j)_{j=1}^n$, then in an arbitrary neighbourhood of G there exists a matrix function the tuple of partial indices of which is $(\varkappa'_j)_{j=1}^n$.

Proof. The assertion is evident if $(\varkappa'_j) = (\varkappa_j)$. If $(\varkappa'_j) \neq (\varkappa_j)$, then, according to Lemma 6.1, there exist tuples $(\varkappa_j^{(k)})_{j=1}^n$, $k=0,\ldots,N$, such that $\varkappa_j^{(0)} = \varkappa_j$, $\varkappa_j^{(N)} = \varkappa'_j$ and the $(k+1)$ th tuple is obtained from the k th one with the aid of one elementary operation. Arguing as in the proof of the necessity in Theorem 6.3, we conclude that in any neighbourhood of the given (factorable) matrix function there exists a matrix function whose tuple of partial indices is obtained from the tuple of partial indices of the initial one with the help of one elementary operation previously chosen. Specifying $\varepsilon > 0$, we choose in the $\varepsilon/2$-neighbourhood of G a matrix function G_1 with the tuple of partial indices $(\varkappa_j^{(1)})_{j=1}^n$, in the $\varepsilon/4$-neighbourhood of G_1 a matrix function G_2 the tuple of partial indices of which is $(\varkappa_j^{(2)})_{j=1}^n$ and so on. After all, we obtain a matrix function G_N from the ε-neighbourhood of G with the tuple of partial indices $(\varkappa'_j)_{j=1}^n$, which proves Theorem 6.5. ≡

COROLLARY 6.1. In any neighbourhood of a factorable matrix function there exists a matrix function with a stable tuple of partial indices.

With the help of Theorem 3.7' all results derived in the present section can be transferred to the case of right-factorization.

COROLLARY 6.2. In any neighbourhood of a matrix function factorable from the left and the right there exists a matrix function with a stable tuple of left as well as right indices.

Proof. For given $\varepsilon > 0$, we take in the $\varepsilon/2$-neighbourhood of the original matrix function G a matrix function G_1 with a stable tuple of partial indices (due to Corollary 6.1, this is always possible). If ε is sufficiently small, then the matrix function G_1 will be factorable not only from the left but also from the right. In view of the stability of its left indices there is a δ-neighbourhood of G_1 to which belong only matrix functions with the same tuple of partial indices. Setting $\delta_1 = \min\{\delta, \varepsilon/2\}$ and using an analogue of Corollary 6.1 related to right-factorization, we find in the δ_1-neighbourhood of G_1 a matrix function G_2 with a stable tuple of right indices. This is just the desired one. ≡

The results of the present section admit a topological interpretation. Denote by $\Phi_{(\varkappa_j)}$ the set of factorable matrix functions[1)] the tuple of partial indices of which is $(\varkappa_j)$. The representation $\Phi = \bigcup_{(\varkappa_j)\in\sigma_n} \Phi_{(\varkappa_j)}$ yields a decomposition of the class of all factorable matrix functions into disjoint non-empty sets, where:

1) if $(\varkappa_j) \succ (\varkappa'_j)$, then $\Phi_{(\varkappa_j)} \subseteq \operatorname{clos} \Phi_{(\varkappa'_j)}$,

2) if $(\varkappa_j) \not\succ (\varkappa'_j)$, then $\Phi_{(\varkappa_j)} \cap \operatorname{clos} \Phi_{(\varkappa'_j)} = \emptyset$.

---

1) Recall that in this section the factorization is always understood as Φ-factorization.

Properties 1) and 2) are reformulations of Theorem 6.5 and 6.4, respectively. Theorem 6.3 means that

3) the set $\Phi_{(\varkappa_j)}$ is open if and only if the tuple $(\varkappa_j)$ satisfies condition (6.2).

Since a tuple $(\varkappa_j)$ having the property (6.2) is comparable with all tuples with the same sum and is minimal (in the sense of the ordering $\succ$), then property 1) means, especially, that

4) under the condition (6.2), the set $\Phi_{(\varkappa_j)}$ is dense in

$$\Phi_{\varkappa} = \cup\{\Phi_{(k_j)} : (k_j) \in \sigma_n , \sum_{j=1}^{n} k_j = \sum_{j=1}^{n} \varkappa_j \ (= \varkappa)\} .$$

If the tuple $(\varkappa_j)$ fails to have property (6.2), then the set $\Phi_{(\varkappa_j)}$ lies in the complement (with respect to $\Phi_{\varkappa}$) of an open dense set. Therefore, it is nowhere dense. Because, for different $\varkappa$, the sets $\Phi_{\varkappa}$ are open and disjoint, $\Phi_{(\varkappa_j)}$ is nowhere dense in their union either. Thus:

5) if $\varkappa_1 - \varkappa_n \geq 2$, then $\Phi_{(\varkappa_j)}$ is a nowhere dense subset of Φ.

Finally, from property 2) we can conclude that

6) the sets $\Phi_{(\varkappa_j)}$ and $\Phi_{(\varkappa'_j)}$ are separated iff the tuples $(\varkappa_j)$ and $(\varkappa'_j)$ are not comparable.

With Theorem 3.20 in mind we can imagine the set Φ as the union of a countable number of separated open sets $\Phi_{\varkappa}$. Now we can every of the $\Phi_{\varkappa}$'s divide into an open set $\Phi_{(\varkappa_j)}$ $(\sum_{j=1}^{n} \varkappa_j = \varkappa , \ \varkappa_1 - \varkappa_n \leq 1)$ and a nowhere dense set $\Phi_{(\varkappa'_j)}$ $(\sum_{j=1}^{n} \varkappa'_j = \varkappa , \ \varkappa'_1 - \varkappa'_n \geq 2)$.

6.3. ESTIMATES FOR THE PARTIAL INDICES OF MEASURABLE BOUNDED MATRIX FUNCTIONS

In Section 6.1 it was shown that inequality (6.2) is a necessary and sufficient condition of stability of partial indices. However, if we want to use this criterion, we must know the value of the largest and smallest partial index. Therefore, we are interested in assertions in which besides Φ-factorability of a certain matrix function also estimates of its partial indices are established.

Sufficient criteria of non-negativity and non-positivity of partial indices are also of some significance, since the defect numbers of the vector-valued Riemann boundary value problem with matrix coefficient G are stable if and only if all its partial indices have the same sign. The present section is devoted to a result of precisely such a type. Owing to Theorem 6.2, every $n \times n$ matrix function which is Φ-factorable in L_p with zero partial indices has a neighbourhood entirely consisting of matrix functions which have the same property. In particular, the matrix function identically equal to I possesses such a neighbourhood. Denote by $\delta_{p,n}(\Gamma)$ the largest possible value of the radius of such a neighbourhood. In other words, $\delta_{p,n}(\Gamma)$ is a number such that, for

$$\|I - G_0\| < \delta_{p,n}(\Gamma) , \tag{6.6}$$

the operator $P + G_0Q$ is invertible and there exists an $n \times n$ matrix function F for which $\|I - F\| = \delta_{p,n}(\Gamma)$ and the operator $P + FQ$ is non-invertible.

We denote by A_p the interval with ends p and $q = p/(p-1)$.

<u>**LEMMA 6.2.**</u> The following assertions related to $\delta_{p,n}(\Gamma)$ are true:

1) the numbers $\|Q\|^{-1}_{L^n_p}$ and $\dfrac{2\|S\|_{L^n_p}}{1 + \|S\|^2_{L^n_p}}$ are estimates from below for $\delta_{p,n}(\Gamma)$; if $\Gamma \in \widetilde{\mathfrak{R}}$, then $\delta_{p,n}(\Gamma) \leq \sin \pi/\max\{p,q\}$;

2) $\delta_{p,n}(\Gamma) = \delta_{q,n}(\Gamma)$;

3) $\delta_{r,n}(\Gamma) \geq \delta_{p,n}(\Gamma)$ for $r \in A_p$; in particular, $\delta_{2,n}(\Gamma) = \max\{\delta_{p,n}(\Gamma): 1 < p < \infty\}$.

<u>Proof.</u> 1) The estimate $\delta_{p,n}(\Gamma) \geq \|Q\|^{-1}_{L^n_p}$ follows from the contracting mapping principle, since, for $\|I - G\| < \|Q\|^{-1}_{L^n_p}$, the operator $P + GQ = I + (G - I)Q$ differs from the unity operator by a summand with norm smaller than one. To prove the second estimate from below for $\delta_{p,n}(\Gamma)$, we consider an arbitrary matrix function G from the $2\|S\|/(1+\|S\|^2)$-neighbourhood of the identity matrix function. Then

$$\|I - G\| \leq \frac{2\sigma}{1 + \sigma^2} \tag{6.7}$$

for some $\sigma > \|S\|_{L_p^n}$ (≥ 1). We introduce the matrix functions

$G_1 = \frac{\sigma^2 + 1}{\sigma^2 - 1} G$ and $\hat{G} = (I-G_1)(I+G_1)^{-1}$. Obviously, G and G_1 are Φ-factorable only simultaneously and the tuple of their partial indices coincide. Moreover, due to condition (6.7) and the identity

$$I + G_1 = \frac{2\sigma^2}{\sigma^2 - 1}\left(I + \frac{\sigma^2+1}{2\sigma^2}(G-I)\right),$$

the matrix function $I + G_1$ is invertible in L_∞. Consequently, the invertibility if the operator $P + G_1Q = \frac{1}{2}(I+\hat{G}S)(I+G_1)$ is equivalent to the invertibility of the operator $I + \hat{G}S$. In this way, the desired estimate will be obtained, if we clarify that the inequality $\|\hat{G}\| \leq \sigma^{-1}$ holds, provided that condition (6.7) is valid. In other words, we have to show that the matrix function $Y = \sigma\hat{G}$ will be contracting as soon as the matrix function $X = \frac{1 + \sigma^2}{2\sigma}(I-G)$ is contracting. But X and Y are connected via the relation

$$Y = (I - \varepsilon X)^{-1}(X - \varepsilon I), \tag{6.8}$$

where $\varepsilon = \sigma^{-1}$. It remains to use the well-known fact from matrix theory that, for any $\varepsilon \in [0,1]$, transformation (6.8) maps a contraction into a contraction. The latter can easily been checked by showing that the spectrum of Y lies inside the unit disk.

Finally, let the condition $\Gamma \in \tilde{\mathcal{R}}$ be fulfilled. Consider a piecewise continuous function f given on Γ which takes only the two values $\cos^2\pi/\bar{p} + \frac{i}{2}\sin 2\pi/\bar{p}$ and $\cos^2\pi/\bar{p} - \frac{i}{2}\sin 2\pi/\bar{p}$ with $\bar{p} = \max\{p,q\}$. Clearly, $\|f - 1\| = \sin \pi/\bar{p}$. Furthermore, at the half of the points of discontinuity of the function f the argument jump is equal to $2\pi/\bar{p}$ and at the other half it equals $2\pi - 2\pi/\bar{p} = 2\pi/\min\{p,q\}$. If at the discontinuities of f there exist the tangents to the contour Γ (which can always be effected), then we are able to use Theorem 5.15, according to which this function is not Φ-factorable in L_p (and L_q). Hence $\delta_{p,1}(\Gamma) \leq \sin \pi/\bar{p}$. From here and the obvious relation $\delta_{p,n}(\Gamma) \leq \delta_{p,1}(\Gamma)$ we deduce the inequality to be proved.

2) It is not hard to see that the inverse transformation to (6.8) is this transformation itself. Therefore, it maps the 1-neighbourhood of the zero matrix function onto itself (and not only into itself as mentioned above). Moreover, it is a one-to-one mapping. Setting $Y_1 = I + \varepsilon Y$, $X_1 = (I - \varepsilon X)'$, from (6.8) we observe that, for arbitrary $\varepsilon < 1$, the transformation

$$Y_1 = (1 - \varepsilon^2)(X_1')^{-1}$$

maps the ε-neighbourhood of the identity matrix function one-to-one onto itself. At the same time, Theorem 3.7' and Corollary 3.6 imply that the matrix function X_1 is Φ-factorable in L_p with zero partial indices if and only if the matrix function Y_1 is Φ-factorable in L_q with zero partial indices. Thus, for some $\varepsilon < 1$, the ε-neighbourhood of the identity matrix function consists completely of matrix functions which are Φ-factorable in L_r with zero partial indices for $r = p$ if and only if they possess this property for $r = q$. This as well as the evident relation $\delta_{r,n}(\Gamma) \leq 1$ (which is true for all $r \in (1,\infty)$) imply assertion 2) of the lemma.

3) Now we consider the $\delta_{p,n}(\Gamma)$-neighbourhood of the identity matrix function I . According to assertion 2), every matrix function G from this neighbourhood is Φ-factorable with zero partial indices not only in L_p but also in L_q . Owing to Theorem 3.9 and 2.2, the matrix function G is Φ-factorable with zero partial indices in all L_r with $r \in A_p$. Hence, the $\delta_{p,n}(\Gamma)$-neighbourhood of the identity matrix function is a part of its $\delta_{r,n}(\Gamma)$-neighbourhood, i.e. $\delta_{p,n}(\Gamma) \leq \delta_{r,n}(\Gamma)$ $(r \in A_p)$. Since, for all $p \in (1,\infty)$, the interval A_p contains the point $r = 2$, we deduce that the last of the relations to be proved is also correct. ≡

<u>THEOREM 6.6.</u> Let the matrix functions G and $X_{\pm} \in L_\infty$ be such that

$$\| I - X_+ G X_- \|_{L_p^n} < \delta_{p,n}(\Gamma) . \qquad (6.9)$$

Then:

1) if $X_+ \in M_\infty^+$, $X_+^{-1} \in L_\infty^+$, $X_- \in L_\infty^-$, $X_-^{-1} \in M_\infty^-$, then $A_p \subseteq \Phi(G)$ and, for all $r \in A_p$, the partial r-indices of G are non-negative,

2) if $X_+ \in L_\infty^+$, $X_+^{-1} \in M_\infty^+$, $X_- \in M_\infty^-$, $X_-^{-1} \in L_\infty^-$, then $A_p \subseteq \Phi(G)$ and, for all $r \in A_p$, the partial r-indices of G are non-positive.

Proof. Under the condition (6.9), the matrix function $G_0 = X_+GX_-$ satisfies relation (6.6). Therefore, from assertion 3) of Lemma 6.2 we conclude that the matrix function G_0 is Φ-factorable in L_r for all $r \in A_p$ and its partial r-indices are equal to zero. With regard to Theorem 3.10 we can claim that not only in case 1) but also in case 2) the matrix function G. is Φ-factorable in L_r $(r \in A_p)$ simultaneously with G_0, i.e. $A_p \subseteq \Phi(G)$. The non-negativity of the partial r-indices of G in case 1) and their non-positivity in case 2) follow from Theorem 2.11 by setting $B_\pm = X_\pm^{-1}$. Let us check this, say, in case 2). Taking into consideration that the partial indices of G_0 are equal to zero, we find that, in view of Theorem 2.11, the value $\mathcal{P}(B_-) + (n-1) \times \mathcal{P}(B_+) + \text{ind}_+\det B_+$ serves as an upper bound of the partial indices of G. But $B_- = X_-^{-1} \in L_\infty^-$, so that $\mathcal{P}(B_-) \leq 0$. We represent the matrix function B_+ in the form $B_+ = f^{-1}A_+$, where $A_+ \in L_\infty^+$ and f is a rational function with index equal to $\mathcal{P}(B_+)$. Then $\text{ind}_+\det B_+ =$ $= \text{ind}_+\det A_+ - n\mathcal{P}(B_+)$, thus, $\text{ind}_+\det B_+ + (n-1)\mathcal{P}(B_+) =$ $= \text{ind}_+\det A_+ - \mathcal{P}(B_+)$. The value standing on the right-hand side of the last equation is opposite to the number $\text{ind}_+ B_+^{-1}$ and, with regard to the condition $B_+^{-1} \in L_\infty^+$, non-positive. Consequently, the upper estimate of the partial indices of the matrix function G is non-positive and, therefore, also the partial indices themselves. The proof of Theorem 6.6 is complete. ≡

On the strenght of Theorem 6.6 we can get sufficient conditions under which the partial indices are contained in a given interval.

In particular, we may obtain stability criteria for partial indices.

THEOREM 6.7. Let the matrix functions $G \in L_\infty$ and $X_{1,2} \in L_\infty^+$, $Y_{1,2} \in L_\infty^-$ be such that $X_{1,2}^{-1} \in M_\infty^+$, $Y_{1,2}^{-1} \in M_\infty^-$ and

$$\begin{aligned} &\|I - t^{-a}X_1^{-1}GY_1\| < \delta_{p,n}(\Gamma), \\ &\|I - t^{-b}X_2GY_2^{-1}\| < \delta_{p,n}(\Gamma), \end{aligned} \qquad (6.9')$$

where a and b $(a \leq b)$ are integers. Then for all $r \in A_p$, we have $r \in \Phi(G)$ and the partial r-indices of G are included between a and b.

<u>Proof.</u> The first of the inequalities (6.9') ensures that assertion 1) of Theorem 6.6 can be applied to the matrix function $t^{-a}G(t)$. The second one enables us to apply assertion 2) of the same theorem to the matrix function $t^{-b}G(t)$. Hence it follows that the matrix function G is Φ-factorable in L_r for $r \in A_p$, the partial r-indices of the matrix function $t^{-a}G(t)$ are non-negative and the partial r-indices of $t^{-b}G(t)$ are non-positive, which is equivalent to the assertion of the theorem. ≡

<u>THEOREM 6.8.</u> Suppose $G \in L_\infty$. For the matrix function G to be Φ-factorable in L_p and for its partial p-indices to be stable, it is sufficient that there exist an integer k as well as matrix functions $X_{1,2} \in L_\infty^+$ and $Y_{1,2} \in L_\infty^-$ such that $X_{1,2}^{-1} \in M_\infty^+$, $Y_{1,2}^{-1} \in M_\infty^-$ and $\|I - t^{-k}X_1^{-1}GY_1\| < \delta_{p,n}(\Gamma)$, $\|I-t^{-k-1}X_2GY_2^{-1}\| < \delta_{p,n}(\Gamma)$.

<u>Proof.</u> Setting $a = k$, $b = k+1$, we meet the conditions of Theorem 6.7, according to which $p \in \Phi(G)$ and the partial p-indices of G can only attain the values k and $k+1$. In virtue of Theorem 6.3, the partial p-indices of G are stable, which proves Theorem 6.8. ≡

The obtained stability criterion for partial indices can be simplified, if the total index $\varkappa$ is assumed to be known. This supposition is justified, for example, for piecewise continuous matrix functions (under the additional restrictions on the contour Γ mentioned in Section 5.3), but also for matrix functions of class $L_\infty^\pm + C$. In the general case $G \in L_\infty$, too, the calculation of the total index is, in virtue of its stability, a simpler problem than the determination of all partial indices.

<u>THEOREM 6.9.</u> Let the matrix function $G \in L_\infty$ admit a factorization in L_p with total index $\varkappa$. If

1) $\varkappa \equiv 0 \pmod n$ or $\varkappa \equiv 1 \pmod n$ and

   a) there exist matrix functions $X_\pm \in L_\infty^\pm$ such that $X_\pm^{-1} \in M_\infty^\pm$ and $\|I - t^{-[\varkappa/n]}X_+^{-1}GX_-\| < \delta_{n,p}(\Gamma)$ or

2) $\varkappa \equiv 0 \pmod n$ or $\varkappa \equiv -1 \pmod n$ and

b) there exist matrix functions $X_{\pm} \in L_{\infty}^{\pm}$ such that $X_{\pm}^{-1} \in M_{\infty}^{\pm}$ and $\|I - t^{-[\varkappa/n]-1} X_{+} G X_{-}^{-1}\| < \delta_{p,n}(\Gamma)$,

then the tuple of partial p-indices of G is stable.

Proof. In the first case, $\varkappa = kn + r$, where $r = 0$ or $r = 1$, and k is an integer. Due to Theorem 6.6, $p \in \Phi(G)$, so that a factorization of G is, properly, a Φ-factorization. Moreover, the inequalities $\varkappa_j \geq k$, $j=1,\ldots,n$ are fulfilled. Since in this case $\sum_{j=1}^{n} \varkappa_j \leq kn+1$, we have, actually, $\varkappa_j = k$ $(j=2,\ldots,n)$, $\varkappa_1 = k$ for $r = 0$ and $\varkappa_1 = k+1$ if $r = 1$. Consequently, the tuple of partial indices of the matrix function G is stable. The second case may be considered analogously. Theorem 6.9. has been proved. ≡

COROLLARY 6.3. If $n = 2$, then either of the conditions a) and b) mentioned in Theorem 6.9 is sufficient for the stability of the partial indices. If $n = 3$, then at least one of these conditions is sufficient.

6.4. SUFFICIENT CONDITIONS FOR COINCIDENCE OF PARTIAL INDICES OF MATRIX FUNCTIONS WHOSE NUMERICAL DOMAIN IS SEPARATED FROM ZERO

Theorems 6.6 - 6.9 are based on Φ-factorability in L_p and equality to zero of partial p-indices of a matrix function G_0 satisfying condition (6.6). Now we are going to obtain another sufficient conditions of coincidence of partial indices which rest upon some additional suppositions on algebraic properties of the matrix function G. Using them and acting as in Theorems 6.6 - 6.9, one can get criteria of non-negativity, non-positivity and stability of partial indices. This is left to the reader.

Below we shall need the following result on the behaviour of partial indices of a matrix function on multiplication by a scalar continuous function.

LEMMA 6.3. Let the matrix functions F and G $(\in L_{\infty})$ be connected by the equation $F = \chi G$, where χ is a continuous non-vanishing function. Then F and G are Φ-factorable in some L_p only simultaneously and the tuple $(\hat{\varkappa}_j)$ of partial p-indices of F is obtained from the tuple $(\varkappa_j)$ of partial p-indices of G by the formula $\hat{\varkappa}_j = \varkappa_j + k$, $j=1,\ldots,n$, where k is the index of the function χ .

<u>Proof.</u> The assertion concerning the simultaneous Φ-factorability of the matrix functions F and G results from Corollary 5.10. Since any function continuous on Γ may be uniformly approximated by rational ones with arbitrary accuracy, then, for every $\varepsilon > 0$, there exists a representation of F of the form $F = \chi_1 G_1$, where $\|\chi-\chi_1\| < \varepsilon$, $\|G-G_1\| < \varepsilon$ and the function χ_1 is rational. For sufficiently small ε, the function χ_1 does not vanish, its index coincides with the index of the function χ , the matrix function G_1 is Φ-factorable in L_p and its tuple of partial p-indices $(\varkappa_j^{(1)})$ is majorized by the tuple $(\varkappa_j)$ in the sense of the ordering $\succ$ (Theorem 6.4).

If $G_1 = G_+^{(1)} \Lambda^{(1)} G_-^{(1)}$ provides a Φ-factorization of the matrix function G_1 in L_p and $\chi_1 = \chi_+^{(1)} t^k \chi_-^{(1)}$ is a factorization of the rational function χ_1 in which, by Theorem 2.9, $(\chi_+^{(1)})^{\pm 1} \in L_\infty^+$, $(\chi_-^{(1)})^{\pm 1} \in L_\infty^-$, then, setting $F_+ = \chi_+^{(1)} G_+^{(1)}$, $F_- = \chi_-^{(1)} G_-^{(1)}$ and $\Lambda(t) = t^k \Lambda^{(1)}(t)$, we get the Φ-factorization $F = F_+ \Lambda F_-$ of F in L_p . Consequently, $\hat{\varkappa}_j = \varkappa_j^{(1)} + k$, $j=1,\ldots,n$ and, due to what was proved above, $(\varkappa_j) \succ (\hat{\varkappa}_j - k)$. Applying similar arguments to the equation $G = \chi^{-1} F$, we can show that $(\hat{\varkappa}_j - k) \succ (\varkappa_j)$. Owing to assertion 3) of Lemma 4.1, we thus deduce that $\hat{\varkappa}_j = \varkappa_j + k$, $j=1,\ldots,n$, which proves Lemma 6.3. ≡

Recall that the matrix A is said to be normal, if it commutes with its adjoint: $AA^* = A^*A$. The norm of a normal matrix coincides with the largest absolute value of its eigenvalues (GANTMAKHER [1]).

<u>THEOREM 6.10.</u> Let F be a measurable $n\times n$ matrix function taking normal values a.e. on Γ and having the following properties: for every point $t \in \Gamma$, there are an arc γ_t $(\ni t)$ on Γ and a disk Δ_t which is visible from the origin at an angle of $\alpha_t < \arcsin \delta_{p,n}(\Gamma)$

such that the eigenvalues of almost all matrices $F(\tau)$, $\tau \in \gamma_t$ lie in the disk Δ_t.

Then the interval A_p is completely contained in some component of the domain of Φ-factorability of F and all partial r-indices of F are equal to each other ($r \in A_p$).

Proof. Without loss of generality, the arcs γ_t may be assumed to be open. From the covering $\{\gamma_t\}_{t\in\Gamma}$ we choose a finite subcovering $\{\gamma_{t_j}\}_{j=1}^{N}$. Cancelling, if necessary, some of the arcs γ_{t_j}, we may assume that every point $t \in \Gamma$ is contained in at most two elements of the covering. Furthermore, we take the largest (say, α) of the numbers α_{t_j}, $j=1,\dots,N$, and set $\delta = \sin \alpha/2$. Evidently $\delta < \delta_{p,n}(\Gamma)$.
By z_j we denote the centre of the disk Δ_{t_j} and by l_{kj} the interval connecting the points z_k^{-1} and z_j^{-1} $(k,j=1,\dots,N)$. Finally, on the contour Γ we define a continuous function ψ with the following properties:

1) if $t \in \gamma_{t_j}$, $t \notin \gamma_{t_k}$ for $k \neq j$, then $\psi(t) = z_j^{-1}$,

2) if $t \in \gamma_{t_j} \cap \gamma_{t_k}$, $j \neq k$, then $\psi(t) \in l_{kj}$.

The function ψ does not vanish on Γ. Indeed, if $\gamma_{t_j} \cap \gamma_{t_k} \neq \emptyset$, then, by assumption, the eigenvalues of $F(\tau)$ lie in the intersection of Δ_{t_j} and Δ_{t_k} a.e. on $\gamma_{t_j} \cap \gamma_{t_k}$.
Therefore, $\Delta_{t_j} \cap \Delta_{t_k} \neq \emptyset$. Since neither of the disks Δ_{t_j} and Δ_{t_k} contains zero, then the interval connecting their centres z_j and z_k does not contain zero either. Consequently, zero does not belong to the interval l_{kj}.

Now we are going to show that the matrix function $G_0 = \psi F$ satisfies the condition (6.6). For this purpose, note that, if the eigenvalues of $F(t)$ lie in the disk Δ_{t_j}, then the eigenvalues of the matrix $z_j^{-1}F(t)$ lie in the disk Δ_0 with centre at the point 1 and radius $\delta = \sin \alpha/2$ $(\geq \sin \alpha_j/2)$. If the eigenvalues of $F(t)$ lie both in Δ_{t_j} and Δ_{t_k}, then we recognize that the eigenvalues of both the matrices $z_j^{-1}F(t)$ and $z_k^{-1}F(t)$ lie in Δ_0 and, therefore, also of all matrices of the form $\zeta F(t)$ with $\zeta \in l_{kj}$. Hence, the eigenvalues of the matrix $G_o(t)$ lie in Δ_o a.e. on Γ, and those of $I - G_o(t)$ in the disk $\{z:|z| \leq \delta\}$. Taking into account that $\delta < \delta_{p,n}(\Gamma)$ and the matrices $I - G_o(t)$ are normal together with $F(t)$, we find that condition (6.6) is fulfilled.

In view of statement 3) of Lemma 6.2, the matrix function G_o is Φ-factorable in L_r for all $r \in A_p$ and its partial indices are equal to zero. Applying now Lemma 6.3 with $\chi = \psi^{-1}$, we conclude that the matrix function F is also Φ-factorable in L_r, $r \in A_p$ and its partial indices are equal to each other and to the index of the function χ.
Thus Theorem 6.10 is proved. ≡

<u>COROLLARY 6.4.</u> Let the matrix $U(t)$ be unitary a.e. on Γ. Furthermore, suppose that, for any point $t \in \Gamma$, we are able to indicate an arc γ_t $(\ni t)$ of the contour Γ and a sector S_t with vertex at the origin of apex angle $\alpha_t < 2 \arcsin \delta_{p,n}(\Gamma)$ such that all eigenvalues of the matrices $U(\tau)$ lie in S_t for almost all $\tau \in \gamma_t$. Then the interval A_p is entirely contained in some component of the domain of Φ-factorability of the matrix function U and all partial p-indices of U are equal to each other.

In fact, a unitary matrix is a normal matrix the eigenvalues of which have absolute value 1. Therefore, the fact that the eigenvalues of the matrices $U(\tau)$, $\tau \in \gamma_t$ lie in the sector S_t implies that they lie in the intersection l_t of this sector with the unit circle. It remains to note that the arc l_t may be included into a disk Δ_t inscribed in the sector S_t. After that, Theorem 6.10 will be applied. ≡

The normality condition occurring in Theorem 6.10 is very incisive. Thus, we try to find an analogue of these propositions related to matrix functions which do not necessarily take normal values. In looking for such an analogue, the following considerations are useful.
The eigenvalues of the normal matrix $G(t)$ belong to some disk if and only if the convex hull of the set of eigenvalues of $G(t)$ belongs to this disk. It is well-known that the convex hull of the set of eigen-

values of the normal matrix $A = (a_{ij})_{i,j=1}^{n}$ coincides with the set

$$H(A) = \{ \sum_{i,j=1}^{n} a_{ij}\xi_i\bar{\xi}_j : \sum_{j=1}^{n} |\xi_j|^2 = 1\} . \tag{6.10}$$

However, formula (6.10) continues to make sense for a matrix A not necessarily normal.

<u>DEFINITION 6.2.</u> The set $H(A)$ is called the <u>numerical domain</u> (or Hausdorff set) of the matrix A.

For any matrix A, $H(A)$ is a convex compact subset of the complex plane containing all eigenvalues of the matrix A (GLAZMAN, LYUBICH [1], MARCUS, MINC [1]).

It is quite natural to suspect that there exist sufficient conditions of Φ-factorability of the matrix function G and estimates for its partial indices formulated in terms of $H(G)$.

<u>DEFINITION 6.3.</u> By $d(A)$ we denote the smallest value of the real part of points from $H(A)$.

It is well-known (cf. MARCUS, MINC [1]) that $d(A)$ is the smallest eigenvalue of the Hermitian matrix $\operatorname{Re} A = \frac{1}{2}(A+A^*)$.

<u>LEMMA 6.4.</u> Assume the matrix function $G \in L_\infty$ to satisfy the condition

$$\operatorname*{ess\,inf}_{t\in\Gamma} d(G(t)) > \sqrt{1-\delta^2_{p,n}(\Gamma)}\ \|G\| . \tag{6.11}$$

Then $A_p \subseteq \Phi(G)$ and all partial r-indices of G are equal to zero for $r \in A_p$.

<u>Proof.</u> We shall show that the matrix function G differs from the matrix function G_0 satisfying inequality (6.6) only by a constant multiplier $\mu > 0$. For this purpose, using the equation $\|A\|^2 = \|AA^*\|$ and choosing the point $t \in \Gamma$ arbitrarily, we write $\|I-\mu G(t)\|^2 =$
$= \|(I-\mu G(t))(I-\mu G(t))^*\| = \|I+\mu^2 G(t)G(t)^* - 2\mu \operatorname{Re} G(t)\|$.
Inside the norm sign we have obtained a Hermitian non-negative matrix N the norm of which, as well-known, is the maximum of the quantity (Nx,x) taken over all vectors x of the unit sphere. Since $(Ix,x) = \|x\|^2 = 1$, $(\mu^2 G(t)G(t)^*x,x) = \mu^2\|G(t)^*x\|^2 \le \mu^2\|G(t)^*\|^2 = \mu^2\|G(t)\|^2$ and

$(-2\mu \operatorname{Re} G(t)x,x) = -2\mu(\operatorname{Re} G(t)x,x) \le -2\mu d(G(t))$, we deduce $\|I-\mu G(t)\|^2 \le 1 + \mu^2\|G(t)\|^2 - 2\mu d(G(t))$. Therefore, the inequality $\|I-\mu G(t)\|^2 \le 1 + \mu^2\|G\|^2 - 2\mu \operatorname{ess\,inf}_{t\in\Gamma} d(G(t))$ holds a.e. on Γ .

Choosing $\mu = \|G\|^{-2} \operatorname{ess\,inf}_{t\in\Gamma} d(G(t))$, we obtain that $\|I-\mu G(t)\|^2 \le 1 - \|G\|^{-2} \operatorname{ess\,inf}_{t\in\Gamma} d(g(t))$ a.e. on Γ. Thus, due to condition (6.11), $\|I-\mu G\| < \delta_{p,n}(\Gamma)$. From this and assertion 3) of Lemma 6.2 we conclude the Φ-factorability in L_r , $r \in A_p$, of the matrix function μG . Moreover, we see that its partial r-indices are equal to zero. Clearly, the initial matrix function G is, together with μG , also Φ-factorable and the tuples of partial indices of these matrix functions coincide. $\equiv$

Acting as in the proof of Theorem 6.10, from Lemma 6.4 one can get the following result.

THEOREM 6.11. Let the matrix function $G \in L_\infty$ be such that, for every point $t \in \Gamma$, there exist an arc $\gamma_t\ (\ni t)$ and a straight line l_t whose distance from zero is greater than $\|G\| \sqrt{1-\delta^2_{p,n}(\Gamma)}$ and which has the property: for almost all $\tau \in \gamma_t$, the set $H(G(\tau))$ is separated from zero by the straight line l_t . Then the matrix function G is Φ-factorable in L_r for $r \in A_p$ and all its partial indices coincide.

Supposing, in addition, that the contour Γ satisfies the Lyapunov condition, we are able to strengthen Lemma 6.4. To this end, we need the following result related to factorization of positive functions.

LEMMA 6.5. Let α be a function given on the Lyapunov contour Γ which takes only non-negative values and for which $\alpha^{\pm 1} \in L_\infty$. Then there exists a representation $\alpha = \alpha_+\alpha_-$ with $\alpha_+^{\pm 1} \in L^+_\infty$, $\alpha_-^{\pm 1} \in L^-_\infty$.

Proof. We set $\alpha_+ = \exp(P \ln \alpha)$, $\alpha_- = \exp(Q \ln \alpha)$. Evidently, $\alpha = \alpha_+\alpha_-$ and the function α_+ (α_- , respectively) is analytic and nondegenerate in $\mathcal{D}^+$ ($\mathcal{D}^-$) . Also, $|\alpha_+^{\pm 1}(z)| = \exp(\pm \operatorname{Re}(P \ln \alpha)(z))$. Furthermore, if we prove that $\operatorname{Re}(P \ln \alpha)$ is a function bounded in $\mathcal{D}^+$, then the inclusion $\alpha_+^{\pm 1} \in L^+_\infty$ will be proved.

We have

$$(P \ln \alpha)(z) = \frac{1}{2\pi i} \int_\Gamma \ln \alpha(\tau) \frac{d\tau}{\tau - z},$$

so that

$$|\mathrm{Re}(P \ln \alpha)(z)| = \left|\frac{1}{2\pi} \int_\Gamma \ln \alpha(\tau) \,\mathrm{Im} \frac{d\tau}{\tau - z}\right| \leq \frac{M}{2\pi} \int_\Gamma \left|\mathrm{Im} \frac{d\tau}{\tau-z}\right|,$$

where $M = \operatorname{ess\,sup}_{t\in\Gamma} |\ln \alpha(\tau)|$ ($< \infty$, because $\alpha^{\pm 1} \in L_\infty$). In this way, the crucial fact to be verified is the finiteness of $\sup_{z \in \mathcal{D}^+} \int_\Gamma \left|\mathrm{Im} \frac{d\tau}{\tau-z}\right|$.

Obviously, it is sufficient to conduct the proof for a connected contour. By ω we denote a function realizing a conformal mapping of the open unit disk onto the domain $\mathcal{D}^+$. Setting $\tau = \omega(e^{i\theta})$, $z = \omega(\zeta)$, we obtain

$$\left|\int_\Gamma \mathrm{Im} \frac{d\tau}{\tau-z}\right| = \int_T \left|\mathrm{Im} \frac{\omega'(e^{i\theta}) de^{i\theta}}{\omega(e^{i\theta}) - \omega(\zeta)}\right| = \int_{-\pi}^{\pi} \left|\mathrm{Re} \frac{\omega'(e^{i\theta}) e^{i\theta}}{\omega(e^{i\theta}) - \omega(\zeta)}\right| d\theta$$

$$\leq \int_{-\pi}^{\pi} \left|\frac{\omega'(e^{i\theta})}{\omega(e^{i\theta})-\omega(\zeta)} - \frac{e^{i\theta}}{e^{i\theta}-\zeta}\right| d\theta + \int_{-\pi}^{\pi} \left|\mathrm{Re} \frac{e^{i\theta}}{e^{i\theta}-\zeta}\right| d\theta .$$

Since the contour Γ is Lyapunov, the function ω' satisfies a Hölder condition (GOLUZIN [1]). Thus, the kernel of the first integral obtained is weakly polar (see e.g. LITVINCHUK [3], p.35). From this we deduce the boundedness of this integral by a certain quantity not depending on ζ. The second integral is an integral with Poisson kernel and therefore equal to 2π. Consequently, we have proved the boundedness of the function $\mathrm{Re}(P \ln \alpha)$ and, thus, also the condition $\alpha_+^{\pm 1} \in L_\infty^+$. The inclusion $\alpha_-^{\pm 1} \in L_\infty^-$ can be verified in a similar way. ≡

<u>LEMMA 6.4'.</u> Assume that Γ is a Lyapunov contour. If, for some $\varepsilon > 0$, the matrix function $G_0 \in L_\infty$ satisfies the inequality

$$d(G_0(t)) \geq \sqrt{1-\delta^2_{p,n}(\Gamma)}\, \|G_0(t)\| + \varepsilon \qquad (6.11')$$

a.e. on Γ, then G_0 is Φ-factorable in L_r ($r \in A_p$) and all its partial r-indices are equal to zero.

<u>Proof.</u> The inequality (6.11') means, in particular, that $d(G_0(t)) \geq \varepsilon$ a.e. on Γ. Consequently, $\|G_0(t)^{-1}\| \leq \varepsilon^{-1}$ a.e. on Γ [1], so that

$G_0^{-1} \in L_\infty$. Since $\|G_0(t)\|^{-1} \leq \|G_0(t)^{-1}\|$, then the function $\alpha(t) = \|G_0(t)\|^{-1}$ meets the assumptions of Lemma 6.5. Therefore, the matrix functions G_0 and αG_0 are Φ-factorable only simultaneously and the tuples of their partial indices coincide.

Multiplying inequality (6.11') termwise by $\alpha(t)$, we get $d(G(t)) \geq \sqrt{1-\delta^2_{p,n}(\Gamma)}\,\|G\| + \varepsilon\|G_0\|^{-1}$, so that G fulfils the assumptions of Lemma 6.4. Now the assertion to be proved follows immediately. ≡

Lemma 6.4' enables us to strengthen Theorem 6.11 in the following manner:

> <u>THEOREM 6.11'.</u> Let the contour Γ be Lyapunov, and assume the matrix function G to belong to L_∞. Moreover, let, for all $t \in \Gamma$, exist an arc γ_t ($\ni t$) of the contour Γ and a straight line l_t with distance to zero greater than $\sqrt{1-\delta^2_{p,n}(\Gamma)}$ such that, for almost all $\tau \in \gamma_t$, the set $\|G(\tau)\|^{-1}H(G(\tau))$ is separated from zero by the straight line l_t. Then G is Φ-factorable in L_r for $r \in A_p$ and all its partial indices are equal to each other.

Finally, note that the conditions of all the theorems in the present section are insensitive to transposition of the matrix function under study. Consequently, a matrix function satisfying these theorems is Φ-factorable not anly from the left but also from the right. Furthermore, its right partial indices coincide with the left ones (and are also equal to each other).

6.5. ON PARTIAL INDICES OF MATRIX FUNCTIONS OF CLASSES $L^{\pm}_\infty + C$

The results obtained in Sections 6.3 and 6.4 provided information both about the existence of a Φ-factorization and about the values of the partial indices. In case when the matrix function G belongs to one of the classes considered in Chapter 5, the question of existence of a

---

[1] Here we used the fact from matrix theory that, if $d(A) > 0$, then $\|A^{-1}\| \leq d(A)^{-1}$. It is a straightforward matter to prove this: on the one hand, $|(Ax,x)| \geq \operatorname{Re}(Ax,x) \geq d(A)\|x\|^2$, on the other hand, $|(Ax,x)| \leq \|Ax\|\cdot\|x\|$, which implies $\|Ax\| \geq d(A)\|x\|$. Hence A is invertible and $\|A^{-1}\| \leq d(A)^{-1}$.

Φ-factorization is already well understood. This allows for making more precise the results on the values of partial indices.

First of all, suppose that $G \in L_\infty^+ + C$ $(L_\infty^- + C)$. According to Theorem 5.2, from the Φ-factorability of such a matrix function in some L_p we deduce the Φ-factorability in all L_p, $1 < p < \infty$, moreover, with the same tuple of partial indices. Hence, in the formulations of Theorems 6.6 - 6.11' and Lemmas 6.4, 6.4' we may replace $\delta_{p,n}(\Gamma)$ by its maximal value $\delta_{2,n}(\Gamma)$, which will be denoted by $\delta_n(\Gamma)$, and claim the existence of a Φ-factorization with non-negative (non-positive, coinciding, etc.) partial indices irrespective of the parameter value p. In addition, Theorems 6.6 - 6.9 can be simplified as follows: if $G \in L_\infty^+ + C$, then in assertion 2) of Theorems 6.6 and 6.9 the requirement $X_+^{-1} \in M_\infty^+$ can be omitted, and in Theorems 6.7 and 6.8 - the demand that $X_2^{-1} \in M_\infty^+$. If $G \in L_\infty^- + C$, then in assertion 1) of Theorems 6.6 and 6.9 we may cancel the requirement $X_-^{-1} \in M_\infty^-$, but in Theorems 6.7 and 6.8 - the demand $Y_1^{-1} \in M_\infty^-$.

Now we are going to state the exact formulation and to conduct the proof of the restated assertion 1) of Theorem 6.6 in the case $G \in L_\infty^- + C$. The other variants can be considered analogously, which is left to the interested reader.

<u>THEOREM 6.12.</u> Suppose that $G \in L_\infty^- + C$, $X_+ \in M_\infty^+$, $X_- \in L_\infty^-$ satisfy the conditions: 1) $\|I - X_+ G X_-\| < \delta_n(\Gamma)$, 2) $X_+^{-1} \in L_\infty^+$.
Then (i) $X_-^{-1} \in M_\infty^-$, (ii) the matrix function G admits a Φ-factorization in all spaces L_p $(1 < p < \infty)$ with a non-negative tuple of partial indices.

<u>Proof.</u> The relation $X_- \in L_\infty^-$ implies that $PX_-Q = 0$. Moreover, the operator PGQ is compact by Lemma 5.1. Consequently,

$$P + X_+GX_-Q = (P+X_+Q)(P+GQ)(P+X_-G) + T, \tag{6.12}$$

where T is compact.

From condition 1) of the theorem we conclude the invertibility of the operator $P+X_+GX_-Q$ in the space L_2^n, and from condition 2) and Theorem 5.2 we get the Fredholmness of the operator $P+X_+Q$ (in all spaces L_p^n).

Therefore, equation (6.12) and Theorem 1.9 imply the Fredholmness of the product $(P+GQ)(P+X_-Q)$ in the space L_2^n. In turn, from the Fredholmness of this product and the same Theorem 1.9 there follows that $P+GQ$ is a Φ_--operator.

Due to Corollary 5.1, if $P+GQ$ is a Φ_--operator and $G \in L_\infty^- + C$, then the operator $P+GQ$ is Fredholm. In this case the second factor $P+X_-Q$ of the product $(P+GQ)(P+X_-Q)$ will also be Fredholm. Since $X_- \in L_\infty^-$, then, in virtue of Theorem 3.15', we may conclude that $X_-^{-1} \in M_\infty^-$. In this way, assertion (i) has been proved. As soon as this condition is fulfilled, we can apply assertion 1) of Theorem 6.6, according to which the partial 2-indices of the matrix function G are non-negative (the Φ-factorability of this matrix function in L_2 results from the Fredholmness of the operator $P+GQ$ just proved). The result obtained can be transferred to all parameter values p, as was mentioned at the beginning of the section. ≡

If the considered matrix function G is continuous, then we can apply the strengthenings of Theorems 6.6 - 6.11 for the case $G \in L_\infty^+ + C$ as well as for the case $G \in L_\infty^- + C$ mentioned at the beginning of the section. In the case of a continuous matrix function G, a particularly clear form can be given to Theorem 6.11.

<u>DEFINITION 6.3.</u> By $h(A)$ we denote the distance from the numerical domain of matrix A to zero. The matrix A will be called strictly non-degenerate, if $h(A) > 0$, i.e. if $H(A)$ does not contain zero.

This definition is justified by the fact that a strictly non-degenerate matrix is clearly non-degenerate.

<u>THEOREM 6.13.</u> If the matrix function G is continuous and, for all $t \in \Gamma$,

$$h(G(t)) > \sqrt{1-\delta_n^2(\Gamma)}\, \|G(t)\| , \tag{6.13}$$

then all the partial indices of G are equal to each other.

<u>Proof.</u> The continuity of G implies the continuity of the function $\alpha(t) = \|G(t)\|$. In accordance with inequality (6.13) this function does not vanish.

From relation (6.13) we deduce an analogous relation for the matrix function $F = \alpha^{-1}G$. In addition, for all $t \in \Gamma$, we have $\|F(t)\| = 1$. Consequently, for all $t \in \Gamma$,

$$h(F(t)) > \sqrt{1-\delta_n^2(\Gamma)}\,\|F\| . \tag{6.14}$$

The closed convex set $H(F(t))$ contains a unique point nearest to zero, which we denote by $z(t)$. Due to inequality (6.14), $z(t) \neq 0$. Let l_t be the straight line passing through the point $\frac{1}{2}(h(F(t)) + \sqrt{1-\delta_n^2(\Gamma)}\,\|F\|)z(t)$ and orthogonal to the radius vector of this point. Utilizing inequality (6.14) again, we find that l_t has a distance to zero greater than $\sqrt{1-\delta_n^2(\Gamma)}\,\|F\|$ and separates the set $H(F(t))$ from zero. Since the matrix functions F and G are continuous, the straight line l_t separates from zero all sets $H(F(\tau))$, where τ is sufficiently close to t.
Summarizing, for the matrix function F, the assumptions of Theorem 6.11 are fulfilled (for $p = 2$). In view of this theorem, the partial indices of F are equal to each other. Applying Lemma 6.3 for $\chi = \alpha^{-1}$, we observe that the partial indices of G are also equal to each other, which proves Theorem 6.13. ≡

For the total index of a continuous matrix function, Muskhelishvili's formula is true. Therefore, it is not hard to evaluate the partial indices of a matrix function G which meets the assumptions of Theorem 6.13: each of them is equal to $\frac{1}{2\pi n}\{\arg\det G(t)\}_\Gamma$ (thus, the value $\frac{1}{2\pi}\{\arg\det G(t)\}_\Gamma$ automatically proves to be divisible by n). Notice, however, that in the suppositions of Theorem 6.13 the values of the partial indices of G are closely connected with the behaviour of its numerical domain. To illuminate this connection, we first remark that the index of a function ψ continuous on Γ is uniquely defined by the condition

$$\psi(t) \in H(G(t)) . \tag{6.15}$$

In fact, if ψ_0 is any other function continuous on Γ and satisfying this condition, then, for every $\lambda \in [0,1]$, the function $\psi_\lambda(t) = \lambda\psi(t) + (1-\lambda)\psi_0(t)$ is continuous and fulfils condition (6.15), therefore, it does not vanish on Γ . The mapping $\lambda \to \psi_\lambda$ maps the interval [0,1] continuously into the functional space $C(\Gamma)$. Consequently, the indices of all functions ψ_λ coincide. Especially, the indices of ψ_0 and $\psi_1 = \psi$ are the same.

It is quite natural to call the common value of the indices of all functions ψ continuous on Γ and satisfying condition (6.15) the <u>winding number of the numerical domain</u> around zero. The partial indices of G coincide with precisely this number as soon as Theorem 6.13 applies. Indeed, the function ψ may be chosen in such a way that the matrix function $\psi^{-1}(t)\|G(t)\|^{-1}$ satisfies the hypothesis of Lemma 6.4 (it suffices to set, e.g., $\psi(t) = z(t)$) . But then the partial indices of F are equal to zero, and, as follows from Lemma 6.3, the partial indices of G coincide with the index of the function ψ , i.e., with the winding number of rotations of the numerical domain around zero.

Clearly, the remark made at the end of the previous section can be applied to Theorem 6.13 in full. Thus, not only the left but also the right indices of an arbitrary matrix function G satisfying the conditions of this theorem coincide with the winding number of the numerical domain around zero.

6.6. ON THE STABILITY OF THE FACTORIZATION FACTORS $G_\pm$

Up to now, the question of stability of the factors $G_\pm$ from the Φ-factorization (2.1) of the matrix function G in L_p has not been studied in full detail. Hence, we restrict ourselves to the explanation of results associated with the factorization problem in a (C,R)-algebra $\mathfrak{A}$. Following the agreement at the beginning of the chapter, instead of "factorization in $\mathfrak{A}$" we shall simply write "factorization", under "norm" we shall understand the norm in $\mathfrak{A}$, etc.

Closeness of the factorization factors $G_\pm$ of the matrix function G and $F_\pm$ of the matrix function

$$F(t) = F_+(t)\Lambda_1(t)F_-(t) \tag{6.16}$$

is characterized by the quantities

$$\|G_+-F_+\| \; , \; \|G_--F_-\| \; , \; \|G_+^{-1}-F_+^{-1}\| \; , \; \|G_-^{-1}-F_-^{-1}\| \; . \tag{6.17}$$

THEOREM 6.14. Let G be a certain matrix function admitting a factorization (2.1). Then there exists a $\delta > 0$ such that, if $\|G-F\| < \delta$, (6.16) is a factorization of the matrix function F and at least one of the values (6.17) is smaller than δ, then $\Lambda_1 = \Lambda$.

Proof. For the sake of definiteness, let $\|G-F\| < \delta$ and $\|G_-^{-1}-F_-^{-1}\| < \delta$. Since $G_+\Lambda - F_+\Lambda_1 = GG_-^{-1} - FF_-^{-1} = G(G_-^{-1}-F_-^{-1}) + (G-F)F_-^{-1}$, then

$$\begin{aligned}\|G_+\Lambda - F_+\Lambda_1\| &\le \|G\|\,\|G_-^{-1}-F_-^{-1}\| + \|G-F\|\,\|F_-^{-1}\| \\ &\le \delta(\|G\| + \|F_-^{-1}\|) \\ &\le \delta(\|G\| + \|G_-^{-1}-F_-^{-1}\| + \|G_-^{-1}\|) \\ &\le \delta(\|G\| + \|G_-^{-1}\| + \delta)\end{aligned}$$

so that the difference of corresponding columns of the matrix functions $G_+\Lambda$ and $F_+\Lambda_1$ can be made arbitrarily small, if δ is chosen in a suitable manner.

Consequently, if we choose δ small enough, then the norm of the vector function $g_j - f_j t^{\varkappa_j^{(1)}-\varkappa_j}$ can also be made arbitrarily small, where g_j and f_j, $j=1,\dots,n$ are the j th columns of the matrix functions G_+ and F_+, respectively. Assume that, for some j, $\varkappa_j^{(1)} > \varkappa_j$. Then $g_j(0) = \frac{1}{2\pi i}\int_\Gamma (g_j(\tau)-f_j(\tau)\tau^{\varkappa_j^{(1)}-\varkappa_j})\,\tau^{-1}d\tau$. Letting $\delta \to 0$, the right-hand side becomes arbitrarily small and the left one does not depend on δ. Consequently, $g_j(0) = 0$, which is impossible, as the matrix function G_+ is non-degenerate in $\mathfrak{D}^+$. The contradiction obtained shows that, for sufficiently small δ, $\varkappa_j^{(1)} \le \varkappa_j$, $j=1,\dots,n$. Since, for δ small enough, the total indices of G and F agree, we conclude that $\varkappa_j^{(1)} = \varkappa_j$, $j=1,\dots,n$, which proves Theorem 6.14. ≡

In this way, for close matrix functions with different tuples of partial indices, the factorization factors $G_\pm$ cannot be too close. It turns out that the converse statement, the formulation of which must somewhat be strengthened because of the lack of uniqueness of the factors $G_\pm$, is also true.

THEOREM 6.15. Let G be a factorable matrix function. For any $\varepsilon > 0$, we can specify a $\delta > 0$ such that, if $\|F-G\| < \delta$ and the partial indices of the matrix function F coincide with the partial indices of the matrix function G, then, for every factorization of G, there can be found a factorization of F for which any of the values in (6.17) are smaller than ε.

In order to prove this theorem, first of all remark that $\|F_+-G_+\| = \|F(F_-^{-1}-G_-^{-1})\Lambda^{-1} + (F-G)G_-^{-1}\Lambda^{-1}\| \leq (\|F\|\|F_-^{-1}-G_-^{-1}\|+\|G_-^{-1}\|\|F-G\|)\|\Lambda^{-1}\|$. Thus, it can be guaranteed that the quantity $\|F_+-G_+\|$ is small as soon as $\|F-G\|$ and $\|F_-^{-1}-G_-^{-1}\|$ are small. Since in every Banach algebra the operation of taking the inverse element is continuous for fixed G, $\|F_+^{-1}-G_+^{-1}\|$ will be small, if $\|F_+-G_+\|$ is small. Analogously, the smallness of $\|F_--G_-\|$ results form the smallness of $\|F_-^{-1}-G_-^{-1}\|$.

In summary, it suffices to prove a simplified form of Theorem 6.15 in which it is only stated that $\|G_-^{-1}-F_-^{-1}\| < \varepsilon$. We shall show that the columns of the matrix function G_-^{-1} may be interpreted as vectors from the kernel of an operator of the form $P+Qt^{-k}G$. The Fredholmness of such operators results from the following proposition.

LEMMA 6.6. The kernels of the operators $P+GQ$ and $T_Q(G)$ have equal dimensions and the images are closed only simultaneously and have equal codimensions.

Lemma 6.6 is close to assertion 1) of Lemma 3.5 with respect to its formulation and way of proving. Thus, its proof is omitted here. We only wish to explain that $\ker(P+GQ) = 0 \dotplus \ker T_Q(G)$, i.e., $x \in \ker(P+GQ)$ if and only if $x^+ (= Px) = 0$ and $x^- \in \ker T_Q(G)$.

LEMMA 6.7. Let G be a factorable matrix function the partial indices of which are $\varkappa_1 \geq \dots \geq \varkappa_n$. Then

1) the j th column of the matrix function G_-^{-1} is a solution of the equation

$$(Pt^{-1} + Qt^{-\varkappa_j}G)x = 0 \; ; \qquad (6.18)$$

2) if X is a matrix function which is non-degenerate at least at one point of Γ and the j th column of which solves the equation (6.18), then X^{-1} may play the role of a right factor in some factorization of G.

Proof. Relation (6.18) is equivalent to both equations $Pt^{-1}x = 0$ and $Qt^{-\varkappa_j}Gx = 0$. The first of the latter two equations means that $t^{-1}x \in (\mathring{A}^-)^n$ or, what is the same, $x \in (A^-)^n$; the second amounts to saying that $t^{-\varkappa_j}Gx \in (A^+)^n$. If we take $x = x_j$, where x_j is the j th column of the matrix function G_-^{-1}, then obviously $x_j \in (A^-)^n$. Moreover, $t^{-\varkappa_j}Gx_j = t^{-\varkappa_j}G_+\Lambda G_-x_j$ is the j th column of the matrix function G_+. Thus, assertion 1) has been proved. Now we proceed to the proof of 2). If the vector function x_j is the solution of equation (6.18), then, as mentioned above, $x_j \in (A^-)^n$ and $t^{-\varkappa_j}G_+\Lambda G_-x_j \in (A^+)^n$, therefore, we deduce that $G_-x_j \in (A^-)^n$ and $t^{-\varkappa_j}\Lambda G_-x_j \in (A^+)^n$. Hence, for the regarded matrix function X, we have $G_-X \in A^-$, $H = \Lambda G_-X\Lambda^{-1} \in A^+$. From this we obtain (compare with the considerations in Section 2.2) that the matrix function H is block-lower triangular, its (i,j) th element is equal to zero for $\varkappa_i < \varkappa_j$, constant for $\varkappa_i = \varkappa_j$, and a polynomial of degree not higher than $\varkappa_i - \varkappa_j$, if $\varkappa_i > \varkappa_j$. In particular, $\det H$ is constant. Since $\det H = \det G_- \cdot \det X$, then $\det G_-$ is different from zero everywhere on Γ, and because $\det X$ is non-zero at least at one point of Γ, then H is non-degenerate. Thus, H satisfies the condition 2) of Theorem 2.2. According to this theorem, setting $\hat{G}_+ = G_+H$, $\hat{G}_- = X^{-1} = \Lambda^{-1}H^{-1}\Lambda G_-$, we get the factorization $\hat{G}_+\Lambda\hat{G}_-$ of the matrix function G, which proves Lemma 6.7. ≡

A similar result holds for the left factorization factor, too. Now we are able to prove Theorem 6.15.

Proof of Theorem 6.15. Due to Lemma 6.7, the j th column x_j of the matrix function G_-^{-1} is a solution of equation (6.18), i.e., it belongs to the kernel of the operator $A = Pt^{-1} + Qt^{-\varkappa_j}G = (P+Qt^{1-\varkappa_j}G)t^{-1}$. Besides of A we consider the operator $B = (P+Qt^{1-\varkappa_j}F)t^{-1}$, where F is a matrix function with the same tuple of partial indices as G. Since the operator of multiplication by t^{-1} is invertible, the operator A is Fredholm simultaneously with $P+Qt^{1-\varkappa_j}G$ and has the same defect numbers. In virtue of Lemma 6.6 the dimension of the kernel of the operator $P-Qt^{1-\varkappa_j}G$ $(P+Qt^{1-\varkappa_j}F)$ is equal to the sum of positive partial indices of $t^{1-\varkappa_j}G$ ($t^{1-\varkappa_j}F$) . Consequently, the dimensions of the kernels of the operators $P+Qt^{1-\varkappa_j}G$ and $P+Qt^{1-\varkappa_j}F$ (and, together with them, also of the operators A and B) are equal. Due to a theorem from GOHBERG, KRUPNIK [4], p. 170, for any vector x from $\ker A$ there exists a vector $y \in \ker B$ such that $\|x-y\| < c\,\|A-B\|$, where the constant c is determined only by the operator A . Since $\|A-B\| \leq \|Q\| \cdot \|t^{1-\varkappa_j}\| \cdot \|G-F\|$, for any $\varepsilon_j > 0$, one can find a $\delta_j > 0$ such that, for $\|G-F\| < \delta_j$, there exists a vector $y_j \in \ker B$ with $\|y_j - x_j\| < \varepsilon_j$. Now we form the matrix function $Y = (y_1 \dots y_n)$. Since the matrix function $(x_1 \dots x_n)$ $(=G_-^{-1})$ is non-singular, for sufficiently small ε_j , the matrix Y will also be non-singular at least at one point of Γ . Owing to assertion 2) of Lemma 6.7, there exists a factorization of the matrix function F such that $F_- = Y^{-1}$. Clearly, for given $\varepsilon > 0$, one can choose ε_j , $j=1,\dots,n$ such that the inequalities $\|y_j - x_j\| < \varepsilon_j$ $(j=1,\dots,n)$ imply $\|X-Y\| < \varepsilon$, i.e., $\|F_-^{-1} - G_-^{-1}\| < \varepsilon$.

In order to finish the proof of the simplified form of Theorem 6.15, it remains only to set $\delta = \min\{\delta_j : j=1,\dots,n\}$. At the same time. we have proved Theorem 6.15 in full. ≡

<u>COROLLARY 6.5.</u> Assume the matrix function G to be factorable and the tuple of its partial indices to be stable. Then, for every $\varepsilon > 0$, there exists a $\delta > 0$ such that, for $\|F-G\| < \delta$, the matrix function F admits a factorization in which each of the quantities (6.17) is less then ε .

In fact, if the tuple of partial indices of the matrix function G is stable, then, for $\delta > 0$ small enough, the condition $\|F-G\| < \delta$ implies the coincidence of the tuples of partial indices of F and G. Thus, Theorem 6.15 can be applied. ≡

In brief, the latter result can be formulated in the following way: If the middle factorization factor Λ is stable, then the factors $G_\pm$ are also stable.

Now we intend to formulate Theorem 6.15 in terms of sequences.

THEOREM 6.15'. Let $\{G^{(k)}\}_{k=1}^{\infty}$ be a sequence of matrix functions admitting a factorization with one and the same tuple of partial indices and converging to the matrix function G which is also assumed to be factorable and to have the same tuple of partial indices. Then one may choose factorizations $G^{(k)} = G_+^{(k)}\Lambda G_-^{(k)}$ in such a manner that $G_\pm^{(k)} \to G_\pm$, $(G_\pm^{(k)})^{-1} \to G_\pm^{-1}$ (in the norm (5.8), where $G_\pm$ are factorization factors of G.

COROLLARY 6.5'. Let $\{G^{(k)}\}_{k=1}^{\infty}$ be a sequence of matrix functions which converge to the factorable matrix function G with a stable tuple of partial indices. Then, beginning with some element of the sequence, the conclusion of Corollary 6.5 is valid.

For a factorable matrix function all partial indices of which are equal to each other, Corollary 6.5 can be made more precise. This refinement amounts for stating that the closeness of the factorization factors of close matrix functions F and G may be ensured by the requirement $G_-(\infty) = F_-(\infty)$.

Up to now, in the present chapter we have been concerned with the following problem: knowing the properties of a given matrix function, determine the properties of matrix functions sufficiently close to it. Now we turn to the "inverse" problem consisting in the determination of the factorization properties for the limit of a sequence of matrix functions, if we know the factorization properties of the terms of this sequence.

Theorem 6.5 shows that any factorable matrix function G with the tuple of partial indices $(\varkappa_j)_{j=1}^{n}$ may be represented as the limit of a sequence $\{G^{(k)}\}_{k=1}^{\infty}$ of factorable matrix functions the tuples of partial indices $(\varkappa_j^{(k)})_{j=1}^{n}$ of which obey the condition $(\varkappa_j) \succ (\varkappa_j^{(k)})$, but, in other respect, they are not interconnected or joined with the tuple $(\varkappa_j)$. Vice versa, if the sequence of factorable matrix functions $\{G^{(k)}\}_{k=1}^{\infty}$ converges to the factorable matrix function G, then, according to Theorem 6.4, for sufficiently large k, the condition $(\varkappa_j)_{j=1}^{n} \succ (\varkappa_j^{(k)})_{j=1}^{n}$ is fulfilled. Finally, the limit of a sequence of factorable matrix functions may not permit any factorization at all. However, we are able to indicate conditions under which the passage to the limit does not go beyond the class of factorable matrix functions and does not influence on the values of partial indices.

<u>THEOREM 6.16.</u> Let $\{G^{(k)}\}_{k=1}^{\infty}$ be a sequence of factorable matrix functions converging to the matrix function G. Then the following assertions are equivalent:

1) the factorization factors of the matrix functions $G^{(k)}$ may be chosen in such a way that $\|(G_{\pm}^{(k)})^{-1}\| \leq M$, where the constant $M < \infty$ does not depend on k;

2) the matrix function G can be factored and all matrix functions $G^{(k)}$, beginning with some number k, have one and the same tuple of partial indices which agrees with that of the matrix function G.

<u>Proof.</u> Suppose that 2) is fulfilled. Due to Theorem 6.15', the factorizations of $G^{(k)}$ can be chosen in such a fashion that the sequences $\{(G_+^{(k)})^{-1}\}$ and $\{(G_-^{(k)})^{-1}\}$ converge (to G_+^{-1} and G_-^{-1}, respectively). In this case these sequences will be bounded either. Thus, 1) follows from 2).

Conversely, let 1) be valid. As it was mentioned in the proof of Theorem 5.6, the operator $(G_-^{(k)})^{-1}\Lambda_k^{-1}Q(G_+^{(k)})^{-1}$ considered in $(\mathring{R}^-)^n$ is a regularizer for $T_Q(G^{(k)})$. Hence, there exists an equibounded sequence of regularizers of the operators $T_Q(G^{(k)})$. In accordance with Theorem 1.11 this means that the limit of the sequence $T_Q(G^{(k)})$, i.e., the operator $T_Q(G)$, is also Fredholm. The Fredholmness of the operator $T_P(G')$ can be shown in a similar way. With regard to Theorem 5.6, the matrix function G can be factored. Its tuple of partial indices will be denoted by $(\varkappa_j)_{j=1}^n$. In virtue of Theorem 6.4, in a certain neighbourhood of G we encounter only such matrix functions whose tuples of partial indices $(\varkappa'_j)$ fulfil the condition $(\varkappa_j) \succ (\varkappa'_j)$. Since there exists only a finite number of such tuples, one of the following two situations always applies: a) there is a neighbourhood of G such that all matrix functions $G^{(k)}$ contained in it have one and the same tuple of partial indices $(\varkappa_j)$; b) it is possible to choose a sequence $\{G^{(k_s)}\}_{s=1}^{\infty}$ all terms of which have one and the same tuple of partial indices $(\hat{\varkappa}_j)$, where $(\hat{\varkappa}_j) \neq (\varkappa_j)$.

Assume that situation b) applies. As the sequence $G^{(k_s)} = G_+^{(k_s)} \hat{\Lambda} G_-^{(k_s)}$ converges to G and $\| \hat{\Lambda} - (G_+^{(k_s)})^{-1} \Lambda (G_-^{(k_s)})^{-1} \| =$

$= \| (G_+^{(k_s)})^{-1} (G^{(k_s)} - G)(G_-^{(k_s)})^{-1} \| \leq M^2 \| G^{(k_s)} - G \|$, then the sequence $F_s = (G_+^{(k_s)})^{-1} \Lambda (G_-^{(k_s)})^{-1}$ converges to $\hat{\Lambda}$. Consequently, in any neighbourhood of the matrix function $\hat{\Lambda}$ there exists a matrix function the tuple of partial indices of which is just the same as that of the matrix function G . According to Theorem 6.4, $(\hat{\varkappa}_j) \succ (\varkappa_j)$, and this is a contradiction to the conditions $(\varkappa_j) \succ (\hat{\varkappa}_j)$ and $(\varkappa_j) \neq (\hat{\varkappa}_j)$.

Summarizing, situation a) holds by all means, i.e., assertion 2) is true. The proof of Theorem 6.16 is complete. ≡

6.7. COMMENTS

The question of stability of partial indices was first considered in MANDZHAVIDZE [2] in connection with the problem of the approximate solution of the Riemann boundary value problem. Theorems 6.1 and 6.2 as well as the criterion of stability of partial indices (Theorem 6.3) were obtained by GOHBERG and KREĬN [2,3] and BOJARSKI [1]. The proof of Theorems 6.1 and 6.2 conducted here is close to the one suggested in MANDZHAVIDZE [5]. The same applies to the proof of the sufficiency in Theorem 6.3. The proof of the necessity in Theorem 6.3 given above generalizes correspondent arguments from BOJARSKI [1], GOHBERG, KREĬN [3] to the case of matrix functions with unbounded factorization factors. In Section 10.2 of the book CLANCEY, GOHBERG [3] another way of proving the necessity is explained, which is based on the following general result of WIDOM [3] related to perturbations of Fredholm operators: Let T be a Fredholm operator in the Banach space X , and let S be a subspace of $[X]$ such that, for every non-zero $x \in X$, $f \in X^*$, there exists an $A \in S$ such that $f(Ax) \neq 0$. Then there exist $B \in S$ and a number $r > 0$ such that, for any $\varepsilon \in (0,r)$, at least one of the defect numbers of the operator $T + \varepsilon B$ is equal to zero. The application of this result to the case $T = P+t^{-s}GQ$, $S = \{AP+BQ: A,B \in \in L_\infty\}$, where s is an integer strictly included between the smallest and largest partial index of the matrix function G, proves the necessity of condition (6.2) for the stability of partial indices.

Theorems 6.4 and 6.5 and Lemma 6.1 are due to GOHBERG and KREĬN [3]. The topological interpretation of Theorems 6.3 - 6.5 is given by the authors under the influence of BOJARSKI's paper [3].
Remark, incidentally, that in this paper there was proved the connectedness of the intersection of the sets $\Phi_\varkappa$ with the class of matrix functions continuous on Γ and of the sets $\Phi_{(\varkappa_j)}$ with the class of matrix functions continuous on Γ and factorable in $C(\Gamma)$. The connectedness of the intersection of $\Phi_\varkappa$ and the class of piecewise continuous matrix function was studied in GLEBOV, DEUNDYAK [1]. Furthermore, the question of connectedness of the sets $\Phi_\varkappa$ themselves as also of the density in L_∞ of the set $\bar{\Phi}$ of all Φ-factorable matrix functions remains to be solved.

Theorems 6.6 - 6.13 are obtained by one of the authors (SPITKOVSKIĬ [2,3,5,10]). The idea of transition from the matrix function G_1 to its Cayley transform used in the proof of Lemma 6.2 is borrowed from KRUPNIK, NYAGA [1], where this device was applied to the study of unitary matrix functions.

In connection with Lemma 6.3, note that in the article of SPITKOVSKIĬ [9] one can find the description of the class $\mathfrak{M}$ consisting of all those functions the multiplication by which does not violate the Φ-factorability in L_p. It appeared that this class does not depend on p and, being multiplied by a function $\chi \in \mathfrak{M}$, the partial indices of an arbitrary Φ-factorable matrix function G behave precisely as in case of a continuous function χ, i.e., they are shifted by its index.

Lemma 6.4 is due to DANILYUK [4,5], its precising for the case of a Lyapunov contour (Lemma 6.4') was obtained by one of the authors (SPITKOVSKIĬ [2]). Lemma 6.5, with the help of which the latter lemma was derived, is well-known for a long time and was already used in SIMONENKO [4]. The proof of Lemma 6.5 is borrowed from the book GOHBERG, KRUPNIK [4].

In SHELEPOV [1] it was mentioned that, with regard to RADON's [1] results, Lemma 6.5 may be transferred to the case when Γ is a Radon curve without cusps. Therefore, Theorem 6.11' can also be transferred to this case. Concerning the considerations after Theorem 6.13 related to the values of the partial indices, let us still mention the following.

Generally speaking, starting from a matrix function G continuous on Γ, it is not always possible to construct a tuple of n functions $\lambda_j(t)$, $j=1,\ldots,n$ continuous on Γ and coinciding, for all $t \in \Gamma$, with the tuple of its eigenvalues counted with regard to their algebraic multiplicity (as a matter of fact, in moving along the components of the contour Γ, the continuous branches of the solutions of the equation $\det(G(t)-\lambda I) = 0$ move into another, not necessarily coinciding with the original ones). But if we do succeed in doing so, then it makes sense to speak about indices of eigenvalues of the matrix funktion G.

In GAKHOV's review article [3] among others there was posed the problem of describing all those matrix functions the partial indices of which coincide with the indices of eigenvalues. The complete solution of this problem has not been found yet [1]. The relations $\lambda_j(t) \in H(G(t))$, however, imply that, under the conditions of Theorem 6.13, the indices of eigenvalues coincide with the winding number of the numerical domain of the matrix function G around zero and, thus, with the partial indices of G. In this way, matrix functions satisfying the conditions of Theorem 6.13 form a certain subclass of the desired set.

The application of Theorems 6.6 - 6.13 is impossible without the knowledge of lower estimates for the values $\delta_{p,n}(\Gamma)$. Assertion 1) of Lemma 6.2 allows for replacing the problem of finding such estimates by the problem of evaluating (or obtaining upper estimates for) the norms

---

1) The problem of the description of all triangular matrix functions whose partial indices coincide with the indices of the diagonal elements solved in Chapter 4 is a special case of this problem.

of operators Q and S. In case when the contour Γ is a circle, the norm of the operator S has been calculated in all spaces L_p^n, $1 < p < \infty$ (VERBITSKIĬ, KRUPNIK [1,2], KRUPNIK [4]). This fact will be used in the next chapter. With our definition of the norm in L_p^n (see Section 1.2) the norms of the operators S and Q, generally speaking, depend on n. However, for $p = 2$, such a dependence fails to exist. In SPITKOVSKIĬ [3] it was shown that the norms of the operators S and Q in L_2 are connected via the relation $\|Q\| = \frac{1}{2}(\|S\|+\|S\|^{-1})$. Therefore, the estimates for $\delta_{2,n}(\Gamma)$ mentioned in assertion 1) of Lemma 6.2 coincide. Unfortunately, we do not know whether these estimates are exact or not. It is also unknown, which of these estimates is the most exact one for $p \neq 2$ and a contour Γ different from the circle.

Theorem 6.15 (on the continuity of the factorization factors for fixed partial indices) is due to SHUBIN [1]. He has also proved there that, if the matrix function G is a k times continuously differentiable (analytic) function of some parameter ω and the partial indices do not depend on ω, then the factorization factors $G_{\pm}$ are also k times continuously differentiable (analytic) functions of ω. The factorization of matrix functions depending on a parameter was also examined by NIKOLENKO [1,2]. In NIKOLENKO [1] he derived the stability criterion for partial indices in case when the parameter ranges over even-dimensional spheres or triangulations. In this paper it was further shown that the triple $(T_+ \times T_-, p_{\overline{\varkappa}}, \mathcal{L}_{\overline{\varkappa}})$ forms a locally trivial fibre bundle, where $T_{\pm}$ is the class of $n \times n$ matrix functions non-singular and satisfying a Hölder condition in $\mathcal{D}^{\pm} \cup \Gamma$ and analytic in $\mathcal{D}^{\pm}$, $\mathcal{L}_{\overline{\varkappa}}$ is the class of Hölder continuous matrix functions with the tuple of partial indices $\overline{\varkappa}$ $(=(\varkappa_1,\dots,\varkappa_n))$, and $p_{\overline{\varkappa}}$ is the mapping from $T_+ \times T_-$ into $\mathcal{L}_{\overline{\varkappa}}$ acting according to formula $p_{\overline{\varkappa}}(F_+,F_-) = F_+\Lambda F_-$. In NIKOLENKO [2] it is shown that this fibre bundle is trivial if and only if $\varkappa_1 = \dots = \varkappa_n$.

Theorem 6.16 was first formulated in LITVINCHUK, NIKOLAĬCHUK, SPITKOVSKIĬ [1].

A series of papers is devoted to the approximate solution of singular integral equations (and, especially, to the vector-valued Riemann boundary value problem). A systematic explanation of the state of the arts (up to 1978) related to methods of their approximate solution the reader can find in the last chapter of the monograph PRÖSSDORF [1]. Concerning later results, see BABESHKO [3], DIDENKO [1-3], DIDENKO, TIKHONENKO [1,2], ZOLOTAREVSKIĬ [1], PRÖSSDORF, SCHMIDT [1], SILBERMANN [1].

In MEUNARGIYA [1,2] several cases are examined in which the solution of the vector-valued Riemann boundary value problem with matrix $G_0 + \lambda G_1$ may be found as a series $\sum_{k=0}^{\infty} \lambda^k F^{(k)}$, where $F_\lambda^{(k)}$ are successively determined as solutions of Riemann problems with matrix G_0 and a free term depending on $F^{(j)}$ $(j=0,\dots,k-1)$.

Now we focus our attention on another results associated with the stability problem and estimates of partial indices. KRAVCHENKO and NIKOLAĬCHUK [1] have obtained formulae for partial indices of second-order matrix functions such that at least one element of them does not degenerate on Γ. These formulae involve the dimension of the kernel of an integral operator which can be effectively constructed by a given matrix function.

CHORIEV [1-3] has studied the relationships between stability of partial indices of sufficiently smooth matrix functions G and the validity of Liouville's theorem for the equation

$$\bar{\partial} U = BU ,$$

where $B(z) = \mathcal{G}(z)^{-1}\bar{\partial}\mathcal{G}(z)$ for $z \in \mathcal{D}^+$, $B(z) = 0$ for $z \in \mathcal{D}^-$, and $\mathcal{G}$ is an extension of the matrix function GD^{-1} into the domain $\mathcal{D}^+$ differentiable in L_p $(p > 2)$ in the generalized sense.

Here $D(t) = \operatorname{diag}[t^{k+1},\dots,t^{k+1},t^k,\dots,t^k]$ is a matrix function annulling the total index of G.

In LAX [1] it was established that the property of the partial indices of a matrix function G given on the circle to be equal to zero is equivalent to the solvability of the Dirichlet problem for the differential equation $G_{z\bar{z}} = G_{z}G^{-1}G_{\bar{z}}$.

In the author's opinion, the problem of finding estimates for the partial indices and of obtaining effectively verifiable conditions of their stability or non-negativity is still far from being completely solved and any new result towards its solution is of considerable interest.

CHAPTER 7. FACTORIZATION ON THE CIRCLE

The results of this chapter may be divided into two groups, which are associated with each other in a natural way. The first of them is formed by problems of factorization and Φ-factorization on the circle $\mathbb{T}$ of Hermitian (in particular, of definite) and unitary matrix functions. The properties in which these matrix functions are peculiar enable for obtaining more complete and exact theorems on factorization. Some of these theorems are useful, especially, for the study of relationships between necessary and exact conditions of factorability, a problem, which was already stated in Chapters 2 and 3.

The second group represents essential strengthenings for the space $L_2(\mathbb{T})$ of a series of important results from the preceding chapter related to Φ-factorization with non-negative, non-positive, stable, in particular, with equal partial indices of arbitrary matrix functions from the classes L_∞ , $L_\infty^{\pm} + C$ and C . This group of results is based on the utilization of exact estimates for the norms of the operators S, P, Q, PGQ, QGP, which hold in $L_2(\mathbb{T})$, but also on the application of results of the first group concerning Φ-factorization of matrix functions with the algebraic properties mentioned above.

In Section 7.1 a definition of factorization specific for the circle $\mathbb{T}$ will be introduced, from which we immediately shall obtain the connection between factorizations of the matrix function G in $L_p(\mathbb{T})$ and the matrix function G^* in $L_q(\mathbb{T})$, $q = p/(p-1)$, which is of some significance to what follows.

In Section 7.2 we shall be concerned with factorization and Φ-factorization of Hermitian matrix functions, where the structure of the factorability domain will be revealed, relations for partial indices will be derived and a necessary condition for the domain of factorability to be non-empty will be established. A representation for a Hermitian matrix function factorable in $L_2(\mathbb{T})$ the total index of which is equal to zero will be described. This representation enables us to estimate the number of zero partial indices of a given matrix function G with the aid of the absolute value of the signature of G.

Section 7.3 is dedicated to the factorization problem in L_2 of those Hermitian matrix functions for which the corresponding quadratic form preserves the sign. The main result of this section is a theorem which asserts that, for a positively definite matrix function G, the necessary condition $G^{\pm} \in L_1$ of its factorability is also sufficient. The proof of this theorem rests upon the claim on the existence of a representation of a positively definite matrix function G as the product of a certain outer matrix function A and its adjoint A^*. This proposition is of independent interest and significance for applications.

The aim of Section 7.4 consists in making more precise various results of 6.3 and 6.4. The most general result of this section is the theorem on the existence of a Φ-factorization in L_p with coinciding partial indices for those matrix functions the numerical domain of which may be locally (i.e., for any sufficiently small arc of the contour $\mathbb{T}$) included into a sector whose apex angle is less than $2\pi/\max(p,q)$. A special case of this theorem is provided by the assertion that every uniformly positively definite matrix function admits a Φ-factorization with zero partial indices. It appears that the factorization factors of such matrix functions are necessarily bounded. This assertion, which is of independent interest (cf. Theorem 5.2 from Section 5.1), allows for reducing in Section 7.5 the problem of Φ-factorability of an arbitrary matrix function $G \in L_\infty$ to the same problem for a certain unitary matrix function. Moreover, this assertion permits us with the help of this reduction to obtain criteria of Φ-factorability in L_2 with non-negative, non-positive and stable (in particular, zero) partial indices. These criteria are the conversions (and that even in a stronger form) of corresponding sufficient conditions from 6.3. Thus, for the problem of Φ-factorization in $L_2(\mathbb{T})$ they are also necessary. Note that for unitary matrix functions the criterion of existence of a Φ-factorization in L_2 with definite partial indices will be derived from the sophisticated result on the calculation of the norms of the Hankel operators PGQ and QGP , which has important applications beyond the factorization theory for matrix functions either.

Furthermore, we intend to specify the results of 7.4 and 7.5 for the case of a matrix function G which satisfies certain analytic conditions, mainly for the case of a continuous matrix function G . This topic is pursued in Section 7.6. Here, in particular, it is established that the partial indices of a continuous strictly non-singular (i.e., satisfying the condition $0 \notin H(G(t))$) matrix function G are equal to each other. This result is a strengthening but also a simplification of Theorem 6.13 for the case when the contour Γ is a circle.

Section 7.7 contains information on the history of the subject and comments related to current literature.

7.1. DEFINITION OF FACTORIZATION ON THE CIRCLE

In the present chapter we deal with the factorization problem in case when the contour Γ is a circle. Without loss of generality, we may suppose that Γ is the unit circle $\mathbb{T} = \{\zeta : |\zeta| = 1\}$.
In this case complex conjugation realizes a one-to-one correspondence between the classes L_p^- and L_p^+ $(= H_p)$. Indeed, the function f defined in the unit disk belongs to the Hardy class H_p if and only if the function $\tilde{f}$ defined outside the unit disk by the equation $\tilde{f}(\zeta) = \overline{f(\bar{\zeta}^{-1})}$ lies in the class E_p^- , where the boundary values of the functions f and $\tilde{f}$ are connected via the relation $\tilde{f}(t) = \overline{f(t)}$.

Redenoting the matrix function G_+ from the representation (2.1) by A and the matrix function G_- by B^* , in view of the remark made above, we may claim that, in the case $\Gamma = \mathbb{T}$, Definition 2.1 is equivalent to the following one.

<u>DEFINITION 7.1.</u> A representation

$$G = A\Lambda B^* \qquad (7.1)$$

will be called a <u>factorization in</u> L_p of the matrix function G defined on the unit circle, if the matrix functions A and B^{-1}

(A^{-1} and B) belong to the class H_p (H_q, $q = p/(p-1)$) and Λ is a diagonal matrix function of the form (2.2).

From Theorems 1.17 and 1.18 it is clear that the requirements $A \in H_p$, $A^{-1} \in H_q$ are fulfilled if and only if A is an outer matrix function[1], $A \in L_p$ and $A^{-1} \in L_q$. A similar assertion is true for the matrix function B.

THEOREM 7.1. The matrix function G defined on $\mathbb{T}$ can be factored in L_p if and only if the matrix function G^* can be factored in L_q. In this case the tuple $(\varkappa_j^*)$ of partial q-indices of G^* and the tuple $(\varkappa_j)$ of partial p-indices of G are connected via the relations

$$\varkappa_j^* = -\varkappa_{n-j+1}, \quad j = 1,2,\dots,n. \tag{7.2}$$

In particular, the total q-index of G^* is opposite to the total p-index of G.

Proof. We introduce the constant matrix

$$T = \begin{pmatrix} 0 & \cdots & 1 \\ & \cdot^{\cdot^{\cdot}} & \\ 1 & \cdots & 0 \end{pmatrix}. \tag{7.3}$$

The matrix T is both Hermitian and unitary, i.e. $T = T^* = T^{-1}$. Taking these properties into consideration, the expression $G^* = B\Lambda^*A^*$, which is obtained from (7.1), may be rewritten in the equivalent form

$$G^* = BT(T\Lambda^*T)(AT)^*. \tag{7.4}$$

Since $T\Lambda^*T = T\Lambda^{-1}T = \mathrm{diag}[t^{-\varkappa_n},\dots,t^{-\varkappa_1}]$, the representation (7.1) is a factorization of the matrix function G in L_p if and only if equation (7.4) provides a factorization of G^* in L_q (therefore, the factorizations exist only simultaneously) and the corresponding partial indices of G and G^* are connected by the relations (7.2), which proves the theorem.

THEOREM 7.1'. A matrix function G of class L_∞ defined on $\mathbb{T}$ is Φ-factorable in L_p if and only if the matrix function G^* is Φ-fac-

---

[1] A n×n matrix function A is called outer, if its entries belong to the class D and det A is an outer function.

torable in L_q. Moreover, the assertions concerning the values of total and partial indices formulated in Theorem 7.1 continue to be valid.

<u>Proof.</u> If the matrix function G is not factorable in L_p, then, owing to the previous theorem, the matrix function G^* is not factorable in L_q, which is consistent with the assertion to be proved. It remains to consider the case of a matrix function G which can be factored in L_p. According to Corollary 3.3 and the definition of Φ-factorization, the representation (7.1) is a Φ-factorization of G in L_p if and only if the operator $K_G = (B^*)^{-1}\Lambda^{-1}QA^{-1}$ is bounded in L_p^n, and the representation (7.4) is a Φ-factorization of G^* in L_q iff the operator $K_{G^*} = [(AT)^*]^{-1}(T\Lambda^*T)^{-1}Q(BT)^{-1}$ is bounded in L_q^n.

But the operator $K_{G^*} = (A^*)^{-1}\Lambda TQTB^{-1} = (A^*)^{-1}\Lambda QB^{-1}$ may be viewed as adjoint (in the sense of Definition 1.6) to the operator $(B^*)^{-1}Q\Lambda^{-1}A^{-1}$, if we identify L_q^n with the dual of the space L_p^n with the aid of the bilinear form

$$(f,g) = \int_{\mathbb{T}} g(t)^*f(t)|dt| .^{1)}$$

In virtue of assertions 2) and 3) of Lemma 3.3 the operators K_G and $(B^*)^{-1}Q\Lambda^{-1}A^{-1}$ are densely defined and closed. Therefore, assertion 4) of the same lemma may be applied, owing to which these operators are bounded only simultaneously.

Thus, the operators K_G and K_{G^*} considered in L_p^n and L_q^n, respectively, are bounded only simultaneously. From here and Theorem 3.8 we deduce the assertion to be proved. ≡

<u>COROLLARY 7.1.</u> The matrix functions G and G^* are factorable (and, provided that $G \in L_\infty$, also Φ-factorable) in L_2 only simultaneously. In addition, their total 2-indices are opposite and the partial indices are connected by relation (7.2).

Corollary 3.6 taken together with Theorem 7.1' leads to the following conclusion.

---

1) This can be established just as the corresponding part from the proof of Lemma 3.4.

COROLLARY 7.2. The matrix function $G \in L_\infty$ is left Φ-factorable in L_p if and only if the matrix function $(G^*)^{-1}$ is right Φ-factorable in L_q, where the corresponding tuples of partial indices coincide.

Due to Corollary 7.2, a unitary matrix function G (i.e., a function for which $G(t)^* = G(t)^{-1}$ a.e. on $\mathbb{T}$) is left Φ-factorable in L_p if and only if it is right Φ-factorable in L_q.

7.2. FACTORIZATION OF HERMITIAN MATRIX FUNCTIONS

From Theorems 7.1 and 7.1' it is clear that Hermitian matrix functions (i.e. such that $G(t) = G(t)^*$ a.e. on $\mathbb{T}$) play a special role in the factorization problem with respect to $\mathbb{T}$.

THEOREM 7.2. Under the mapping $p \to p/(p-1)$, the domain of factorability (and, if $G \in L_\infty$, then also the domain of Φ-factorability) of a Hermitian matrix function G is transformed into itself. In addition, for its partial p-indices, the relations

$$\begin{aligned} \varkappa_j + \varkappa_{n-j+1} &\leq 0 \qquad \text{if } p > 2, \\ \varkappa_j + \varkappa_{n-j+1} &= 0 \qquad \text{if } p = 2, \\ \varkappa_j + \varkappa_{n-j+1} &\geq 0 \qquad \text{if } p < 2 \end{aligned} \tag{7.5}$$

are valid.

The assertion of Theorem 7.2 follows from Corollary 7.1 and Theorem 2.1 (on the monotony of partial p-indices).

COROLLARY 7.3. The components of the factorability domain of a Hermitian matrix function which correspond to positive (negative) values of the total index are situated strictly to the left (right) of the point $p = 2$. The component $\mathfrak{F}_0(G)$, provided that it is not empty, necessarily contains the point $p = 2$.

Consider the Hermitian matrix function G with non-empty factorability domain. In accordance with Theorem 7.2 G can be factored in L_p for some $p \geq 2$. If (7.1) provides a factorization of G in L_p, then, since G is Hermitian, equation (7.4) yields a factorization of this matrix function in L_q. According to formulae (2.5), which express the connection between factorization factors from different factorizations of one and the same matrix function, we have

$$A = BTH \,, \tag{7.6}$$

where H is a polynomial matrix function the structure of which is described in Chapter 2 immediately after formulae (2.5). In particular, there was also shown that $\det H$ is a polynomial whose degree is equal to the difference of the total indices of G in the considered factorizations. In the case under study, thanks to Theorem 7.1 and Corollary 7.3, we may assert that this degree is equal to $-2\varkappa$, where $\varkappa$ (≤ 0) is the total p-index of G.

Utilizing equations (7.1) and (7.6), we may represent the matrix function G in the form

$$G = BTH\Lambda B^* \,. \tag{7.7}$$

Consequently, for almost all $t \in \mathbb{T}$, the quadratic forms associated with the matrices $G(t)$ and $TH(t)\Lambda(t)$ are congruent. Hence, the signatures (i.e. the differences of the numbers of positive and negative squares of the corresponding quadratic forms) of these matrices also coincide a.e. on $\mathbb{T}$. Since the matrix function $TH\Lambda$ is continuous on $\mathbb{T}$, its signature is a piecewise constant function of t the discontinuities of which are contained in the set of zeros of the function $\det(TH\Lambda) = (-1)^n t^{\varkappa} \det H$, i.e. in the set of zeros of the polynomial $\det H$. However, the latter vanishes at no more than $2|\varkappa|$ points of the circle $\mathbb{T}$. In this way, the following statement has been proved.

<u>THEOREM 7.3.</u> Let G be a Hermitian matrix function. For the set $\mathfrak{F}'_{\varkappa}(G)$ to be non-empty, it is necessary that there exist points $t_j \in \mathbb{T}$, $j = 1,\dots,N$ which partition $\mathbb{T}$ into $N \leq 2|\varkappa|$ arcs such that the signature of the matrix $G(t)$ is constant on a each of them a.e..

<u>COROLLARY 7.4.</u> For $\mathfrak{F}_0(G)$ to be non-empty, it is necessary that the signature of the matrix $G(t)$ is constant a.e..

<u>COROLLARY 7.5.</u> For $\mathfrak{F}(G)$ to be non-empty, it is necessary that there exists a partition of $\mathbb{T}$ into a finite number of arcs such that on each of them the signature of the matrix $G(t)$ is constant a.e..

With the help of Corollary 7.5 one can easily construct an example of a scalar function taking the values 1 and -1 which does not admit a factorization in any of the spaces L_p. In particular, one may set $f(t) = 1$ if $[1/|t-1|]$ is even, and $f(t) = -1$ otherwise.

The conditions found in Theorem 7.3 are not sufficient for the domain of factorability of the matrix function G to be non-empty, even in case we require $G^{\pm 1} \in L_\infty$ and $N = 0$. As an example we consider the matrix function $G(t) = \mathrm{diag}[f(t),-f(t)]$, where the function f is real-valued, changes its sign infinitely many times and $f^{\pm 1} \in L_\infty$. For all $t \in \mathbb{T}$, the matrix $G(t)$ is Hermitian and its signature is equal to zero. At the same time, the matrix function in question differs from the Hermitian matrix function $\mathrm{diag}[f,f]$, the signature of which changes infinitely many times, by the constant multiplier $\begin{pmatrix} 1 & 0 \\ 0 & -1 \end{pmatrix}$. Consequently, the factorability domain of G is empty.

Now we intend to consider in more detail an Hermitian matrix function G for which the component $\mathfrak{F}_0(G)$ is not empty. According to Corollary 7.3, in this case the inclusion $2 \in \mathfrak{F}_0(G)$ necessarily holds and the partial indices of G are connected by relations (7.5). Thus, the tuple of partial indices of G is uniquely determined by the numbers $k_m > \dots > k_1 > 0$ $(m \geq 0)$ of different positive indices and their multiplicities $l_m, \dots, l_1$.

THEOREM 7.4. Assume that for the Hermitian matrix function G the set $\mathfrak{F}_0(G)$ is non-empty. Then G admits a representation in the form

$$G = A_0 \Lambda_0 A_0^* , \tag{7.8}$$

where $A_0^{\pm 1} \in H_2$,

$$\Lambda_0(t) = \begin{pmatrix} & & & & & & t^{k_m} I_{l_m} \\ & & & & & \cdot^{\cdot^{\cdot}} & \\ & & & & t^{k_1} I_{l_1} & & \\ & & & J & & & \\ & & t^{-k_1} I_{l_1} & & & & \\ & \cdot^{\cdot^{\cdot}} & & & & & \\ t^{-k_m} I_{l_m} & & & & & & \end{pmatrix}, \tag{7.9}$$

$J = \begin{pmatrix} I_{l_+} & \\ & -I_{l_-} \end{pmatrix}$, and l_+ and l_- are integers such that their sum is equal to $l_0 = n-2\sum_{j=1}^{m} l_j$ and their difference equals the signature σ of G.

Before proving the theorem, let us remark that by σ one has to understand that value of the signature which is taken by the matrix function G a.e. on $\mathbb{T}$. Because of Corollary 7.4, such a value exists.

Proof. Starting with an arbitrary factorization (7.1) of the matrix function G in L_2 and arguing as in the proof of Theorem 7.3, we obtain equation (7.7) in which $B^{\pm 1} \in H_2$ and H is a transition matrix from one factorization of G in L_2 to another. Consequently, the structure of H is described by assertion 2) of Theorem 2.2. With regard to relations (7.5), we deduce the following description of the structure of the matrix function $W = TH$.

Let the matrix function W be partitioned into blocks in the same way as the matrix function Λ_0 defined by equation (7.9), and let these blocks be numbered in lines and columns by the indices from $-m$ to m from bottom to top and from the left to the right, respectively. Then:

1) if $r < s$, then $W_{r,s} = 0$,

2) $W_{r,r}$ is a constant non-singular matrix,

3) if $r > s$, then $W_{r,s}$ is a polynomial matrix function such that the degrees of its elements do not exceed $k_r - k_s$ (for the sake of uniformity, set $k_s = -k_{-s}$ for $s < 0$, and $k_0 = 0$).

Now we want to show that there exists a polynomial matrix function Ψ with a constant non-zero determinant such that the equation

$$W(t)\Lambda(t) = \Psi(t)\Lambda_0(t)\Psi(t)^* \tag{7.10}$$

is valid.

Partitioning the matrix function Ψ into blocks in the same manner as W and imposing on it the additional condition

$$\Psi_{r,s} = 0 \quad \text{for} \quad r+s < 0, \tag{7.11}$$

we get a system which is equivalent to (7.10):

$$t^{-k_s} W_{r,-s}(t) = \sum_{\nu} \Psi_{r,-\nu}(t)\, J_\nu \Psi_{s,\nu}(t)^* t^{k_\nu}, \tag{7.12}$$

where the summation on the right-hand side is carried out for ν lying between $-s$ and r, $J_\nu = I$ for $\nu \neq 0$ and $J_o = J$.

Since the matrix function $W\Lambda$ is Hermitian, the system (7.12) is transformed into itself after conjugation and changing r and s with each other. Consequently, those equations with $r < s$ may be omitted. The remaining equations of the system (7.12) are partitioned into $2m+1$ groups, assigning to the j th group those of the equations for which $r+s = j$ $(j=0,\dots,2m)$. The zero group consists of the equations

$$\begin{aligned} W_{r,r}(t) &= \Psi_{r,-r}(t)\Psi_{-r,r}(t)^*, \quad r=1,\dots,m, \\ W_{o,o} &= \Psi_{o,o}\, J\Psi^*_{o,o}. \end{aligned} \tag{7.13}$$

For the solvability of the last equation, it is necessary and sufficient to equate the integers l_+ and l_- to the number of positive and negative squares, respectively, of the matrix $W_{o,o}$. The remaining equations are always solvable. The solutions of all the equations (7.13) may be chosen in the class of constant matrices.

Now we take advantage of the fact that the equations of the j th group involve only those unknown matrices $\Psi_{r,s}$ for which $0 \le r+s \le j$. Suppose that, for $j \le j_o$ $(< 2m+1)$, the equations of the j th group are solved, where the matrix functions $\Psi_{r,s}$ already obtained are polynomial with degree of its elements not exceeding k_r+k_s. Each equation of the (j_o+1) th group may be rewritten in the form

$$\begin{aligned} &\Psi_{r,s}(t) J_s \Psi^*_{s,-s} + \Psi_{r,-r} J_r \Psi_{s,r}(t)^* t^{k_r+k_s} = \\ &W_{r,-s}(t) - \sum_{\nu} \Psi_{r,-\nu}(t) J_\nu \Psi_{s,\nu}(t)^* t^{k_\nu+k_s}. \end{aligned} \tag{7.14}$$

The summation on the right-hand side is carried out for ν lying between $-s+1$ and $r-1$.

In view of the assumption stipulated the right-hand side of (7.14) is a polynomial matrix function the degrees of the entries of which are not greater than k_r+k_s. Setting $\Psi_{s,r}=0$ for $s \neq r$, from (7.14) we get $\Psi_{r,s}$ in the form of a polynomial matrix function the degrees of the entries of which are also less than or equal to k_r+k_s.

For $s=r$, equation (7.14) reads as follows:

$$2\,\mathrm{Re}(t^{-k_r}\,\Psi_{r,r}(t)\Psi^*_{r,-r}) =$$

$$t^{-k_r}W_{r,-r}(t) - \sum_{|\nu|\leq r-1} \Psi_{r,-\nu}(t)J_\nu\Psi_{r,\nu}(t)^*t^{k_\nu}\,.$$

Starting from the fact that $W\Lambda$ is a Hermitian matrix function and changing the summation index ν by $-\nu$, it is not hard to verify that the right-hand side of the latter relation is a Hermitian matrix function and its elements are of the form $\sum_{|j|\leq k_r} c_j t^j$. Therefore, it can be represented in the form $2\,\mathrm{Re}\,Z(t)$, where Z is a polynomial matrix function such that the degrees of its entries are not greater than k_r. It remains to set $\Psi_{r,r}(t) = t^{k_r}Z(t)(\Psi^*_{r,-r})^{-1}$.

Concequently, under the assumption made above, all the equations of the (j_0+1) th group can be solved and the obtained solutions $\Psi_{r,s}$ $(r+s = j_0+1)$ are polynomial matrix functions of degree less than or equal to k_r+k_s.

We obtained that there exists a polynomial matrix function Ψ satisfying equation (7.10). Its non-degeneracy results from condition (7.11) and the non-degeneracy of the diagonal blocks $\Psi_{r,-r}$.

The constructed matrix function Ψ satisfies the requirements $\Psi^{\pm 1} \in H_\infty$. Hence, the matrix function $A_0 = B\Psi$ belongs, together with its inverse, to the class H_2. At the same time, successively applying equations (7.10) and (7.7), we get $A_0\Lambda_0A_0^* = B(\Psi\Lambda_0\Psi^*)B = BW\Lambda B^* = BTH\Lambda B^* = G$. Thus, the existence of the desired representation (7.8) has been proved. Equation (7.8) implies that the signature of the matrices $G(t)$ and $\Lambda_0(t)$ coincide a.e. on $\mathbb{T}$. Since the signature of the latter is equal to $l_+ - l_-$, we obtain the equation $l_+ - l_- = \sigma$, whereas the equation $l_+ + l_- = l_0$ is evident. The proof of Theorem 7.4 is complete. ≡

As a consequence from the relations for the numbers $l_{\pm}$ established in Theorem 7.4 and the obvious inequality $l_{+} + l_{-} \geq |l_{+} - l_{-}|$ we derive the following statement.

COROLLARY 7.6. The number of zero 2-indices of a Hermitian matrix function is not less than the absolute value of its signature (to be more precise, of that value of the signature which is attained on a set of complete measure).

COROLLARY 7.7. Let G be a Hermitian matrix function defined on $\mathbb{T}$ which admits a factorization in a certain (L_∞, R)-algebra $\mathfrak{A}$. Then this matrix function admits also a representation (7.8) with $A_0^{\pm 1} \in \mathfrak{A}^+$.

Proof. A factorization of the matrix function G in $\mathfrak{A}$ may be chosen as the factorization (7.1) being the initial one in the argument of Theorem 7.4. Then we have $B^{\pm 1} \in \mathfrak{A}$. Taking into account that the multiplication by polynomial matrix functions does not go beyond the class $\mathfrak{A}$, we see that along with B the matrix function $A_0 = B\Psi$ is an invertible element of $\mathfrak{A}^+$. $\equiv$

COROLLARY 7.8. A second-order Hermitian matrix function G which can be factored in L_2 and has a negative determinant admits a representation

$$G(t) = A(t) \begin{pmatrix} 0 & t^{\varkappa_1} \\ t^{-\varkappa_1} & 0 \end{pmatrix} A(t)^* ,$$

where $A^{\pm 1} \in H_2$, $\varkappa_1$ (≥ 0) and $-\varkappa_1$ $(= \varkappa_2)$ are its partial 2-indices.

In fact, if $\varkappa_1 > 0$, then the mentioned representation exists by Theorem 7.4. If $\varkappa_1 = 0$, then Theorem 7.4 ensures the existence of the representation $G = A_0 \begin{pmatrix} 1 & 0 \\ 0 & -1 \end{pmatrix} A_0^*$. It remains only to remark that

$$\begin{pmatrix} 1 & 0 \\ 0 & -1 \end{pmatrix} = \frac{1}{2} \begin{pmatrix} 1 & 1 \\ -1 & 1 \end{pmatrix} \begin{pmatrix} 0 & 1 \\ 1 & 0 \end{pmatrix} \begin{pmatrix} 1 & -1 \\ 1 & 1 \end{pmatrix}$$

and to set

$$A = \frac{1}{\sqrt{2}} A_0 \begin{pmatrix} 1 & 1 \\ -1 & 1 \end{pmatrix} .$$

7.3. FACTORIZATION OF DEFINITE MATRIX FUNCTIONS

An important subclass of Hermitian matrix functions are positively (negatively) definite matrix functions, i.e. such that the signature of the matrix $G(t)$ is equal to n $(-n)$ a.e. on $\mathbb{T}$. Since negatively definite matrix functions become positively definite after multiplication by -1, it is sufficient to study only positively definite matrix functions.

If a positively definite matrix function is factorable in L_2, then, by Corollary 7.6, all its partial 2-indices are equal to zero. Moreover, $l_+ = n$ and $l_- = 0$, since only in this case the equation $l_+ - l_- = n$ may be guaranteed. Thus, for positively definite matrix functions, the factor Λ_0 is the unit matrix. Consequently, if a positively definite matrix function can be factored in L_2, then it admits also a representation

$$G = AA^* \tag{7.15}$$

with $A^{\pm 1} \in H_2$.

The necessary condition for $\mathfrak{F}_0(G)$ to be non-empty established in Corollary 7.4 (i.e. of factorability in L_2) is fulfilled for a positively definite matrix function. If we complete this condition by the requirements $G^{\pm 1} \in L_1$, which are obviously necessary for factorability, then the obtained collection of conditions proves already to be sufficient. In other words, the following result holds.

<u>THEOREM 7.5.</u> A positively definite matrix function G defined on $\mathbb{T}$ can be factored in L_2 if and only if $G^{\pm 1} \in L_1$.

Theorem 7.5 may be derived as a consequence of a more general result on the representation of a positively definite matrix function G in the form of the product (7.15) of an outer matrix function and its adjoint[1]. Let us begin with the scalar case.

[1] The reader who is only interested in the problem of Φ-factorization of matrix functions from the class L_∞ may omit all the material up to the end of the present section. The question of Φ-factorability of positively definite matrix functions of the mentioned class will be solved in 7.4 irrespective of Theorem 7.5.

LEMMA 7.1. Let g be a non-negative function defined on $\mathbb{T}$. The representation $g(t) = |f(t)|^2$, where f is a certain outer function, is possible if and only if $\log g \in L_1$. If the desired function f exists, then it is defined up to a constant multiplier with absolute value 1.

Proof. From Definition 1.12 of an outer function and Theorem 1.15 it is clear that the relation $g = |f|^2$ holds if and only if $g(t) = k(t)^2$ a.e. on $\mathbb{T}$. Since the role of the function k in (1.7) may be played by any non-negative function with summable logarithm, it follows that the condition $\log g \in L_1$ is necessary and sufficient for the existence of the desired representation.

If this condition is fulfilled, then the corresponding function f is defined up to the constant ν from relation (1.7), which proves Lemma 7.1. ≡

In particular, Lemma 7.1 allows for solving the question of factorability in L_2 of a real-valued function to the very end.

THEOREM 7.6. Let G be a real-valued scalar function. The inclusion $2 \in \mathfrak{F}(G)$ holds if and only if

1) $G^{\pm} \in L_1$ and

2) G takes values of one and the same sign a.e.

Proof. The necessity of condition 1) is obvious; the necessity of condition 2) results from Corollary 7.4. If condition 1) and 2) are satisfied, then $G = \varepsilon g$, where g is a non-negative (a.e.) function such that $g^{\pm 1} \in L_1$ and ε is a constant equal to 1 or -1. Due to Lemma 7.1, $g = |f|^2$, where f is an outer function belonging with its inverse to the class H_2. Setting $A = f$, $B = \varepsilon f$, we get a factorization of G in L_2. Thus Theorem 7.6 is proved. ≡

Now let $G \in L_1$ be a positively definite matrix function of n th order. We introduce the Hilbert space $\mathfrak{h}$ of n-dimensional vector functions (rows) f defined on $\mathbb{T}$ for which the value $\frac{1}{2\pi} \int f(e^{i\theta})G(e^{i\theta})f(e^{i\theta})^* d\theta$[1]

---

[1] All the integrals in the present section are taken over the interval $[-\pi,\pi]$.

makes sense and is finite. The scalar product in $\mathfrak{H}$ is defined by the formula $(f,g) = \frac{1}{2\pi} \int f(e^{i\theta})G(e^{i\theta})g(e^{i\theta})^* d\theta$. We denote by $\mathcal{L}$ the closure (with respect to the norm in $\mathfrak{H}$) of the space L consisting of all vector functions of the form

$$f(e^{i\theta}) = \sum_{k=0}^{N} e^{ik\theta} f^{(k)}, \tag{7.16}$$

where $f^{(k)}$ $(k=0,\dots,N;\ N=0,1,\dots)$ are constant row vectors.

Let $\mathcal{L}_0$ be the closure (again with respect to the norm in $\mathfrak{H}$) of the space L_0 of those vector functions $f \in L$ for which $f^{(0)} = 0$. Clearly, $\mathcal{L} \supset \mathcal{L}_0$ and $\dim \mathcal{L} | \mathcal{L}_0 \leq n$.

<u>LEMMA 7.2.</u> Let $G \in L_1$ be a positively definite matrix function of n th order such that $\log \det G \in L_1$. Then $\dim \mathcal{L} | \mathcal{L}_0 = n$.

<u>Proof.</u> Suppose at first $n = 1$. According to Lemma 7.1, the condition $\log G \in L_1$ implies that $G = |G_+|^2$, where G_+ is an outer function. Since $G \in L_1$, then $G_+ \in H_2$. For an arbitrary function f, instead of $\int f(e^{i\theta})G(e^{i\theta})\overline{f(e^{i\theta})}d\theta$ we may write $\int |G_+(e^{i\theta})f(e^{i\theta})|^2 d\theta$. From this it is clear that, for all $f \in L$, $(f,f) = \frac{1}{2\pi} \int |G_+(e^{i\theta})f(e^{i\theta})|^2 d\theta \geq |G_+(0)f(0)|^2$. Setting $f = 1-g$ with $g \in L_0$, we thus obtain $\|1-g\|_{\mathfrak{H}} = (1-g,1-g)^{1/2} \geq |G_+(0)|$. Consequently, the distance from 1 to L_0 (and, therefore, also to the subspace $\mathcal{L}_0$) is not less than $|G_+(0)|$. Thus, $\mathcal{L}_0 \neq \mathcal{L}$ and $\dim(\mathcal{L} | \mathcal{L}_0) = 1$.
Returning to the general case, we denote by $\lambda_1(e^{i\theta}) \geq \dots \geq \lambda_n(e^{i\theta})$ the collection of all eigenvalues of the matrix $G(e^{i\theta})$. Since

$$\log \lambda_n(e^{i\theta}) \leq \frac{1}{n} \sum_{j=1}^{n} \log \lambda_j(e^{i\theta}) =$$

$$\frac{1}{n} \log \prod_{j=1}^{n} \lambda_j(e^{i\theta}) = \frac{1}{n} \log \det G(e^{i\theta})$$

and

$$\log \det G(e^{i\theta}) = \sum_{j=1}^{n} \log \lambda_j(e^{i\theta}) \leq \log \lambda_n(e^{i\theta}) + (n-1)\log \lambda_1(e^{i\theta})$$

$$< \log \lambda_n(e^{i\theta}) + (n-1)\lambda_1(e^{i\theta}) < \log \lambda_n(e^{i\theta}) + (n-1)\sum_{j=1}^{n} \lambda_j(e^{i\theta})$$

$$= \log \lambda_n(e^{i\theta}) + (n-1)\operatorname{tr} G(e^{i\theta}),$$

the conditions $\log \det G \in L_1$ and $\log \lambda_n \in L_1$ are equivalent for a positively definite matrix function $G \in L_1$. By $\operatorname{tr} G(e^{i\theta})$ we denote the trace of the matrix $G(e^{i\theta})$[1]. We shall make use of its summability. Choosing a vector function $f \in L$ and denoting its j th component by f_j, we have

$$(f,f) = \frac{1}{2\pi} \int f(e^{i\theta})G(e^{i\theta})f(e^{i\theta})^* d\theta$$

$$\geq \frac{1}{2\pi} \int f(e^{i\theta})\lambda_n(e^{i\theta})f(e^{i\theta})^* \, d\theta = \sum_{j=1}^{n} \frac{1}{2\pi} \int \lambda_n(e^{i\theta})|f_j(e^{i\theta})| d\theta .$$

The inclusion $\log \lambda_n \in L_1$ established above permits us to assert that $\lambda_n = |\lambda_+|^2$, where λ_+ is an outer function. In addition, due to the inequality $0 \leq \lambda_n(e^{i\theta}) \leq \operatorname{tr} G(e^{i\theta})$, the function λ_n is summable and, thus, $\lambda_+ \in L_2$. By what was proved in the first part of the lemma,

$$\frac{1}{2\pi} \int \lambda_n(e^{i\theta})|f_j(e^{i\theta})|^2 d\theta \geq |\lambda_+(0)f_j(0)|^2 .$$

Hence

$$(f,f) \geq \sum_{j=1}^{n} |\lambda_+(0)f_j(0)|^2 = |\lambda_+(0)|^2\|f(0)\|^2 .$$

Setting $f = a - g$, where a is a fixed constant vector and g ranges over L_0, we find that the distance from a to g is not less than $|\lambda_+(0)|\|a\|$. Since L_0 is dense in $\mathcal{L}_0$, every constant non-zero vector does not belong to $\mathcal{L}_0$. Consequently, $\dim(\mathcal{L}|\mathcal{L}_0) = n$, which proves Lemma 7.2. ≡

<u>THEOREM 7.7.</u> The matrix function G defined on $\mathbb{T}$ admits a representation $G = AA^*$ with an outer matrix function A if and only if

1) G is positively definite,
2) $\log \det G \in L_1$,
3) $\log\|G\| \in L_1$.

If the indicated representation exists, then the matrix function A is defined up to a constant right unitary multiplier.

<u>Proof.</u> Assume that $G = AA^*$, where $A = (a_{jk})_{j,k=1}^{n}$ is an outer matrix function. Then, for an arbitrary n-dimensional vector x, $(G(t)x,x) = (A(t)A(t)^*x,x) = \|A(t)^*x\|^2 \geq 0$ holds a.e. on $\mathbf{T}$, i.e.,

[1] $\operatorname{tr} G(e^{i\theta})$ is the sum of the diagonal elements of $G(e^{i\theta})$, where $G \in L_1$.

the matrix function G is positively definite and condition 1) is fulfilled. Moreover, from Lemma 7.15 and the relation $\det G = |\det A|^2$ we deduce that condition 2) is also satisfied.

Furthermore, we denote by $\lambda_1(t) \geq \dots \geq \lambda_n(t)$ the collection of eigenvalues of the matrix $G(t)$ at those points $t \in \mathbb{T}$ where this matrix is positively definite. Then $\|G(t)\| = (\lambda_{max}(G(t)G(t)^*))^{1/2} = \lambda_{max}(G(t)) = \lambda_1(t)$. We have, on the one hand, $\lambda_1(t) \geq (\lambda_1(t)\dots\lambda_n(t))^{1/n} = (\det G(t))^{1/n}$. On the other hand, $\lambda_1(t) \leq \lambda_1(t)+\dots+\lambda_n(t) = \operatorname{tr} G(t) = \sum_{j=1}^{n} g_{jj}(t) = \sum_{j,k=1}^{n} |a_{jk}(t)|^2 \leq n^2 \max_{j,k}|a_{jk}(t)|$. Thus

$$\frac{1}{n}\log\det G(t) \leq \log\|G(t)\| \leq 2\log n + \max_{j,k}\log|a_{jk}(t)| \quad \text{a.e. on } \mathbb{T}.$$

With regard to the definition of an outer function each of the functions a_{jk} belongs to the class D. Therefore, it is either identically equal to zero or, taken absolutely, coincides with a certain outer function. In any case, $\int \log|a_{jk}(e^{i\theta})|\,d\theta < +\infty$ and, consequently, $\int \max_{j,k}\log|a_{jk}(e^{i\theta})|\,d\theta < +\infty$. From the inequalities obtained for $\log\|G(t)\|$ and the summability of the function $\log\det G$, we thus conclude that $\log\|G\| \in L_1$. In this way, condition 3) is also necessary for the existence of a representation (7.15).

It is obvious that, replacing in (7.15) the matrix function A by AU, where U is a constant unitary multiplier, we do not violate equation (7.15) and do not go beyond the class of outer matrix functions. Vice versa, assume that, besides of the representation (7.15), the equation $G = BB^*$ with an outer matrix function B is valid. Then

$$A^{-1}(t)B(t) = (B^{-1}(t)A(t))^*$$

a.e. on $\mathbb{T}$.

The latter equation means that the matrix function $U = A^{-1}B$ is unitary and, in particular, belongs to the class L_∞. Moreover, this matrix function is outer along with A and B. Therefore, on the basis of Theorem 1.18, from the condition $U \in L_\infty$ we can conclude that $U \in H_\infty$. To verify that the matrix function U^* $(= B^{-1}A)$ lies in the class H_∞ can be done in a similar way. But then $U \in H_\infty \cap L^-$, so that, by Theorem 1.28, the matrix function U is constant. Thus $B = AU$, where U is a constant unitary matrix.

It remains to prove the existence of a representation (7.15) in case the conditions 1) - 3) are fulfilled. Replacing the matrix function G by $f^{-1}G$, where $f(t) = \|G(t)\|$, we get a summable positively definite matrix function with summable logarithm of the determinant. If the existence of a representation $f^{-1}G = BB^*$ with an outer matrix function B will be proved, then setting $A = Bf_+$ (where f_+ is an outer function with the property $|f_+|^2 = f$ existing in view of Lemma 7.1), we shall obtain a representation (7.15) of the original matrix function G. Hence, without loss of generality, we may assume the matrix function G to be summable.

If we stipulate the mentioned assumption and use the notations introduced before Lemma 7.2, we observe that, due to this lemma, the orthogonal complement of $\mathcal{L}_0$ with respect to $\mathcal{L}$ is n-dimensional. Now we choose an orthonormal basis $\{\varphi_j\}_{j=1}^n$ in $\mathcal{L} \ominus \mathcal{L}_0$ [1]. Since, for arbitrary $f \in \mathcal{L}$ and $k = 1,2,\ldots$, $e^{ik}f \in \mathcal{L}_0$, then for all $f \in \mathcal{L}$:

$$\int f(e^{i\theta})G(e^{i\theta})\varphi_j(e^{i\theta})^* e^{ik\theta}d\theta = 0, \qquad (7.17)$$
$$j = 1,\ldots,n;\ k = 1,2,\ldots .$$

In particular, taking $f = \varphi_l$, we have

$$\int \varphi_l(e^{i\theta})G(e^{i\theta})\varphi_j(e^{i\theta})^* e^{ik\theta}d\theta = 0,$$
$$j,l = 1,\ldots,n;\ k = 1,2,\ldots .$$

Changing in the last equation j and l with each other and passing to the adjoint, we find that this equation continues to be true also for $k = -1,-2,\ldots$. But this means that the measure $\varphi_l(e^{i\theta})G(e^{i\theta})\varphi_j(e^{i\theta})^*d\theta$ is orthogonal to all trigonometric polynomials vanishing at zero. In its turn, this allows for concluding that the function $\varphi_l(e^{i\theta})G(e^{i\theta})\varphi_j(e^{i\theta})^*$ is equal to a certain constant c_{lj} $(l,j = 1,\ldots,n)$ a.e. .

Furthermore, $c_{lj} = \frac{1}{2\pi}\int \varphi_l(e^{i\theta})G(e^{i\theta})\varphi_j(e^{i\theta})^*d\theta = (\varphi_l,\varphi_j) = \delta_{lj}$ (here the orthonormality of the system has been used).

---

1) $\mathcal{L} \ominus \mathcal{L}_0$ denotes the orthogonal complement of $\mathcal{L}_0$ in $\mathcal{L}$.

In this way, if we denote by Φ that matrix function the j th row of which is φ_j , then $\Phi(e^{i\theta})G(e^{i\theta})\Phi(e^{i\theta})^* = I$ a.e. on $\mathbb{T}$. From here we conclude that equation (7.15) holds, for instance, for $A = G\Phi^*$ $(= \Phi^{-1})$. It only remains to show that the matrix function A defined in this manner is outer.

Relation (7.15) and condition 2) of the theorem imply the inclusion $A \in L_2$. Equation (7.17) means (if we choose in it the rows $(0,\dots,0,1,0,\dots,0)$ as f , where the 1 stands on the l th place) that the Fourier coefficients with negative numbers of the (l,j) th element of $A\ (= G\Phi^*)$ are equal to zero and, thus, $A \in H_2$. Furthermore, the rows of the matrix function $A^{-1}\ (= \Phi)$ are elements of $\mathfrak{L}$. Consequently, they can be approximated (in the metric of $\mathfrak{H}$) by vector functions of the form (7.16). In other words, there exists a sequence of polynomial matrix functions $\{P_k\}_{k=1}^{\infty}$ such that

$$\int (P_k(e^{i\theta})-(A(e^{i\theta}))^{-1})G(e^{i\theta})\,(P_k(e^{i\theta})-(A(e^{i\theta}))^{-1})^*d\theta =$$

$$\int (P_k(e^{i\theta})A(e^{i\theta})-I)(P_k(e^{i\theta})A(e^{i\theta})-I)^*d\theta \to 0 .$$

This means that the sequence $\{P_kA\}_{k=1}^{\infty}$ converges to I in L_2 . But then the sequence $\{\det A \det P_k\}_{k=1}^{\infty}$ converges to 1 in the metric of the space $L_{2/n}$ [1] . If $\det A = f^{(i)}f^{(e)}$ is a representation of $\det A$ in the form of the product of an inner function and an outer one, then the sequence $\{f^{(e)}\det P_k\}_{k=1}^{\infty}$ lying in $H_{2/n}$ converges to the function $(f^{(i)})^{-1}$ in the metric of the space $L_{2/n}$. Consequently, $(f^{(i)})^{-1}$ belongs to the class $H_{2/n}$, i.e., $f^{(i)}$ is an invertible element of $H_{2/n}$ and thus an outer function. Since $f^{(i)}$ is an outer and inner function simultaneously, it reduces to a constant. Hence, the function $\det A$ and, by definition, also the matrix function A is outer. The proof of Theorem 7.7 is complete. ≡

Now we are prepared to prove Theorem 7.5.

---

[1] Notice that, for $p\in(0,1)$, L_p is a complete metric space with the metric $\varrho(f,g) = \int |f(e^{i\theta})-g(e^{i\theta})|^p d\theta$, and H_p is its closed subspace (see, for example, DUREN [1], p. 37).

Proof of Theorem 7.5. The inequality $-nx^{-1/n} < \log x < nx^{1/n}$, which is true for all $x > 0$, shows that the condition $G^{\pm 1} \in L_1$ implies the summability of $\log \det G$. Therefore, Theorem 7.7 can be applied to the matrix function G, which consequently, admits a representation (7.15) with an outer matrix function A. In virtue of equation (7.15), the condition $G \in L_1$ $(G^{-1} \in L_1)$ is equivalent to $A \in L_2$ $(A^{-1} \in L_2)$ Thus, the representation (7.15) yields a factorization of G in L_2. ∎

COROLLARY 7.9. The positively definite matrix function G defined on $\mathbb{T}$ can be factored in L_{p_0} with a zero total index if and only if $G^{\pm 1} \in L_r$, where $r = \frac{1}{2}\max\{p_0, p_0/(p_0-1)\}$.

Proof. If $G^{\pm 1} \in L_r$, then in any case $G^{\pm 1} \in L_1$ and the matrix function G admits a representation (7.15) with $A^{\pm 1} \in H_2$. The inclusion $G \in L_r$ implies that $A \in L_{2r}$ and, consequently, $A \in H_{2r}$. Analogously, it can be verified that $A^{-1} \in H_{2r}$, i.e., (7.15) is also a factorization of G for all $p \in [2r/(2r-1), 2r]$ and, in particular, for $p = p_0$.

Conversely, let the matrix function G be factorable in L_{p_0} with a zero total index. Then $\mathfrak{F}_0(G)$ is not empty and, thus, $2 \in \mathfrak{F}_0(G)$ as mentioned in Corollary 7.3. Owing to Theorem 7.2, the component $\mathfrak{F}_0(G)$ contains along with the point p_0 the point $q_0 = p_0/(p_0-1)$. As was shown at the beginning of the present section, a factorization of the matrix function G in L_2 may be chosen in the form (7.15). Thanks to Corollary 2.2, it will also be a factorization of G in all L_p for $p \in \mathfrak{F}_0(G)$, especially, for $p = p_0$ and $p = q_0$. From this the inclusion $A^{\pm 1} \in H_{2r}$ and thus $G^{\pm 1} \in L_r$ results. ≡

Corollary 7.9 illustrates that every connected subset of the ray $(1,\infty)$ which, under the mapping $p \to p/(p-1)$, is transferred into itself, is a component of the factorability domain corresponding to the zero total index of a certain positively definite matrix function defined on $\mathbb{T}$.

7.4. CRITERIA FOR EXISTENCE OF Φ-FACTORIZATION AND COINCIDENCE OF PARTIAL INDICES ASSOCIATED WITH THE BEHAVIOUR OF THE NUMERICAL DOMAIN

For the factorization problem on the circle the results of Sections 6.3 - 6.5 enable for being made more precise. One of the reasons guaranteeing these strengthenings is the exact equation

$$\|S\|_{L_p^n} = \operatorname{ctg} \frac{\pi}{2 \max(p,q)} \tag{7.18}$$

for the norm of the operator S in the space L_p^n proved in VERBITSKIĬ, KRUPNIK [1], KRUPNIK [4]. In view of relation (7.18), the lower estimate $2\|S\|/(1+\|S\|^2)$ for the quantity $\delta_{p,n}(\Gamma)$ from assertion 1) of Lemma 6.2 coincides, for $\Gamma = \mathbb{T}$, with its upper estimate $\sin(\pi/\max(p,q))$. Consequently, the equations

$$\delta_{p,n}(\mathbb{T}) = \sin \frac{\pi}{\max(p,q)}, \quad \delta_n(\mathbb{T}) = 1 \tag{7.19}$$

are true, which allows us to give the formulations of the statements from 6.3 - 6.5 a more concrete form.

In particular, Lemma 6.4 (and its strengthening, Lemma 6.4') can, for $p = 2$, be reformulated in the following manner.

THEOREM 7.8. Let G $(\in L_\infty)$ be a matrix function with a uniformly positive real part, i.e., the value $d(G(t))$ in Definition 6.3 is bounded from below a.e. on $\mathbb{T}$ by a positive constant which does not depend on t. Then G is Φ-factorable in L_2 and all its partial 2-indices are equal to zero.

From Theorem 7.8 we can easily derive the analogue of Theorem 7.5 related to the special case when $G^{\pm 1} \in L_\infty$, which is of particular interest. To be more exact, the following result is true.

THEOREM 7.9. Let $G \in L_\infty$ be a positively definite matrix function. The domain of Φ-factorability of G is either empty (if $G^{-1} \notin L_\infty$) or consists of the only component $\Phi_0(G)$ filling up the whole ray $(1,\infty)$ (if $G^{-1} \in L_\infty$). In the latter case the matrix function G admits a representation (7.15) in which $A^{\pm 1} \in H_\infty$.

Proof. The necessity of the condition $G^{-1} \in L_\infty$ for the set $\Phi(G)$ to be non-empty is of a general character and has already been proved in Chapter 3 (Corollary 3.5). Now we assume that this condition is fulfilled. For a positively definite matrix function G, the value $d(G(t))$ is the smallest eigenvalue of the matrix $G(t)$. Hence, it coincides with $\|G(t)^{-1}\|^{-1}$. Consequently, the condition $G^{-1} \in L_\infty$ allows us to apply Theorem 7.8 to the matrix function G. According to this theorem, there exists a factorization of G in L_2. But then, as was pointed out at the beginning of Section 7.3, there exists also a representation of G in the form (7.15) with an outer matrix function A. From the relations $G^{\pm 1} \in L_\infty$ we deduce that $A^{\pm 1} \in H_\infty$. Therefore, the representation (7.15) is a Φ-factorization of the matrix function G in all L_p, $1 < p < \infty$ and, thus, $\Phi_0(G) = (1,\infty)$, which completes the proof of Theorem 7.9. ≡

Theorem 7.9, which is specific for the circle, is the second circumstance (after equations (7.19)) which permits us to make the results of the preceding chapter more precise. Now we are going to show how the assertion of Corollary 6.4 (after substitution of the set of eigenvalues by the numerical domain) can be transferred from unitary matrix functions to arbitrary ones with the aid of this theorem as well as Theorem 6.10.

THEOREM 7.10. Let the matrix function $F \in L_\infty$ defined on $\mathbb{T}$ have the property: for every point $t \in \mathbb{T}$ there exists an arc γ_t ($\ni t$) of the contour $\mathbb{T}$, a scalar $\varepsilon_t > 0$ and a sector S_t with vertex at the origin of apex angle $\alpha_t < 2\pi/\max(p,q)$ $(q=p/(p-1))$ such that $h(F(\tau)) \geq \varepsilon_t$ and $H(F(\tau)) \subseteq S_t$ a.e. on γ_t. Then the interval with the end points p and q lies entirely inside a certain component of the Φ-factorability domain of the matrix function F and all partial p-indices of F are equal to each other.

Proof. Arguing as in the proof of Theorem 6.10, we establish the existence of a continuous and non-vanishing function χ such that $F = \chi G$, where G is a matrix function the numerical domain of which is contained in the sector $S_\alpha = \{z : |\arg z| \leq \alpha/2\}$ a.e. on $\mathbb{T}$, $\alpha < 2\pi/\max(p,q)$ and $h(G(\tau)) \geq \varepsilon > 0$ a.e. on $\mathbb{T}$. Since $H(G(\tau))$ lies outside the disk of radius ε with centre at 0 and in the sector S_α, we conclude that $H(G(\tau))$ is located to the right of the vertical line passing through the point $(\varepsilon \cos \frac{\alpha}{2}, 0)$. Consequently, a.e. on $\mathbb{T}$ we have $d(G(\tau)) \geq \varepsilon \cos \alpha/2$, so that Theorem 7.9 may be applied to the matrix function $M = \mathrm{Re}\, G$. Owing to this theorem, $M = M_+ M_+^*$, where $M_+^{\pm 1} \in H_\infty$. Denoting $K = \mathrm{im}\, G$, we obtain $G = M + iK = M_+ M_+^* + iK = M_+(I + iM_+^{-1}K(M_+^*)^{-1})M_+^* = M_+(I+iN)M_+^*$, where $N = M_+^{-1}K(M_+^*)^{-1}$. The matrix function N is Hermitian together with K. Since $H(G(\tau)) \subseteq S_\alpha$, then $|(K(\tau)x,x)| \leq \mathrm{tg}\, \frac{\alpha}{2} \|M_+^*(\tau)x\|^2$. Denoting $M_+^*(\tau)x$ by y and using the equation $(K(\tau)x,x) = (M_+(\tau)N(\tau)M_+^*(\tau)x,x) = (N(\tau)y,y)$, we find that $|(N(\tau)y,y)| \leq \mathrm{tg}\, \frac{\alpha}{2}\|y\|^2$. This means that the numerical domain of the matrix $N(\tau)$ is situated on the interval $[-\mathrm{tg}\, \frac{\alpha}{2}, \mathrm{tg}\, \frac{\alpha}{2}]$ a.e. on $\mathbb{T}$. Consequently, the numerical domain of the matrix $I + iN(\tau)$ is located on the interval which connects the points $1-i\cdot\mathrm{tg}\, \frac{\alpha}{2}$ and $1+i\cdot\mathrm{tg}\, \frac{\alpha}{2}$. This interval lies in the disk with centre at the point $(\sec^2 \frac{\alpha}{2}, 0)$ of radius $\sin(\alpha/2)/(\cos^2(\alpha/2))$, which is visible from the origin at the angle α. In addition, the matrix $I + iN(\tau)$ is obviously normal. Therefore, Theorem 6.10 can be applied to the matrix function $I + iN$. Due to this theorem, the interval A_p with the end points p and q lies entirely in a certain component of the domain of Φ-factorability and all partial indices are equal to each other. Since the interval A_p passes through the point 2 and the matrix function $I + iN$ has a uniformly positive real part, we may claim that, according to Theorem 7.8, all its partial 2-indices are equal to zero. Consequently $A_p \subseteq \Phi_0(I+iN)$. Apparently, this implies that $A_p \subseteq \Phi_0(G)$ and all partial r-indices of G, $r \in A_p$, are equal to zero. Utilizing Lemma 6.3, we conclude that the interval A_p lies inside a certain component of the domain of Φ-factorability of the matrix function F and all partial r-indices $(r \in A_p)$ of F are equal to each other. $\equiv$

Theorem 7.10 is an essential generalization of Theorem 6.10, because the sector S_t contains the disk Δ_t and the rigorous requirement of normality is no longer imposed upon the matrix $F(t)$ in Theorem 7.10. Moreover, Theorem 7.10 generalizes Theorem 6.11', too. In fact, under the condition of Theorem 6.11', the sets $\|F(\tau)\|^{-1}H(F(\tau))$, $\tau \in \gamma_t$ are separated from zero by the straight line l_t which is distant from zero by more than $\cos(\pi/2\max(p,q))$. At the same time, these sets are concentrated in the unit disk, as the numerical domain of an arbitrary matrix A lies in the disk with centre at zero and radius $\|A\|$. Thus, the sets $\|F(\tau)\|^{-1}H(F(\tau))$ are located in the circular segment which is cut off from the unit disk by the line l_t , and this segment may be inscribed into a sector S_t the apex angle of which satisfies the restrictions of Theorem 7.10. Consequently, the numerical domain of all matrices $F(\tau)$, $\tau \in \gamma_t$ lie in the same sector, i.e., the conditions of Theorem 7.10 are fulfilled for the matrix function F .

For $p = 2$, however, Theorem 7.10 is equivalent to Theorem 6.11' reformulated with regard to equations (7.19). This reformulation looks as follows.

<u>THEOREM 7.11'.</u> Assume the matrix function F of class L_∞ defined on $\mathbf{T}$ to be such that, for every point $t \in \mathbf{T}$, there exist an arc γ_t ($\ni t$) and a straight line l_t which does not pass through zero and has the following property: for almost all $\tau \in \gamma_t$, the set $H(F(\tau))$ is separated from zero by l_t . Then the matrix function G is Φ-factorable in L_2 and all its partial 2-indices are equal to each other.

Finally, note that, as also in Section 6.4, under the conditions of Theorem 7.8 - 7.11, there exists not only a left but also a right Φ-factorization of the matric function G (and F) , where the right partial indices coincide with the left ones and with each other.

7.5. DEFINITENESS CRITERIA AND STABILITY OF PARTIAL INDICES OF BOUNDED MEASURABLE MATRIX FUNCTIONS

One more circumstance (besides equations (7.19) and Theorem 7.9) thanks to which, ultimately, we succeed in proving assertions converse to Theorems 6.6 - 6.9 for $p = 2$, is the existence of exact formulae for the norms of the operators QGP and PGQ in the case $\Gamma = \mathbf{T}$.

<u>LEMMA 7.3.</u> Let G be a matrix function from L_∞ defined on $\mathbf{T}$. The norms of the operators QGP and PGQ in the space L_2^n are calculated by the formulae

$$\|QGP\| = \min_{X\in H^+_\infty} \|G-X\| \,, \qquad \|PGQ\| = \min_{Y\in H^-_\infty} \|G-Y\| \,. \tag{7.20}$$

Proof. We shall consider the spaces L_1 and L_2 of $n\times n$ matrix functions as Banach spaces the norm in which is defined by the formulae

$$\|A\|_1 = \frac{1}{2\pi} \int \operatorname{tr}(A(e^{i\theta})A(e^{i\theta})^*)^{1/2} d\theta$$

and

$$\|A\|_2 = \left(\frac{1}{2\pi} \int \operatorname{tr} A(e^{i\theta})A(e^{i\theta})^* d\theta\right)^{1/2} ,$$

respectively.

The space dual to L_1 may be identified with the space of measurable essentially bounded matrix functions (supplied with the norm (3.19)) by assigning to the matrix function $G \in L_\infty$ the functional

$$\psi_G(A) = \frac{1}{2\pi} \int \operatorname{tr} A(e^{i\theta})G(e^{i\theta})^* d\theta \,. \tag{7.21}$$

With regard to the mentioned correspondence the subspace H^-_∞ is an orthogonal complement to the subspace $\overset{\circ}{H}{}^+_1$ of those $n\times n$ matrix functions from H^+_1 which vanish at zero.

The space L_2 supplied with the scalar product

$$(A,B) = \frac{1}{2\pi} \int \operatorname{tr}(A(e^{i\theta})B(e^{i\theta})^*) d\theta$$

becomes a Hilbert space, where for all $A,B \in L_2$, $\|AB\|_1 \le \|A\|_2 \|B\|_2$.

In this way, the set of matrix functions of the form

$$F = AB \tag{7.22}$$

with $A,B \in H^+_2$, $\|A\|_2$, $\|B\|_2 < 1$ lies in the open unit ball of the subspace H^+_1 of the space L_1 . Let us show that it is dense in this ball.

For this purpose, above all, note that an arbitrary matrix function $F \in H^+_1$, $\|F\|_1 < 1$ for which $\log|\det F| \in L_1$ admits a representation (7.22). Indeed, in this case the matrix function $(FF^*)^{1/2}$ satisfies the conditions of Theorem 7.7. Thus, $(FF^*)^{1/2} = AA^*$, where $A \in H^+_2$ is an outer matrix function. Besides,

$$\|A\|_2^2 = \frac{1}{2\pi} \int \operatorname{tr} A(e^{i\theta})A(e^{i\theta})^* d\theta = \frac{1}{2\pi} \int \operatorname{tr}(F(e^{i\theta})F(e^{i\theta})^*)^{1/2} d\theta$$

$$= \|F\|_1 < 1 \,.$$

In order to satisfy (7.22), we set $B = A^{-1}F$. Since

$$\|B\|_2^2 = \frac{1}{2\pi}\int tr(B(e^{i\theta})B(e^{i\theta})^*)d\theta$$

$$= \frac{1}{2\pi}\int tr(A(e^{i\theta})^{-1}F(e^{i\theta})F(e^{i\theta})^*(A(e^{i\theta})^*)^{-1})d\theta$$

$$= \frac{1}{2\pi}\int tr(A(e^{i\theta})^*A(e^{i\theta}))d\theta$$

$$= \frac{1}{2\pi}\int tr(A(e^{i\theta})A(e^{i\theta})^*)d\theta$$

$$= \|F\|_1 < 1 ,$$

the matrix function B is situated in the open unit ball of the space L_2. Furthermore, since B is of class D, then it also lies in the open unit ball of the space H_2^+. Hence the desired representation has been constructed.

In turn, the set of matrix functions $F \in H_1^+$, $\|F\|_1 < 1$ for which $\log|\det F| \in L_1$ is dense in the unit ball of H_1^+. In fact, with the aid of an arbitrarily small perturbation of the matrix function F we may guarantee that $\det F(0) \neq 0$. Applying now Jensen's inequality (GOLUZIN [1], PRIVALOV [2]) to the function $\det F(\in H^+_{1/n})$, we get

$$\frac{1}{2\pi}\int \log|\det F(e^{i\theta})|d\theta \geq \log|\det F(0)| > -\infty .$$

Multiplying the matrix functions F and B by $e^{i\theta}$, it can easily be seen that the set of matrix functions of the form AB with $A \in H_2^+$, $B \in \mathring{H}_2^+$, $\|A\|_2$, $\|B\|_2 < 1$ is dense in the ball $\{F \in \mathring{H}_1^+ : \|F\|_1 < 1\}$.

Now we intend to show that the norm of the operator PGQ in L_2^n coincides with the norm of the functional ψ_G defined by formula (7.21) on $\mathring{H}_1^+$.

By the very definition,

$$\|\psi_G|_{\mathring{H}_1^+}\| = \sup\{\frac{1}{2\pi}\int tr(F(e^{i\theta})G(e^{i\theta})^*)d\theta : \|F\|_1 < 1 , F \in \mathring{H}_1^+\} ,$$

where sup may be taken not over the whole of the unit ball, but only over its dense subset consisting of matrix functions of the form $F_1F_2^*$ with $\|F_j\|_2 < 1$ $(j=1,2)$, $F_1 \in H_2^+$, $F_2 \in \mathring{H}_2^-$. At the same time,

$$\frac{1}{2\pi}\int \operatorname{tr}(F_1(e^{i\theta})F_2(e^{i\theta})^*G(e^{i\theta})^*)d\theta$$

$$= \frac{1}{2\pi}\int \operatorname{tr}(F_2(e^{i\theta})^*G(e^{i\theta})^*F_1(e^{i\theta}))d\theta$$

$$= \frac{1}{2\pi}\int \operatorname{tr}((G(e^{i\theta})F_2(e^{i\theta}))^*F_1(e^{i\theta}))d\theta$$

$$= \frac{1}{2\pi} \operatorname{tr}\int (G(e^{i\theta})F_2(e^{i\theta}))^*F_1(e^{i\theta})d\theta .$$

Denoting the columns of the matrix functions F_j by $f_1^{(j)},\dots,f_n^{(j)}$ $(j=1,2)$, instead of $\int (G(e^{i\theta})F_2(e^{i\theta}))^*F_1(e^{i\theta})d\theta$ we may write the matrix the (i,k) th entry of which is the scalar product (in L_2^n) of $f_k^{(1)}$ and $Gf_i^{(2)}$. Therefore

$$\frac{1}{2\pi} \operatorname{tr}\int (G(e^{i\theta})F_2(e^{i\theta}))^*F_1(e^{i\theta})d\theta = \frac{1}{2\pi}\sum_{k=1}^{n}(f_k^{(1)},Gf_k^{(2)})$$

$$= \frac{1}{2\pi}\sum_{k=1}^{n}(f_k^{(1)},PGQf_k^{(2)}) .$$

The condition $\|F_j\|_2 < 1$ is equivalent to the inequality

$\frac{1}{2\pi}\sum_{k=1}^{n}\|f_k^{(j)}\|^2 < 1 \quad (j=1,2).$

Thus, it only remains to note that, under the conditions

$\frac{1}{2\pi}\sum_{k=1}^{n}\|f_k^{(j)}\|^2 < 1 \quad (j=1,2)$, $f_k^{(1)}\in(H_2^+)^n$, $f_k^{(2)}\in(\overset{\circ}{H}{}_2^-)^n$, the value

$\frac{1}{2\pi}\sup\sum_{k=1}^{n}(f_k^{(1)}, PGQf_k^{(2)})$ is equal to $\|PGQ\|$.

Summarizing, $\|PGQ\| = \|\psi_G|_{\overset{\circ}{H}{}_1^+}\|$. An arbitrary extension of the functional $\psi_G|_{H_1^+}$ to the whole space L_1 is of the form ψ_{G_1} . Besides, since $\psi_G(F) = \psi_{G_1}(F)$ for all $F\in\overset{\circ}{H}{}_1^+$, then $G-G_1\in H_\infty^-$, i.e., $G_1 = G-Y$ with $Y\in H_\infty^-$. Conversely, for any $Y\in H_\infty^-$, the functional ψ_{G-Y} is an extension of the functional $\psi_G|_{\overset{\circ}{H}{}_1^+}$. Using the fact that in extending a functional the norm does not decrease, we recognize that $\|PGQ\| \le \|\psi_{G-Y}\| = \|G-Y\|$ for all $Y\in H_\infty^-$. At the same time, due to the Hahn-Banach theorem, there exists a[n exte]nsion which preserves the norm. This proves the second of the formu[la]s (7.20).

[U]tilizing the fact that QGP is an operator adjoint to PG^*Q , so that

$$\|QGP\| = \|PG^*G\| = \min_{Y\in H_\infty^-}\|G^*-Y\| = \min_{X\in H_\infty^+}\|G^*-X^*\| = \min_{X\in H_\infty^+}\|G-X\| ,$$

the first formula may be readily derived from the second one. Thus, the proof of Lemma 7.3 is complete. ≡

The criterion of non-negativity (non-positivity) of partial 2-indices, first of all, will be proved for unitary matrix functions. Above all, this is caused by the fact that the proof of the corresponding criterion for arbitrary matrix functions employs the statement related to unitary matrix functions as an auxiliary result. However, there are, in addition, two more important reasons: 1) for unitary matrix functions the definiteness criterion for partial indices can be formulated in the most transparent form, revealing the relationships between the structure of the set of partial indices and the approximation properties of matrix functions, and 2) matrix functions occurring in applied problems are often unitary (see, for instance, KREĬN, MELIK-ADAMYAN [1], TROITSKIĬ [1]).

THEOREM 7.12. Let U be a unitary matrix function defined on $\mathbb{T}$. The operator $P + UQ$ is invertible from the right (left) in L_2^n if and only if there exists a matrix function $X \in H_\infty^+$ (H_∞^-) such that

$$\|U - X\| < 1 . \tag{7.23}$$

Proof. Owing to assertion 1) of Lemma 3.5, the operator $P + UQ$ is invertible from the right only simultaneously with the operator $T_Q(U)$, which, in turn, is invertible from the right only simultaneously with the operator $P + QUQ$. The latter is invertible from the right if and only if the operator

$$\begin{aligned}(P+QUQ)(P+QUQ)^* &= (P+QUQ)(P+QU^*Q) = P+QUQU^*Q = P+QUU^*Q-QUPU^*Q \\ &= P+Q-(QUP)(PU^*Q) = I-(QUP)(QUP)^*\end{aligned}$$

is invertible [1]. Since $\|QUP\| \leq \|U\| = 1$, then for the invertibility of this operator, it is necessary and sufficient that $\|QUP\| < 1$. It remains to apply the first of the formulae (7.20). The case of invertibility from the left can be regarded analogously. Thus Theorem 7.12 is proved. ≡

[1] Here we used the fact that a linear bounded operator T acting in a Hilbert space is invertible from the right if and only if the operator TT^* is invertible.

THEOREM 7.13. Assume that U is a unitary matrix function defined on $\mathbb{T}$. Then the following assertions are equivalent:

1) U is Φ-factorable in L_2 and all its partial indices are non-negative (non-positive),

2) there exists a matrix function $X \in H_\infty^+$ (H_∞^-) such that $X^{-1} \in M_\infty^+$ (M_∞^-) and $\|U-X\| < 1$,

3) there exists a matrix function $Y \in M_\infty^-$ (M_∞^+) such that $Y^{-1} \in H_\infty^-$ (H_∞^+) and $\|U-Y\| < 1$.

In this case the total index of U coincides with the number of zeros of $\det X$ inside the unit disk and with the number of poles of $\det Y$ outside it.

Proof. We shall conduct the proof for the case of non-negative partial indices. Since, for any matrix function $X \in L_\infty$,

$$\|U - X\| = \|I - XU^*\| , \tag{7.24}$$

then the inequality (7.23) guarantees the applicability of assertion 2) from Theorem 6.6 to the matrix function U^* (with $X_+ = X$, $X_- = I$), according to which this matrix function is Φ-factorable in L_2 and its partial indices are non-positive. In view of Corollary 7.1, from here we deduce the Φ-factorability in L_2 of the matrix function U as well as the non-negativity of its partial indices. Thus, 2) implies 1).

Now we are going to prove that 1) implies 2). From the Φ-factorability of the matrix function U and the non-negativity of its partial indices we conclude that the operator $P + UQ$ is invertible from the right (Corollary 3.8). By Theorem 7.12, there exists a matrix function $X \in H_\infty^+$ satisfying the inequality (7.23). We want to show that $X^{-1} \in M_\infty^+$ and calculate the total index of U. To this end, we shall employ the equation $T_Q(XU^*) = QXU^*|_{(\overset{\circ}{H}{}_2^-)^n} = QXQU^*|_{(\overset{\circ}{H}{}_2^-)^n} + QXPU^*|_{(\overset{\circ}{H}{}_2^-)^n} = T_Q(X)T_Q(U^*)$,

which holds because of $QXP = 0$. As follows from relations (7.23) and (7.24), the operator $T_Q(XU)$ is invertible. Due to Corollary 1.2, the operator $T_Q(U^*) = T_Q(U)^*$ is Fredholm simultaneously with the operator $T_Q(U)$. Consequently (see Theorem 1.9), the operator $T_Q(X)$ is also Fredholm and its index is opposite to the index of $T_Q(U^*)$ and, thus, coincides with the index of the operator $T_Q(U)$. On the strength of Theorem 3.15 we conclude from the Φ-factorability of the matrix function $X \in H_\infty^+$ that $X^{-1} \in M_\infty^+$ and, due to Theorem 2.6, the total index of X (i.e., the index of the operator $T_Q(X)$) coincides with the number of zeros of $\det X$ in the unit disk. Thus, the equivalence of assertions 1) and 2) as well as the validity of one of the formulae for the total index U have been proved. The equivalence of assertions 1) and 3) and the second formula for the total index can be established in a similar way. ≡

COROLLARY 7.10. Assume U to be a unitary matrix function defined on $\mathbf{T}$. The following assertions are equivalent:

1) U is Φ-factorable in L_2 and all its partial indices are equal to zero,

2) there exists a matrix function $X \in H_\infty^+$ such that $X^{-1} \in H_\infty^+$ and $\|X-U\| < 1$,

3) there exists a matrix function $X \in H_\infty^-$ such that $X^{-1} \in H_\infty^-$ and $\|X-U\| < 1$.

The transition from unitary matrix functions to arbitrary matrix functions of class L_∞ may be accomplished with the aid of the following auxiliary result.

LEMMA 7.4. Let the matrix function G given on $\mathbf{T}$ satisfy the condition $G^{\pm 1} \in L_\infty$. Then it can be represented in the form

$$G = AU , \tag{7.25}$$

but also as

$$G = VB^* , \tag{7.25'}$$

where $A^{\pm 1}$, $B^{\pm 1} \in H_\infty$ and the matrix functions U and V are unitary.

Proof. The matrix functions GG^* and G^*G meet the conditions of Theorem 7.9. Therefore, the equations $GG^* = AA^*$, $G^*G = BB^*$ with $A^{\pm 1}$, $B^{\pm 1} \in H_\infty$ are valid. Setting $U = A^{-1}G$, $V = G(B^*)^{-1}$, we may guarantee that relations (7.25) and (7.25') are fulfilled. Thus we are only obliged to check that the matrix functions U and V are unitary:

$U(t)U(t)^* = A(t)^{-1}G(t)G(t)^*(A(t)^*)^{-1} = A(t)^{-1}A(t)A(t)^*(A(t)^*)^{-1} = I$

a.e. on $\mathbb{T}$. The unitarity of the matrix $V(t)$ can analogously be verified. ≡

Equations (7.25) and (7.25') mean that the matrix functions G, U and V are factorable (Φ-factorable) only simultaneously, where the tuples of their partial indices coincide.

Now, finally, we are prepared for proving the existence criterion for a Φ-factorization in L_2 with definite partial indices.

THEOREM 7.14. For a matrix function $G \in L_\infty$ defined on $\mathbb{T}$, the following statements are equivalent:

1) $2 \in \Phi(G)$ and all partial indices of G are non-negative,
2) there exists a matrix function $X_- \in H^-_\infty$ such that $X_-^{-1} \in M^-_\infty$ and $\|I-GX_-\| < 1$,
3) there exists a matrix function $X_+ \in M^+_\infty$ such that $X_+^{-1} \in H^+_\infty$ and $\|I-X_+G\| < 1$,
4) $G = Y_+G_0$, where $Y_+ \in H^+_\infty$, $Y_+^{-1} \in M^+_\infty$, $d(G_0(t)) \geq d > 0$ a.e. on $\mathbb{T}$,
5) $G = G_0Y_-$, where $Y_- \in M^-_\infty$, $Y_-^{-1} \in H^-_\infty$, $d(G_0(t)) \geq d > 0$ a.e. on $\mathbb{T}$.

If these conditions hold, then the total index of the matrix function G coincides with the number of poles of $\det X_+$ and with the number of zeros of $\det Y_+$ in the unit disk, but also with the number auf zeros of $\det X_-$ and the number of poles of $\det Y_-$ outside the unit disk.

Proof. The implications 2) ==> 1) and 3) ==> 1) are special cases of statement 1) from Theorem 6.6. If 1) is fulfilled, then due to Corollary 3.4, $G^{-1} \in L_\infty$. Thus, Lemma 7.4 can be applied. The unitary matrix function V defined in this lemma is Φ-factorable in L_2 together with G and its partial indices are non-negative. According to Theorem 7.13, there exists a matrix function $X \in H^+_\infty$ such that $X^{-1} \in M^+_\infty$ and $\|V-X\| < 1$. Setting $X_- = (XB^{-1})^*$, where B is the matrix function from representation (7.25'), we get $X_- \in H_\infty^-$, $X_-^{-1} \in M_\infty^-$ and $\|I-GX_-\| = \|I-VB^*(B^*)^{-1}X^*\| = \|I-VX^*\| = \|V^*-X^*\| = \|V-X\|$. Therefore, 1) ==> 2).

In order to verify the implication 1) ==> 3), we introduce the matrix function U from representation (7.25) and a matrix function $Y \in M_\infty^-$ such that $Y^{-1} \in H_\infty^-$, $\|U-Y\| < 1$ existing by Theorem 7.13. Furthermore, we set $X_+ = Y^*A^{-1}$. Then $X_+ \in M_\infty^+$, $X_+^{-1} = A(Y^*)^{-1} \in H_\infty^+$ and $\|I-X_+G\| = \|I-Y^*A^{-1}AU\| = \|I-Y^*U\| = \|U^*-Y^*\| = \|U-Y\| < 1$.

In this way, we have checked that statements 1), 2) and 3) are equivalent. Now we want to show the equivalence of assertions 2) and 5). If 2) is true, then the matrix function GX_- has a uniformly positive real part, because $\mathrm{Re}(GX_-x,x) = (x,x) - \mathrm{Re}((I-GX_-)x,x) \geq (1-\|I-GX_-\|)\|x\|^2$. Setting $Y_- = X_-^{-1}$, we conclude that 2) ==> 5). Conversely, if 5) is fulfilled, then the condition $\|I-GX_-\| < 1$ is satisfied for $X_- = \mu Y_-^{-1}$, where μ is some positive constant (concerning the choice of μ, see the proof of Lemma 6.4). Thus, 2) ==> 5). The implication 3) ==> 4) can be verified analogously.

The formulae for the total index established in the theorem can be proved in just the same way as in Theorem 7.13. Thus, the proof of Theorem 7.14 is complete. ≡

<u>COROLLARY 7.11.</u> For the matrix function $G \in L_\infty$ given on $\mathbb{T}$, the following statements are equivalent:

1) $2 \in \Phi(G)$ and all partial indices of G are equal to zero,
2) there is a matrix function $X_+ \in H_\infty^+$ such that $X_+^{-1} \in H_\infty^+$ and $\|I-X_+G\| < 1$,
3) there is a matrix function $X_- \in H_\infty^-$ such that $X_-^{-1} \in H_\infty^-$ and $\|I-GX_-\| < 1$,
4) $G = Y_+G_0$, where $Y_+^{\pm 1} \in H_\infty^+$, $d(G_0(t)) \geq d > 0$ a.e. on $\mathbb{T}$,
5) $G = G_0Y_-$, where $Y_-^{\pm 1} \in H_\infty^-$, $d(G_0(t)) \geq d > 0$ a.e. on $\mathbb{T}$.

In this way, statements 1) and 3) of Theorem 6.6 as well as Lemma 6.4 admit an inversion for the factorization problem in L_2 on $\mathbb{T}$.

Apparently, the correctness of the inversion of Theorem 6.7 - 6.9 and of the remaining statements of Theorem 6.6 may be proved in a similar manner. In this case one of the matrix functions by which the matrix function G in the formulations of the indicated theorems is multiplied may be chosen to be the identity.

In summary, for the Φ-factorization problem in the space L_2 on $\mathbb{T}$ we have derived criteria of non-negativity, non-positivity, and equality to zero of partial indices. In view of its particular importance we now still want to formulate the criterion of stability.

THEOREM 7.15. The matrix function $G \in L_\infty$ given on $\mathbb{T}$ is Φ-factorable in L_2 with a stable tuple of partial indices if and only if one can find an integer k and matrix functions $X_+ \in \overset{+}{H}_\infty$ such that $X_\pm^{-1} \in \overset{\pm}{M}_\infty$ and $\|I - t^{-k} G X_-\| < 1$, $\|I - t^{-k-1} X_+ G\| < 1$.

7.6. ON PARTIAL INDICES OF CONTINUOUS MATRIX FUNCTIONS

In Chapter 6 we have already used the fact that the additional assumption $G \in \overset{\pm}{L}_\infty + C$ excludes the dependence of the Φ-factorability property and the values of partial indices on the parameter value p. In this way, for $G \in \overset{\pm}{L}_\infty + C$, Theorems 7.14, 7.15 and Corollary 7.11 proved above for the factorization in L_2 pass into criteria of non-negativity, stability and equality to zero of partial indices related to factorization in arbitrary L_p , $1 < p < \infty$. In particular, this remark applies to continuous matrix functions G .

Supposing the requirement of factorability in C (which is stronger than the condition $G^{\pm 1} \in C$) to be fulfilled, it seems that we may go even further. Namely, one can claim that the conditions of non-negativity (non-positivity, equality to zero, stability) of partial indices mentioned in Section 6.5 will remain necessary even in case the additional demand of rationality is imposed upon the matrix functions $X_\pm$, $Y_\pm$ involved in them. It should be noted that in this case the proof of necessity is not based on the sophisticated Lemma 7.3. Now we are going to present the exact formulation and the proof for the situation when the partial indices are equal to zero.

THEOREM 7.16. Assume the matrix function G defined on $\mathbb{T}$ to be factorable in C and all of its partial indices to be equal to zero. Then there can be found polynomial matrix functions X_1 and X_2 whose zeros of the determinant lie outside the unit disk and such that

$$\|I - X_1 G\| < 1 \, , \; \|I - G X_2^*\| < 1 \, . \tag{7.26}$$

Proof. By assumption, the factorization of the matrix function G is of the form $G = AB^*$, where $A^{\pm 1}$, $B^{\pm 1} \in C^+$. Therefore, the matrix functions $Y_1 = \mu BA^{-1}$ and $Y_2 = \mu AB^{-1}$ also belong to C^+ and are non-degenerate in the unit disk for arbitrarily chosen $\mu \neq 0$. Since $I - Y_1 G = I - \mu BB^*$, $I - GY_2^* = I - \mu AA^*$, then choosing μ from the interval $(0, \min\{\|A\|^{-2} , \|B\|^{-2}\})$, we may ensure the validity of the inequalities $\|I-Y_1 G\| < 1$, $\|I-GY_2^*\| < 1$. The condition $Y_{1,2} \in C^+$ enables us to uniformly approximate the matrix functions $Y_{1,2}$ by matrix polynomials with arbitrary degree of accuracy. Choosing as $X_{1,2}$ matrix polynomials sufficiently close to $Y_{1,2}$ we may guarantee both the validity of inequality (7.26) and the non-degeneracy of $X_{1,2}$ in the unit disk, which proves Theorem 7.16. ≡

Finally, we intend to present a proposition which is obtained from Theorem 7.11 under the additional requirement of continuity of the matrix function G .

THEOREM 7.17. The partial indices of a continuous strictly non-degenerate matrix function are equal to each other.

Proof. With regard to Definition 6.3 the condition $0 \notin H(G(t))$ is fulfilled for all $t \in \mathbb{T}$. Owing to the convexity of the numerical domain, we may claim that there exists a straight line l_t separating $H(G(t))$ from zero. As the matrix function G is continuous, the straight line l_t will separate from zero also all the sets $H(G(\tau))$ with τ close enough to t . In this way, the conditions of Theorem 7.11 are satisfied. Due to this theorem, the partial indices of G (coinciding with its partial 2-indices) are equal to each other, which proves Theorem 7.17. ≡

COROLLARY 7.12. Suppose the matrix function G is continuous on $\mathbb{T}$ and for all $t \in \mathbb{T}$ $G(t)$ is normal, and the convex hull of the spectrum of $G(t)$ does not contain zero. Then the partial indices of G are equal to each other.

The assertion of the corollary is correct, because the numerical domain of a normal matrix coincides with the convex hull of the spectrum. In particular, if G is a unitary matrix function, then $G(t)$ is, for all $t \in \mathbb{T}$, a normal matrix with a spectrum concentrated on the unit circle. Therefore the following statement is true.

COROLLARY 7.13. If U is a unitary matrix function continuous on $\mathbb{T}$ whose eigenvalues, for any $t \in \mathbb{T}$, are concentrated on an arc of the circle the end points of which are seen at an angle less than π, then all partial indices of U coincide.

COROLLARY 7.14. Assume G to be a continuous second-order matrix function given on $\mathbb{T}$, where, for all $t \in \mathbb{T}$, $G(t)$ is a normal matrix with eigenvalues $\lambda_1(t)$ and $\lambda_2(t)$. For the coincidence of the partial indices of G, it is sufficient for the quotient $\lambda_1(t)/\lambda_2(t)$ to be non-negative for any $t \in \mathbb{T}$.

Indeed, the numerical domain of the matrix function $G(t)$ is an interval with the end points $\lambda_1(t)$ and $\lambda_2(t)$. The condition imposed on $\lambda_1(t)/\lambda_2(t)$ just means that this interval does not contain the zero point.

Incidentally, the structure of the numerical domain for second-order matrices can be easily described, even if they fail to be normal (HALMOS [1]). Namely, if λ_1 and λ_2 are eigenvalues of the second-order matrix A and x_1 and x_2 are normed eigenvectors associated with them, then $H(A)$ is an ellipse with focuses λ_1 and λ_2 and a large axis

$$\frac{|\lambda_1-\lambda_2|}{(1-|(x_1,x_2)|^2)^{1/2}},$$

if $\lambda_1 \neq \lambda_2$, and a disk of radius $\frac{1}{2}\|A-\lambda I\|$ and centre λ, if $\lambda_1 = \lambda_2 = \lambda$. From this we derive the next result.

COROLLARY 7.15. Let G be a continuous second-order matrix function. Furthermore, assume that $\lambda_1(t)$ and $\lambda_2(t)$ are the eigenvalues of the matrix $G(t)$ and $x_1(t)$ and $x_2(t)$ the normed eigenvectors associated with them. The following condition is sufficient for the coincidence of partial indices of G :

$$|\lambda_1(t)| + |\lambda_2(t)| > |\lambda_1(t)-\lambda_2(t)|(1-|x_1(t),x_2(t))|^2)^{-1/2}$$

must hold at those points $t \in \mathbb{T}$ at which $\lambda_1(t) \neq \lambda_2(t)$, and

$$2|\lambda(t)| > \|G(t) - \lambda(t)I\|$$

is valid at those points $t \in \mathbb{T}$ at which $\lambda_1(t) = \lambda_2(t)$ $(= \lambda(t))$.

In connection with Corollary 7.14 the following example is of some significance. Let

$$G(t) = \frac{1}{2}\begin{pmatrix} t^k+t^{-k} & -t^k+t^{-k} \\ t^k-t^{-k} & -t^k-t^{-k} \end{pmatrix},$$

where $k \geq 0$ is integer.

The matrix $G(t)$ is simultaneously Hermitian and unitary, its eigenvalues do not depend on t and are equal to $\lambda_1(t) = 1$ and $\lambda_2(t) = -1$. At the same time,

$$G(t) = \frac{1}{2}\begin{pmatrix} 1 & 1 \\ 1 & -1 \end{pmatrix}\begin{pmatrix} 0 & t^k \\ t^{-k} & 0 \end{pmatrix}\begin{pmatrix} 1 & 1 \\ 1 & -1 \end{pmatrix},$$

so that the partial indices of G are equal to k and $-k$. For $k = 0$, they coincide, although the assumption of Corollary 7.14 is not fulfilled; for $k > 0$, the partial indices of the matrix function G do not coincide. Thus, the sufficient condition for coincidence of partial indices established in Corollary 7.14 is essential but not necessary.

7.7. COMMENTS

The relationship between factorizations of the matrix functions G and G^* in case of Hölder matrix functions mentioned in Theorems 7.1 and 7.1' was established by SHMUL'YAN [2]. Relation (7.5) from Theorem 7.2 for the same case was also proved by him.

Theorem 7.3 and its Corollary 7.5 are given in SPITKOVSKIĬ [6]. Theorem 7.4 and Corollary 7.8 for the case of matrix functions with Hölder elements were proved in NIKOLAĬCHUK, SPITKOVSKIĬ [1,2]; in SPITKOVSKIĬ [7] they were extended to the general case. A special kind of factorization of Hermitian matrix functions with zero partial indices is presented in SHMUL'YAN [2]. Corollary 7.6 is also proved there. Theorem 7.4 may be used for the classification of shifts in a space with indefinite metric in case of a finite-dimensional isotropic subspace (concerning this matter, see the note of one of the authors in the review Referativnyĭ Zhurnal Matematika 1983, 3B, 1001 of the article BALL, HELTON [1], where the inverse way of reasoning was gone: Theorem 7.6 was deduced from results related to shift classification).

A Φ-factorability criterion in L_2 as well as a method for constructing a factorization and calculating partial indices of a Hermitian second-order matrix function with negative determinant and definite diagonal elements are described in LITVINCHUK, SPITKOVSKIĬ [1,2]. In particular, in these papers it is established that the general case of such a matrix function can be reduced to the study of a matrix function of the kind

$$\Omega = \begin{pmatrix} |\omega|^2-1 & \omega \\ \overline{\omega} & 1 \end{pmatrix} \qquad (\omega \in L_\infty) .$$

In this case the matrix function Ω is Φ-factorable if and only if 1 does not belong to the limit spectrum of the operator $H(\omega)H(\omega)^*$ ($H(\omega) = Q\omega P$ is a Hankel operator with the symbol ω) and its partial indices are $\pm l$, where l is the multiplicity of 1 as an s-number of the operator $H(\omega)$ (i.e. as an eigenvalue of $H(\omega)H(\omega)^*$).

The proof scheme of Theorem 7.4 given here was used by DIDENKO and CHERNETSKIĬ [1] in studying the factorization problem for orthogonal matrix functions. Namely, in this article it is shown that, for a Hölder orthogonal matrix function G, equations (7.5) are fulfilled and there always exists a factorization (2.1) subjected to the additional conditions $G_+'G_+ = G_-G_-' = T$, where T is the matrix defined by equation (7.3).

Lemma 7.1 is a result of SZEGŐ (see, e.g., HOFFMAN [1]) obtained by him as a generalization of the Fejer-Riesz lemma which asserts that any trigonometric polynomial non-negative on $\mathbb{T}$ coincides with the square of the modulus of some algebraic polynomial. Lemma 7.2 and Theorem 7.7 are reformulations of an important proposition from the theory of stochastic processes (see, for example, ROZANOV [1]). The extension of this statement to the sectorial case and the history of the problem can be found in KREĬN, SPITKOVSKIĬ [1,2]. In the article KREĬN, SPITKOVSKIĬ [3] the results of the papers just mentioned, in particular, the theorems on factorization of sectorial matrix functions were applied for obtaining generalizations of the Szegö limit theorem. In turn, these generalizations can be used for the discussion of some problems of statistical physics in a similar manner as the corresponding results for the scalar case were used in VLADIMIROV, VOLOVICH [1,2].
We would like to remark that methods and results of factorization theory have been used for extending the Szegö limit theorem before the paper of KREĬN, SPITKOVSKIĬ [3], too. Let us refer, for instance, to one result from WIDOM [3] due to which, for a matrix function F from the class $H_\infty + C$, the left and right partial indices are equal to zero,

$$\lim_{n \to \infty} \frac{\Delta_{n+1}}{\Delta_n} = \lim_{\varrho \to 1} \exp \frac{1}{2\pi} \int_0^{2\pi} \log \det G_\varrho(e^{i\theta}) d\theta .$$

Here G_ϱ are the values of the harmonic continuation of G on the circle with radius ϱ (< 1) and Δ_n is the determinant of the block-Toeplitz matrix $(G_{i-j})_{i,j=0}^{\infty}$ constructed by the Fourier coefficients G_j of the matrix function G.

In case of a contracting positively definite matrix function G, the material of Ch. 5 of the book SZ.-NAGY, FOIAS [1] is also related to Theorem 7.7. The special kind (7.15) of factorization of a positively definite matrix function satisfying a Hölder condition was established by SHMUL'YAN [1]. Theorems 7.5 and 7.9 are explicitly formulated (as corollaries of more general propositions) in SPITKOVSKIĬ [2], however, as a matter of fact, they were known even earlier.

The boundedness of the factorization factors of a positively definite matrix function proved in Theorem 7.9 generalizes (for $\Gamma = \mathbb{T}$) the result of Lemma 6.5 from the scalar case. In Chapter 6 the proposition which stated that the multiplication by a positive function α invertible in L_∞ does not go beyond the class of Φ-factorable matrix functions was easily derived from this lemma. The latter result, however, cannot be transferred to the case of a positively definite matrix function G, even if $\Gamma = \mathbb{T}$: in the paper of KRUPNIK and ROZENBERG [1] it was proved that there exist piecewise continuous 2x2 matrix functions A and B such that A is Φ-factorable in L_p, B is uniformly positively definite, but the product AB fails to be Φ-factorable in L_p. An example of two uniformly positively definite piecewise continuous matrix functions of second order the product of which is not Φ-factorable in L_2 was constructed by one of the authors in SPITKOVSKIĬ [4].

Theorem 7.8 is due to SIMONENKO [4]. The equality to zero of the partial indices of a matrix function factorable in C and with uniformly positive real part was shown by GOHBERG and KREĬN [3]. The factorability in L_2 and the equality to zero of the partial indices of matrix functions of the type $I + F$ with $\|F\| < 1$ were established by WIENER and MASANI [1]. Notice that every matrix function with a uniformly positive real part may be transformed into the type just mentioned via multiplication by a positive constant.

In connection with Theorem 7.3 we also note that on every connected rectifiable closed contour different from the circle there exists a rational matrix function with uniformly positive real part and non-zero partial indices (VIROZUB, MATSAEV [1]). Such matrix functions exist even for $n = 2$ (MARKUS, MATSAEV [2]).

Theorem 7.10 has been proved by one of the authors (SPITKOVSKIĬ [10]). A slightly different formulation of this theorem is given in SPITKOVSKIĬ [5]. In case of unitary matrix functions Theorem 7.10 changes over into a result of KRUPNIK, NYAGA [1] (see also KRUPNIK [4]) coinciding with the reformulation of Corollary 6.4, where equations (7.19) are taken into account. This theorem also covers the factorability criterion of a matrix function G with zero partial indices which is discussed in VERBITSKIǏ, KRUPNIK [2]. Furthermore, notice that the questions of Fredholmness, index and one-sided invertibility of the operator

$$Tf(s) = sf(s) + \frac{B^*(s)}{\pi} \int_E \frac{B(t)f(t)}{s-t}\,dt$$

which was examined by another method in FAOUR NAZIH [1] (concerning this topic, see the note of one of the authors in the review Referativnyǐ Zhurnal, Matematika 1980, 10 B, 674 of the paper in question) may be studied with the aid of Theorem 7.10.

The factorization problem of locally sectorial (i.e. satisfying the condition $H(F(\tau)) \subseteq S_t$ $(\tau \in \gamma_t)$ of Theorem 7.10) but, generally speaking, unbounded matrix functions F was studied in SPITKOVSKIǏ [7]. In this paper the notion of canonical factorization was introduced. On the one hand, it generalizes the corresponding notion from KREǏN, SPITKOVSKIǏ [1,2] (especially, the representation of a positively definite matrix function as the product of an outer matrix function and its conjugate) and, on the other hand, the notion of factorization in L_2 with coinciding partial indices. Moreover, it is shown there that, for locally sectorial matrix functions, a canonical factorization is determined up to constant multipliers. In connection with this it should be mentioned that till now a definition of factorization on the circle which comprehends in a certain sense both factorization in L_p and representations of the form (7.15) and is unique in the sense that it could be reconstructed by the factor Λ with the same degree of arbitrariness of a factorization in L_p has not been found yet. From our point of view, the problem of finding such a definition is very important for the needs of a logically well-composed structure of factorization theory.

Theorem 7.11 in the scalar case was proved by DOUGLAS and WIDOM [1]. In this case the conditions of Theorem 7.11 may be formulated in a more elegant fashion. Namely, the function F satisfies the conditions of Theorem 7.11 (for $n = 1$) if and only if the convex hull of the set of its essential limits at every point $t \in \mathbb{T}$ does not contain zero. With this in mind, in the paper DOUGLAS, WIDOM [1] there was posed the following problem: Is the condition of non-degeneracy at every point $t \in \mathbb{T}$ of all matrices from the convex hull of the essential limits of a matrix function F $(\in L_\infty)$ given on $\mathbb{T}$ sufficient for the Φ-factorability of F in L_2 with zero partial indices? This question was positively answered by CLANCEY [1] under the additional supposition that, at every point $t \in \mathbb{T}$, the matrix function F has no more than two limits. In the general case the answer to this question is negative: In the article AZOFF, CLANCEY [1] one can find examples of matrix functions F satisfying the Douglas-Widom demands, but whose partial 2-indices are not equal to zero. Moreover, there are cited examples of matrix functions F satisfying these demands, but in general being not factorable in L_2 at all.

Lemma 7.3 for the scalar case was proved in ADAMYAN, AROV, KREĬN [1], DUKHOVNYĬ [1] and for the matrix case (and the case of an operator function, which we do not touch upon) in ADAMYAN, AROV, KREĬN [2], NEHARI [1]. The main ideas of the proof described here are borrowed from ADAMYAN, AROV, KREĬN [1], POUSSON [1]. In the present book we do not mention the following sophisticated question naturally arising in connection with Lemma 7.3: Under which conditions is the matrix function X_0 (Y_0, respectively) on which the minimum in (7.20) is attained unitary and to which degree are the properties in which the matrix function G is peculiar acquired by the matrix function $G - X_0$ ($G - Y_0$) ? It appears that the last question is closely related to the properties of the factorization factors of G ; concerning this subject, see the paper ADAMYAN, AROV, KREĬN [1] quoted above. In this connection we still refer to another paper (SILBERMANN [1]), where it is shown that the best approximation of a function G $(\in L_\infty)$ analytic at all but one points $t \in \mathbb{T}$ can be everywhere discontinuous. In this way, the local principle does not work here.

Theorem 7.12 in the scalar case was proved by LEE and SARASON [1]

Theorems 7.13 and 7.14 in the scalar case were established by WIDOM [1] and DEVINATZ [1], and in the matrix case but for zero partial indices by POUSSON [1] and RABINDRANATHAN [1] (in other words, in the latter two papers Corollaries 7.10 and 7.11 were shown). In a general form Theorems 7.13 - 7.15 were presented by one of the authors (SPITKOVSKIĬ [2,10]).

The device of transition to a unitary matrix function described in Lemma 7.4 was suggested by RABINDRANATHAN [1]. In the paper SPITKOVSKIĬ [6] a modification of the Rabindranathan method was proposed which consists in the multiplication of the studied matrix function G by A^{-1} on the left and by $(B^*)^{-1}$ on the right, where A and B are outer matrix functions from the representations $AA^* = (GG^*)^{1/2}$ and $BB^* = (G^*G)^{1/2}$. In particular, on such a multiplication Hermitian matrix functions change into matrix functions which are Hermitian and unitary simultaneously (applying a notion usual in matrix theory, such matrix functions are called $\mathfrak{J}$ -matrix functions). There are two situations in which the factorization problem for $\mathfrak{J}$ -matrix functions can be solved trivially. This is the scalar case, where we simply deal with a function having the values ± 1, and the definite case, in which an $\mathfrak{J}$-matrix function may be reduced to a constant one which equals I or $-I$. On the strength of this simple consideration in SPITKOVSKIĬ [6] a generalization of Theorem 7.6 to the case $p \neq 2$ is proved. Moreover, it is clarified that the requirement for the total index to be equal to zero can be omitted in Corollary 7.9. Incidentally, Theorem 7.6 and Corollary 7.9 themselves, as far as the authors know, have not been published previously.

We intend to note that for an arbitrary $\mathfrak{J}$ -matrix function, certainly, there exists a criterion for the partial indices to be equal to zero formulated in terms of the behaviour (dependence on t) of the subspace of its invariant vectors, since the matrix function itself may be reconstructed uniquely by this subspace. One sufficient condition of this type is presented in the article SPITKOVSKIĬ [6]. Unfortunately, a criterion itself has not been obtained yet.

In connection with the theory of canonical differential equations the factorization of unitary matrix functions defined on the real line has been studied by KREǏN and MELIK-ADAMYAN [1,2]. Among other results they proved, in principle, that, for a given matrix function $X \in H_\infty^+$, one can find a matrix function $Y \in H_\infty^-$ such that $X + Y$ is unitary almost everywhere on $\mathbb{T}$ and has zero partial indices if and only if $\|PXQ\| < 1$. A procedure of finding the matrix function Y, which is uniquely defined by X, is also described there.

Concerning other work dedicated to the specific character of factorization of matrix functions having one or another additional algebraic property, we mention the article of FEINSTEIN, SHAMASH [1]. In this paper an algorithm for constructing a factorization of the form $G(s) = X(s)X'(-s)$ for a rational matrix function G considered on the real line with $G(s) = G'(-s)$, $G(is) \geq 0$ is presented.
Theorem 7.16 was proved in SPITKOVSKIǏ [10], cf. also SPITKOVSKIǏ [2,3]; Theorem 7.17 and its Corollaries 7.12 - 7.15 are borrowed from SPITKOVSKIǏ [1], wheras the example inserted after Corollary 7.15 is taken from SHMUL'YAN [1].

In terms of Toeplitz operators (i.e. operators of the form $T_p(G)$) the theory of the factorization problem in L_2 on the circle for the scalar case is explained in detail in Ch. 7 of the book DOUGLAS [2] [1]. In particular, in the book just mentioned one can find an interesting result on the connectivity of the spectrum of the operator $T_p(G)$ for arbitrary $G \in L_\infty$ obtained by WIDOM [2] as well as an analogue of this result for the case of an essential spectrum. When diagonal matrix functions were first dealt with, we already noticed that the property of the spectrum and the essential spectrum just mentioned disappears in passing over to the matrix case. In the examples known to the authors the number of connected components of the (essential) spectrum does not exceed the order of the matrix. It is of some interest to find out whether this law remains true for all matrix functions from L_∞.

---

[1] See also the monographs HALMOS, SUNDER [1] and BÖTTCHER, SILBERMANN [1].

CHAPTER 8. CONDITIONS OF Φ-FACTORABILITY IN THE SPACE L_p. CRITERION OF Φ-FACTORABILITY IN L_2 OF BOUNDED MEASURABLE MATRIX FUNCTIONS

In this comparatively small chapter we shall be concerned with classes of matrix functions from L_∞ defined on a contour of class $\mathcal{R}$ and Φ-factorable in the spaces $L_p(\Gamma)$, where we pursue the aim to select such classes which are as wide as possible. For the construction of classes of matrix functions from the classes studies previously in Chapters 5, 6 and 7 which admit a Φ-factorization we shall use the ε-local principle (Theorem 3.22) and the properties of a matrix function $F \in L_\infty$ to preserve Φ-factorability on multiplication by a matrix function from the classes $L_\infty^{\pm} + C$ (Theorem 5.5). As a matter of fact, this general approach forced us from the very start to give up any attempts to obtain, simultaneously with conditions of Φ-factorability, any conclusions about the values of partial indices. The conditions of Φ-factorability in L_p derived below are sufficient. However, for $p = 2$ in case of a Lyapunov contour Γ, we are also able to substantiate their necessity. A criterion of Φ-factorability of a matrix function G from L_∞ in the spaces L_p with $p \neq 2$ fails to be known hitherto.

Now we intend to describe the contents of the present chapter in more detail. Section 8.1 is of auxiliary character. On the basis of the ε-local principle the operation of transition from a certain class A of matrix functions from L_∞ to a wider class $\mathfrak{X}$ preserving the property of Φ-factorability is introduced in it. In addition, in 8.1 properties and estimates for the radius $\delta'_{p,n}(\Gamma)$ of the largest neighbourhood of a unitary matrix function entirely consisting of matrix functions which are Φ-factorable in L_p will be established.

The preliminary results are used in Section 8.2 in order to formulate a sufficient condition of Φ-factorability in L_p.

This condition consists in the membership of the matrix function G to the class $B_{p,n}(\Gamma)$ which is constructed by some class of matrix functions F from L_∞ with the help of the method described in 8.1 and is determined by one of the conditions ensuring Φ-factorability of F in L_p. In case of a Lyapunov contour Γ the conditions determining the class $B_{p,n}(\Gamma)$ can be weakened. If $p = 2$ and Γ is a Lyapunov contour, then the condition $G \in B_{2,n}(\Gamma)$ proves to be an exact condition of Φ-factorability. This result is explained in Section 8.3. Comments on current literature are given in 8.4.

8.1. AUXILIARY RESULTS

DEFINITION 8.1. Let A be a certain class of matrix functions from L_∞. The class of all matrix functions G satisfying the following requirements will be denoted by $\tilde{A}$: for every point $t \in \Gamma$, there exist matrix functions $F_t \in A$ and $G_{\pm,t} \in L_\infty$ such that

1) $G_{+,t}^{\pm 1} \in L_\infty^+ + C$, $G_{-,t}^{\pm 1} \in L_\infty^- + C$,

2) the matrix functions G and $G_{+,t}F_tG_{-,t}$ are ε-locally equivalent at the point t.

It is clear by the very definition that the class $\tilde{A}$ contains A. The operation "$\sim$" of transition from A to $\tilde{A}$ is idempotent: choosing the class $B = \tilde{A}$ as the initial one and constructing by it the class $\tilde{B}$, we see that it does not contain any new matrix functions (in comparison with $\tilde{A}$), i.e. $\tilde{\tilde{A}} = \tilde{A}$.

The operation "$\sim$" is of interest for us, since it does not violate the Φ-factorability of a matrix function. To be more exact, the following result holds.

THEOREM 8.1. Assume A to be an arbitrary subclass of the class L_∞ and U ($\subseteq (1,\infty)$) a set lying in a certain component of the Φ-factorability domain of every matrix function from A. Then U is a subset of some component of the Φ-factorability domain of all matrix functions from class $\tilde{A}$.

In case of a one-point set U ($U = \{p\}$) Theorem 8.1 simply means that Φ-factorability in L_p of all matrix functions of the class A implies Φ-factorability in L_p of the matrix functions of the class $\tilde{A}$.

Proof. Starting from the matrix function $G \in \tilde{A}$ and the point $t \in \Gamma$, we reconstruct the matrix functions F_t and $G_{\pm,t}$ involved in Definition 8.1.

Let p be any point of the set U. Then the matrix function F_t $(\in A)$ is Φ-factorable in L_p and, according to Theorem 5. , the matrix function $G_t = G_{+,t}F_tG_{-,t}$ is also Φ-factorable in L_p. Since, for all $t \in \Gamma$, the matrix functions G and G_t are ε-locally equivalent at the point t, Theorem 3.22 enables us to make a conclusion about Φ-factorability of the matrix function G in L_p.

In this way, we have verified that the set U lies in the domain of Φ-factorability of every matrix function G from the class $\tilde{A}$. Now we are still obliged to prove that this set belongs entirely to one component of the Φ-factorability domain, i.e., for any $p_1, p_2 \in U$ $(p_1 < p_2)$, the total p_1- and p_2-indices of G coincide. By assumption, such a coincidence holds for all matrix functions of the class A and, especially, for the matrix function F_t. Owing to the relation $G_{\pm,t} \in L^{\pm}_{\infty} + C$, the total index of the matrix function $G_{\pm,t}$ does not depend on p and, according to Theorem 5.5, the total p_1- (p_2-) index of G_t is equal to the sum of the total p_1- (p_2-) index of F_t and the total indices of $G_{\pm,t}$. Thus, the total p_1-index of the matrix function G_t coincides with its total p_2-index. But then, as follows from Corollary 2.2, the matrix function G_t admits one and the same factorization in L_{p_1} and L_{p_2}, where in view of Theorem 3.8 this factorization is a Φ-factorization (in L_{p_1} and L_{p_2}). From Theorem 3.16 we deduce that in this case it is possible to indicate a regularizer of the operator $P + G_tQ$ in the space $L^n_{p_1}$ such that its restriction on $L^n_{p_2}$ is a regularizer of $P + G_tQ$ in this space. Using the method of constructing a regularizer of a local type operator (in this case, the operator $P + GQ$) by its local regularizers (in this case, by the regularizers of the operators $P + G_tQ$, $t \in \Gamma$) described in Theorem 1.14, we find that for the operator $P + GQ$ there exists also a regularizer R in the space $L^n_{p_1}$ the restriction of which on the space $L^n_{p_2}$ is a regularizer for $P + GQ$ in this space.

Evidently, the index of an arbitrary bounded operator does not increase with the growth of p. Since the index of the product $(P + GQ)R$ is equal to zero in either of the spaces $L^n_{p_1}$ and $L^n_{p_2}$, from this and Theorem 1.6 we deduce that the index of the operator $P + GQ$ in $L^n_{p_1}$ and $L^n_{p_2}$ is one and the same.

In other words, the total p_1-index of the matrix function G coincides with its total p_2-index, which proves Theorem 8.1. ≡

For obtaining new concrete criteria of Φ-factorability with the aid of Theorem 8.1, it is necessary to choose as A one of the classes of Φ-factorable matrix functions which are already known.

In particular, we may assume the class A to agree with the class of matrix functions satisfying the conditions of one of the Theorems 6.6 - 6.11'. For the moment, however, only the Φ-factorability of matrix functions from the class A is essential for us, but not any other information about their partial indices, because the latter gets lost in any case in passing to the class $\mathfrak{X}$. Therefore, in describing the class A, without any loss we may replace the values $\delta_{p,n}(\Gamma)$ involved in Theorems 6.6 - 6.11 by, generally speaking, larger values $\delta'_{p,n}(\Gamma)$. The legality of such a substitution will be founded in Section 8.2. Here we state the definition of these values and discuss some of their properties.

<u>DEFINITION 8.2.</u> By $\delta'_{p,n}(\Gamma)$ we denote the radius of the largest spherical neighbourhood of the $n\times n$ matrix function I consisting entirely of matrix functions which are Φ-factorable in L_p.

Recall that we denote by Δ_p the interval of the real line connecting the point p with the point $q\ (= p/(p-1))$.

<u>LEMMA 8.1.</u> 1) For any contour Γ and all $p \in (1,\infty)$, the following relations hold:

$$\delta_{p,n}(\Gamma) \le \delta'_{p,n}(\Gamma)\ , \tag{8.1}$$

$$\delta'_{p,n}(\Gamma) \le \delta'_{p,1}(\Gamma) \le \delta_{p,1}(\Gamma)\ , \tag{8.2}$$

$$\delta'_{q,n}(\Gamma) = \delta'_{p,n}(\Gamma)\ , \tag{8.3}$$

$$\delta'_{r,n}(\Gamma) \geq \delta'_{p,n}(\Gamma) \quad \text{if} \quad r \in A_p \, ; \tag{8.4}$$

2) if the contour Γ is composed of closed disjoint curves Γ_j, then

$$\delta'_{p,n}(\Gamma) = \min_j \delta'_{p,n}(\Gamma_j) \, ;$$

3) for all simple closed smooth contours Γ, the quantity $\delta'_{p,n}(\Gamma)$ takes one and the same value.

Proof. 1) By definition, the $\delta_{p,n}(\Gamma)$-neighbourhood of the matrix function I consists only of matrix functions which are Φ-factorable in L_p with zero partial indices. Hence, this neighbourhood lies in the $\delta'_{p,n}(\Gamma)$-neighbourhood of I, which implies the relation (8.1). The inequality $\delta'_{p,n}(\Gamma) \leq \delta'_{p,1}(\Gamma)$ results from Corollary 4.2, according to which the diagonal $n\times n$ matrix function fI fails to be Φ-factorable in L_p as soon as the scalar function f is not Φ-factorable in L_p.

Now, let us note that the total p-index of every matrix function from the $\delta'_{p,n}(\Gamma)$-neighbourhood of I is equal to zero. Indeed, owing to its stability and integralness, the total p-index is constant on connected open subsets of the set of matrix functions being Φ-factorable in L_p and, in particular, on the $\delta'_{p,n}(\Gamma)$-neighbourhood of I. In case $n = 1$, this fact simply means that the indices of all functions from the $\delta'_{p,1}(\Gamma)$-neighbourhood of the identity matrix are equal to zero, i.e., this neighbourhood belongs to its $\delta_{p,1}(\Gamma)$-neighbourhood. Thus $\delta'_{p,1}(\Gamma) \leq \delta_{p,1}(\Gamma)$, which together with inequality (8.1) completes the proof of relations (8.2) for $n = 1$.

Proceeding to the verification of equation (8.3), we consider the mapping $Y = (1-\varepsilon^2)(X')^{-1}$. In the proof of Lemma 6.2 it was already mentioned that, for every $\varepsilon \in (0,1)$, the ε-neighbourhood of the identity matrix function I remains invariant under this mapping. Furthermore, the Φ-factorability of the matrix function X in L_p is equivalent to the Φ-factorability of the matrix function Y in L_q. Thus we conclude that, for every $\varepsilon < 1$, the inequalities $\varepsilon < \delta'_{p,n}(\Gamma)$ and $\varepsilon < \delta'_{q,n}(\Gamma)$ are equivalent and, therefore, $\min\{1,\delta'_{p,n}(\Gamma)\} = \min\{1,\delta'_{q,n}(\Gamma)\}$.

But relations (8.2) and the inequality $\delta_{p,1}(\Gamma) \leq 1$ imply that the values $\delta'_{p,n}(\Gamma)$ and $\delta'_{q,n}(\Gamma)$ do not exceed one. Hence, relation (8.3) is true.

Finally, consider an arbitrary matrix function G from the $\delta'_{p,n}(\Gamma)$-neighbourhood of I. Due to equation (8.3), this matrix function is Φ-factorable not only in L_p but also in L_q. Besides, as it was shown above, its total p- (q-) index is equal to zero. In this way, the interpolation Theorem 3.9 can be applied, according to which the matrix function G is Φ-factorable in all spaces L_r, $r \in A_p$.

In summary, all matrix functions from the $\delta'_{p,n}(\Gamma)$-neighbourhood of I are Φ-factorable in L_r, $r \in A_p$. Hence $\delta'_{r,n}(\Gamma) \geq \delta'_{p,n}(\Gamma)$, which completes the proof of assertion 1).

2) Let G be a certain matrix function given on Γ, and assume (as in Theorem 3.21) G_j to be the restriction of this matrix function on Γ_j. The inequality $\|I-G\| < \varepsilon$ is fulfilled iff $\|I-G_j\| < \varepsilon$ for any j. Therefore, if $\varepsilon < \min_j \delta_{p,n}(\Gamma_j)$, then every matrix function G_j is left and right Φ-factorable on Γ_j in L_p (here we used equation (8.3) proved above and Theorem 3.7'). With regard to assertion 2) of Theorem 3.21, from this the Φ-factorability of the matrix function G in L_p results.

In this way, the condition $\|I-G\| < \varepsilon$ guarantees the Φ-factorability of G in L_p, i.e. $\varepsilon \leq \delta'_{p,n}(\Gamma)$. Thus we have proved that $\min_j \delta'_{p,n}(\Gamma_j) \leq \delta'_{p,n}(\Gamma)$. Now we choose a contour Γ_{j_0} on which $\min_j \delta'_{p,n}(\Gamma_j)$ is attained. Moreover, on this contour we take a matrix function G_{j_0} which is not Φ-factorable in L_p (on the left if $j_0=0$, and on the right if $j_0 > 0$) but lies at a distance of $\delta'_{p,n}(\Gamma_{j_0})$ from I.
Finally, set $G_j = I$ for $j \neq j_0$. Then we obtain a matrix function G which is not Φ-factorable in L_p (by Theorem 3.21) and for which $\|I-G\| = \delta'_{p,n}(\Gamma_{j_0})$. Therefore $\delta'_{p,n}(\Gamma) \leq \delta'_{p,n}(\Gamma_{j_0})$, which completes the proof of proposition 2).

Proposition 3) may be deduced from proposition 1) of Theorem 3.21 in an analogous manner. ≡

Owing to inequality (8.1), the lower bounds for $\delta_{p,n}(\Gamma)$ indicated in proposition 1) of Lemma 6.2 are simultaneously lower bounds for $\delta'_{p,n}(\Gamma)$. However, it turns out that more precise estimates are available, in which the norms of the operators S and Q are replaced by their factor norms $|S|_{L_p^n}$ and $|Q|_{L_p^n}$ [1].

In other words, the following statement is true.

<u>LEMMA 8.2.</u> Either of the values $|Q|^{-1}_{L_p^n}$ and $\dfrac{2|S|_{L_p^n}}{1+|S|^2_{L_p^n}}$ is a lower bound for $\delta'_{p,n}(\Gamma)$.

<u>Proof.</u> We intend to show, for instance, that $|Q|^{-1}_{L_p^n} \le \delta'_{p,n}(\Gamma)$. Suppose that the condition $\|I-G\| < |Q|^{-1}_{L_p^n}$ is satisfied. By definition of the factor norm, there exists an operator T compact in the space L_p^n such that $\|I-G\| < \|Q+T\|^{-1}_{L_p^n}$.

Thanks to the contracting mapping principle, in this case the operator $V = I-(I-G)(Q+T) = P+GQ-(I-G)T$ is invertible. The operator $P + GQ$ differing from V by the compact summand $(I-G)T$ will be Fredholm in view of Theorem 1.5. In this way, the Φ-factorability in L_p of all matrix functions from the $|Q|^{-1}_{L_p^n}$-neighbourhood of I and, together with it, the desired estimate have been proved.

In connection with the second estimate of Lemma 8.2 we would like to note that the value $|S|_{L_p^n}$ does not change under smooth mappings β of the contour Γ. Indeed, for such mappings $\beta : L \to \Gamma$, in the proof of Theorem 3.21 the existence of an isometric invertible operator $W : L_p^n(\Gamma) \to L_p^n(L)$ was established for which the operator WSW^{-1} differs from the operator S_0 of singular integration along L only by

---

[1] We remind the reader that the definition of the factor norm was given in Section 1.1.

a compact summand and, possibly, by sign. But then $|WSW^{-1}|_{L_p^n(L)} = |S_o|_{L_p^n(L)}$. At the same time, due to the fact that the operator W is isometric, we have $|WSW^{-1}|_{L_p^n(L)} = |S|_{L_p^n(\Gamma)}$, which proves the invariance of the factor norm of the operator S for sufficiently smooth mappings β.

For smooth contours Γ one may obtain a result of final nature concerning the values $\delta'_{p,n}(\Gamma)$ and $|S|_{L_p^n}$.

<u>LEMMA 8.3.</u> Let the contour Γ be smooth. Then

$$\delta'_{p,n}(\Gamma) = \sin(\pi/\max(p,q)) \tag{8.5}$$

and

$$|S|_{L_p^n} = \mathrm{ctg}(\pi/2\max(p,q)) .$$

<u>Proof.</u> Every simply-connected smooth contour Γ is the image of the circle $\mathbb{T}$ under a certain diffeomorphism $\beta : \mathbb{T} \to \Gamma$.

According to proposition 3) of Lemma 8.1, for an arbitrary contour of this type, the equation

$$\delta'_{p,n}(\Gamma) = \delta'_{p,n}(\mathbb{T}) \tag{8.6}$$

holds.

But then, owing to assertion 2) of Lemma 8.1, the equation (8.6) is also correct for a compound contour Γ the components of which are smooth curves.

Furthermore, as it was clarified in Chapter 7, equations (7.19) are valid for any natural n. Thus, owing to relations (8.1) and (8.2), the relation $\delta'_{p,n}(\Gamma_o) = \sin(\pi/\max(p,q))$ follows. Comparing this result with equation (8.6), we obtain the first of the relations to be proved.[1)]

From the estimate $\frac{2|S|}{1+|S|^2} \leq \delta'_{p,n}(\Gamma)$ and equation (8.5) we deduce that $|S|_{L_p^n} \geq \mathrm{ctg}(\pi/2\max(p,q))$. Since the value $|S|_{L_p^n}$ as also $\delta'_{p,n}(\Gamma)$

---

1) Thus, for the case of a circle, the inequalities in relations (8.1) and (8.2) become equalities.

is one and the same for all smooth contours, it suffices to verify the opposite inequality for $\Gamma = \mathbb{T}$. But in this case the inequality $|S|_{L_p^n} \leq \operatorname{ctg}(\pi/2 \max(p,q))$ results from relation (7.18). Thus Lemma 8.3 is completely proved. ≡

Remark that, for any contour Γ of class $\tilde{\mathfrak{R}}$, proposition 1) of Lemma 6.2 along with inequality (8.2) enable us to claim that $\delta'_{p,n}(\Gamma) \leq \sin(\pi/\max(p,q))$. In this way, for smooth contours, the quantities $\delta'_{p,n}(\Gamma)$ attain the largest possible value for contours of class $\mathfrak{R}$. Clearly, at the same time the value $|S|_{L_p^n}$ takes the smallest of all possible values.

8.2. SUFFICIENT CONDITIONS OF Φ-FACTORABILITY

DEFINITION 8.3. To the class $B_{p,n}(\Gamma)$ we assigne those n×n matrix functions G given on Γ for which, for every point $t \in \Gamma$, one can specify an arc γ_t containing it as well as matrix functions $G_{\pm,t} \in L_\infty^{\pm} + C$ such that $G_{\pm,t}^{-} \in L_\infty^{\pm} + C$ and, for $F_t = G_{+,t} G G_{-,t}$, at least one of the following relations is fulfilled:

1) $\operatorname*{ess\,sup}_{\tau\in\gamma_t} \|I - F_t(\tau)\| < \delta'_{p,n}(\Gamma)$,

2) $\operatorname*{ess\,inf}_{\tau\in\gamma_t} d(F_t(\tau)) > \sqrt{1-(\delta'_{p,n}(\Gamma))^2}\ \operatorname*{ess\,sup}_{\tau\in\gamma_t} \|F_t(\tau)\|$,

3) the set $\dfrac{1}{\operatorname{ess\,sup}\|F_t(\tau)\|} \bigcup_{\tau\in\gamma_t} H(F_t(\tau))$ is separated from zero by a straight line whose distance to zero is greater than $\sqrt{1-\delta'_{p,n}(\Gamma)^2}$,

4) there exists a disk Δ_t which is visible from the origin at an angle $\alpha_t < 2 \arcsin \delta'_{p,n}(\Gamma)$ and such that almost everywhere on γ_t the matrix $F_t(\tau)$ is normal and their eigenvalues are located in Δ_t.

It turns out that all matrix functions from the class $B_{p,n}(\Gamma)$ are Φ-factorable in L_p. Moreover, the following result is true.

THEOREM 8.2. The interval A_p is completely contained in a certain component of the domain of Φ-factorability of any matrix function from the class $B_{p,n}(\Gamma)$.

Proof. Consider the class A of those matrix functions F_t (= F) for which at least one of the requirements 1) - 4) of Definition 8.3 is fulfilled, and that on the curve γ_t coinciding with the whole of the contour Γ. Evidently, the class $\mathfrak{A}$ constructed by A in accordance with Definition 8.1 coincides with the class $B_{p,n}(\Gamma)$. With regard to Theorem 8.1 it is possible to conclude that Theorem 8.2 will be proved as soon as we establish the correctness of its proposition with respect to matrix functions from the class A.

However, the Φ-factorability with one and the same total index in all L_r $(r \in A_p)$ of an arbitrary matrix function F satisfying condition 1) of Definition 8.3 for $\gamma_t = \Gamma$ results directly from Definition 8.2 of the values $\delta'_{p,n}(\Gamma)$ and inequality (8.4). The case when F satisfies one of the conditions 2) - 4) of Definition 8.3 can be reduced to the previous one by methods used in the arguments of Lemma 6.4, Theorems 6.11 and 6.10, respectively. Thus Theorem 8.2 is proved. ≡

Still preserving the claim of Theorem 8.2, in case of a smooth contour Γ one can weaken the restrictions on the class $B_{p,n}(\Gamma)$ to a certain degree. Namely, conditions 1) - 4) from Definition 8.3 related to a Lyapunov contour Γ will be replaced by the following:

1') $\operatorname{ess\,sup}_{\tau\in\gamma_t} \|I-F_t(\tau)\| < \sin(\pi/\max(p,q))$,

2') $d(F_t(\tau)) \geq \|F_t(\tau)\|\cos(\pi/\max(p,q))+\varepsilon_t$, $\varepsilon_t > 0$,

3') the set $\bigcup_{\tau\in\gamma_t} \|F_t(\tau)\|^{-1}H(F_t(\tau))$ is separated from zero by some straight line whose distance to zero is greater than $\cos(\pi/\max(p,q))$

4') there exist a sector S_t with vertex at the origin of apex angle $\alpha_t < 2\pi/\max(p,q)$ and a number $\varepsilon_t > 0$ such that $H(F_t(\tau)) \subset S_t$ and $h(F_t(\tau)) \geq \varepsilon_t$ a.e. on γ_t.

In Chapters 6 and 7 it was already mentioned that the conditions of the type 1) - 4) and 1') - 4') are not independent. For instance, 2) can be reduced to 1), and 3) to 2) by multiplying them by a suitable constant. At the same time, 2) is a special case of 3). Conditions 1') - 3') are interrelated in a similar manner. Furthermore, condition 4') is a consequence of any of the conditions 1') - 3'). Making use of this remark, one could relieve the definition of the class $B_{p,n}(\Gamma)$ and in case of a smooth contour Γ leave over only condition 4') (from all the conditions 1') - 4')). However, we do not intend to do this, because in concrete problems the check of one of the conditions 1) - 4) and 1') - 4') may prove to be easier than that of another condition.

THEOREM 8.3. Let the contour Γ be smooth. Then for all matrix functions of the class $B_{p,n}(\Gamma)$ (in the definition of which the conditions 1) - 4) are replaced by conditions 1') - 4')), the interval A_p lies entirely inside a certain component of the Φ-factorability domain.

Proof. As also in Theorem 8.2, it suffices to pay attention to matrix functions of the class A, to which we assign those matrix functions from L_∞ which possess one of the properties 1') - 4'), however, with the change of γ_t by Γ. From what was mentioned above it is clear that the class A consists exactly of those matrix functions F $(=F_t)$ which satisfy condition 4') for $\gamma_t = \Gamma$.

In case when the contour Γ is a circle, the Φ-factorability of such matrix functions F in L_r $(r \in A_p)$ with zero partial (and, thus, also total) indices results from Theorem 7.10. Since condition 4') is exclusively formulated in terms of values of G, it is invariant with respect to mappings $\beta : L \to \Gamma$ of contours for which the matrix function G given on Γ is replaced by the matrix function $G_0 = G \circ \beta$ defined on L. Therefore, using statement 1) of Theorem 3.21, one can assert that matrix functions F given on an arbitrary simple smooth contour Γ are Φ-factorable in L_r $(r \in A_p)$ with zero total index as soon as they satisfy condition 4') for $\gamma_t = \Gamma$. The passage from a simple smooth contour to a composite one can be accomplished with the aid of statement 2) of the same Theorem 3.21. Thus Theorem 8.3 is completely proved. $\equiv$

8.3. CRITERION OF Φ-FACTORABILITY IN L_2

With the aid of Theorems 8.3 and 7.14 the problem of Φ-factorability in L_2 of matrix functions from the class L_∞ given on a smooth contour Γ may be solved to the very end. Before formulating the corresponding result, we note that, for $p = 2$, $\sin(\pi/\max(p,q)) = 1$, $\cos(\pi/\max(p,q)) = 0$ and, therefore, condition 2) from the definition of the class $B_{2,n}(\Gamma)$ is equivalent to the requirement of uniformly positiveness of the real part of the matrix function F on γ_t, and conditions 1'), 3') and 4') change over into 2'), if F_t is multiplied by a certain constant.

<u>THEOREM 8.4.</u> For a matrix function $G \in L_\infty$ defined on the smooth contour Γ, the following statements are equivalent:

1) $2 \in \Phi(G)$,
2) $G = G_{(o)}B$, where $B^{\pm 1} \in M_\infty^-$ and $\|I-G_{(o)}\| < 1$,
3) $G = G_{(o)}B$, where $B^{\pm 1} \in M_\infty^-$ and the real part of the matrix function $G_{(o)}$ is uniformly positive on Γ,
4) $G \in B_{2,n}(\Gamma)$.

<u>Proof.</u> The equivalence of assertions 2) and 3) was already mentioned above. Evidently, 3) ==> 4). The implication 4) ==> 1 is a special case (for $p = 2$) of Theorem 8.3. It remains to prove that 1) implies 2).

Let the matrix function G be Φ-factorable in L_2 and $\varkappa_n$ be the smallest of its partial 2-indices. Then the matrix function $G_1(t) = t^{-\varkappa_n}G(t)$ is Φ-factorable in L_2 and all its partial 2-indices are non-negative. If the contour Γ is the unit circle, then Theorem 7.14 can be applied, according to which $G_1 = A_1G_{(o)}^{(1)} = G_{(o)}^{(2)}B$, where $A_1 \in H_\infty^+$, $A_1^{-1} \in M_\infty^+$, $B_1 \in M_\infty^-$, $B_1^{-1} \in H_\infty^-$ and $G_{(o)}^{(j)}$ $(j = 1,2)$ are matrix functions with uniformly positive real part. Setting $A(z) = z^{\varkappa_n}A_1(z)$, $B(z) = z^{\varkappa_n}B_1(z)$, we find

$$G = A\,G_{(o)}^{(1)} = G_{(o)}^{(2)}\,B, \tag{8.7}$$

where $A^{\pm 1} \in M_\infty^+$, $B^{\pm 1} \in M_\infty^-$. Thus, if $\Gamma = \mathbb{T}$, then the proposition to be proved is correct.

Now, let Γ be a simple closed smooth contour, and assume ω to be a conformal mapping of the unit disk onto the domain $\mathcal{D}^+$ bounded by Γ. According to proposition 1) of Theorem 3.21, the matrix function G given on Γ is Φ-factorable in L_2 only simultaneously with the matrix function $G \circ \omega$ given on $\mathbb{T}$. By what was proved above, Φ-factorability of $G \circ \omega$ in L_2 means that it is possible to represent it in the form $G \circ \omega = AG_{(o)}$, where $A^{\pm 1} \in M_\infty^+(\mathbb{T})$ and $G_{(o)}$ is a matrix function with a uniformly positive real part. From the last equation we obtain that $G = (A \circ \omega^{-1})(G_{(o)} \circ \omega^{-1})$. Since $(A \circ \omega^{-1})^{\pm 1} \in M_\infty^+(\Gamma)$ and the ranges of the matrix functions $G_{(o)}$ and $G_{(o)} \circ \omega^{-1}$ agree, then the existence of the first of the representations (8.7) for G has been proved. The existence of the second representation can be established by means of a conformal mapping of the exterior of the unit disk onto the domain $\mathcal{D}^-$.

Now we proceed to the general case, when Γ is a collection of m connected contours $\Gamma(j)$, $j=0,\ldots,m-1$. Due to assertion 2) of Theorem 3.21 (and in its notations), the matrix function G is Φ-factorable in L_2 if and only if the matrix function G_o is Φ-factorable in L_2 from the left and G_j, $j=1,\ldots,m-1$ from the right. Consequently, Φ-factorability (from the left) of G in L_2 implies Φ-factorability in L_2 (from the left) of the matrix functions G_o and G'_j $(j=1,\ldots, m-1)$. Owing to what was proved above, the relations

$$G_o = F_oB_o, \quad G'_j = A'_jH'_j, \quad j=1,\ldots,m-1 \tag{8.8}$$

are valid, where the matrix functions F_j $(j=0,\ldots,m-1)$ are given on $\Gamma(j)$ and have uniformly positive real parts, $B_o^{\pm 1} \in M_\infty^-(\Gamma(0))$, $A_j^{\pm 1} \in M_\infty^+(\Gamma(j))$. Equations (8.8) mean that $G = FB$, where $F|_{\Gamma_j} = F_j$ $(j=0,\ldots,m-1)$, $B|_{\Gamma(0)} = B_o$, $B|_{\Gamma(j)} = A_j$ $(j=1,\ldots,m-1)$. Moreover, $B^{\pm 1} \in M_\infty^-(\Gamma)$ and F is a matrix function with a uniformly positive real part. Thus, the proof of Theorem 8.4 is complete. ≡

From Theorem 8.4 we obtain a necessary condition of Φ-factorability making the result of Corollary 3.5 more precise for $p = 2$.

COROLLARY 8.1. For the Φ-factorability in L_2 of the matrix function $G \in L_\infty$ given on the smooth contour Γ, it is necessary for G^{-1} to belong to the closed subalgebra $L_\infty^-[G]$ of the algebra L_∞ generated by the algebra $L_\infty^- + C$ and the matrix function G.

Proof. According to Theorem 8.4, from the Φ-factorability of the matrix function G in L_2 we deduce that $G = G_{(o)}B$, where $\|G_{(o)}-I\| < 1$ and $B^{\pm 1} \in M_\infty^-$. But then we have

$$G^{-1} = B^{-1}G_{(o)}^{-1} = B^{-1} \sum_{k=o}^{\infty} (I-G_{(o)})^k = \sum_{k=o}^{\infty} B^{-1}(I-GB^{-1})^k .$$

Every summand of the series obtained is a polynomial of G and B^{-1} and, therefore, lies in the algebra $L_\infty^-[G]$. Taking into account that this series converges in the metric of L_∞, we observe that the inclusion $G^{-1} \in L_\infty^-[G]$ holds, too.

If $G \in L_\infty^- + C$, then obviously $L_\infty^-[G] = L_\infty^- + C$, and the necessary condition of Φ-factorability stated in Corollary 8.1 coincides with the necessary and sufficient condition described in Theorem 5.2. As to the general case, the relation $G^{-1} \in L_\infty^-[G]$ does not yet guarantee Φ-factorability of G in L_2. An appropriate example is provided by every matrix function G not belonging to $L_\infty^- + C$ for which $G^{-1} \in L_\infty^- + C$.

8.4. COMMENTS

The properties of the values $\delta'_{p,n}(\Gamma)$ proved in Lemma 8.1 are analogous to the well-known property of the factor norm of the operators S and Q (see, e.g., Ch. 7 of the book GOHBERG, KRUPNIK [4]). In connection with Lemma 8.3 it should be noted that in the paper KRUPNIK, POLONSKIĬ [1], for $n = 1$, the coincidence of the norm of the operator S on the circle with its factor norm was established. Furthermore, their common value was found using a result of PICHORIDES [1]. The independence of this value of n is proved in KRUPNIK [4] (see also VERBITSKIĬ, KRUPNIK [1,2]). The coincidence of the factor norm of S (and also of P and Q) for all Lyapunov curves was mentioned e.g. in GOHBERG, KRUPNIK [4]. In the article of NYAGA [1] it is shown that, if one permits the existence of corner points on Γ, then the factor norms of the operators P, Q and S do depend on Γ. The problem of exactness of the estimates derived in Lemma 8.2 in case of a nonsmooth contour Γ remains unsolved. Also, the problem of precisely describing the dependence of the quantities $\delta'_{p,n}(\Gamma)$ on n has not been solved yet for such a contour.

Theorems 8.1 - 8.3 are certain extensions of corresponding results of SIMONENKO [4,6], DANILYUK [4,5] (see also [6]), KRUPNIK and NYAGA [1] and one of the authors (SPITKOVSKIǏ [2,10]). Theorem 8.4 for $n = 1$ can be found in SIMONENKO [6] and, for $n > 1$, in SPITKOVSKIǏ [2]. From Theorem 8.3, applied to the scalar case, the following proposition obtained by SIMONENKO [4,6] results. Let Γ be a smooth contour (which, for the sake of simplicity, is assumed to be connected). To the class $S_p(\Gamma)$ we assign those functions f given on Γ which can be represented in the form $f(t) = \varphi(t)t^k h(t)$, where $\varphi^{\pm 1} \in L_\infty^+$, k is an integer, $h^{\pm 1} \in L_\infty$, $|\arg h(t)| \leq \alpha$ almost everywhere on Γ for a certain $\alpha < 2\pi/\max(p,q)$ and $q = p/(p-1)$. Then all the functions of the class $S_p(\Gamma)$ are Φ-factorable in L_p and L_q with the same index value (equal to k).

KRUPNIK [2] showed that the set of functions which are Φ-factorable in L_p and L_q with equal value of the index is exhausted by the functions of class $S_p(\Gamma)$. This implies, in particular, that the class $\mathfrak{S}_p(\Gamma)$ constructed by $S_p(\Gamma)$ in accordance with Definition 8.1 agrees with $S_p(\Gamma)$ and $B_{p,1}(\Gamma)$.

The question whether the mentioned result can be transferred to the matrix case, i.e., whether the sufficient conditions of Φ-factorability in L_p and L_q with the same value of the total index indicated in Theorem 8.3 are necessary, is of some significance. For $p = 2$, Theorems 8.4 yields a positive answer to this question.

Clearly, the class $S_p(\Gamma)$ does not comprehend all functions which are Φ-factorable in L_p, because such functions are not necessarily Φ-factorable in L_q (all the more, they may or may not have coinciding p- and q-indices). Especially, not all piecewise continuous functions which are Φ-factorable in L_p belong to the class $S_p(\Gamma)$. The investigation of the problem of Φ-factorability of products of a piecewise continuous function multiplied by a function of the class $S_p(\Gamma)$ was initiared by DANILYUK [3]. The corresponding results from DANILYUK [3] and their extension obtained by DANILYUK, SHELEPOV [2] and SHELEPOV [1] are explained in detail in Ch. 5 of the monograph DANILYUK [6]. In FROLOV [2] (see also Section 8.6 in GOHBERG, KRUPNIK [4]) it is proved that, if the contour Γ is Lyapunov and the jumps of the argument of the non-degenerate piecewise continuous factor are included between $\min\{0, \frac{2\pi}{p} - \frac{2\pi}{q}\}$ and $\max\{0, \frac{2\pi}{p} - \frac{2\pi}{q}\}$, then the products of the indicated kind form the class $M_p(\Gamma)$ consisting only of functions which are Φ-factorable in L_p.

KOKILASHVILI and PAATASHVILI [2] have extended this result of Frolov to the case of contours from some wider class T (which is, however, a subclass of $\mathfrak{R}$). It is not hard to see that the class $M_p(\Gamma)$ contains, apart from the class $S_p(\Gamma)$, also all piecewise continuous functions being Φ-factorable in L_p.

The nonvoid proper subset of the class $M_p(\Gamma)$ consists of those products for which the points of discontinuity of the piecewise continuous factor are points of continuity of the factor from $S_p(\Gamma)$. The Φ-factorability in L_p of such functions was proved in KOKILASHVILI, PAATASHVILI [1] for contours of the class T and, for contours of an even wider subclass $\mathfrak{J}^*$ of the class $\mathfrak{R}$, in KOKILASHVILI, PAATASHVILI [3,4]. Let us explain that a contour Γ belongs to class $\mathfrak{J}^*$ if it may be partitioned into a finite number of arcs for each of which there exists a simple smooth curve γ such that

$$\operatorname*{ess\,sup}_{0\le\sigma\le 1} \int_0^1 \left| \frac{t'(s)}{t(s)-t(\sigma)} - \frac{\tau'(s)}{\tau(s)-\tau(\sigma)} \right| ds < \infty ,$$

where $\tau(s)$ and $t(s)$, $0 \le s \le 1$ are natural parametrizations of the curves γ and Γ, respectively, and if, in addition, (1.12) is satisfied.

The class T is distinguished from $\mathfrak{J}^*$ by the conditions

$\gamma = \{\zeta : |\zeta| = 1\}$ and $t([0,1]) = \Gamma$.

The result of Frolov has been recently extended to the matrix case by one of the authors. The corresponding result, which simultaneously generalizes Theorem 8.3 (case 4')), reads as follows: Let the matrix function G given on the smooth contour Γ have the property that, for every point $t \in \Gamma$, there exist a matrix A_t , a number $\theta_t \in \in [0, 2\pi(r-p)/p]$ and a sector S_t with vertex at zero and an apex angle less than $2\pi/\max(p,q)$ such that, for all essential limits F of the matrix function G at the point t from the left (from the right), the numerical domain of the matrices $A_t F$ lies in the sector $S_t(e^{i\theta_t} S_t)$. Then G is Φ-factorable in L_p .

In the scalar case the problem of Φ-factorability of functions with discontinuities of almost-periodic (GOHBERG, KRUPNIK [4], GOHBERG, SEMENTSUL [1], PILIDI [1], SEMENTSUL [1] and semi-almost periodic type (SAGINASHVILI [1-3], SARASON [3]) was studied, too. For $p = 2$, there are also investigations concerning multiplicatively periodic weakly oscillating functions (ABRAHAMSE [1]) and functions from certain algebras generated by them (POWER [1]). The problem of Φ-factorability in L_2 of functions from the algebra generated by piecewise continuous (PC) and quasi-continuous (QC) functions, i.e. functions belonging to $H_\infty + C$ along with their complex conjugates, was also examined (SARASON [4]). A corresponding result for $H_\infty + PC$ fails to be known. One of the difficulties arising in the study of the latter class consists in the fact that it, being a subspace, is not an algebra (BONSALL, GILLESPIE [1]). In the paper ROCHBERG [1] (see also SPITKOVSKIĬ [6,14]) a criterion of Φ-factorability of functions G with p-index $\varkappa$ which is formulated in terms of condition (3.21) was obtained for $\varrho = \exp(-p\tilde{w}/2)$, where $\tilde{w}$ is the harmonic conjugate to a branch w of the argument of the function $G(t)t^{-\varkappa}$ chosen in a suitable manner. Analogues of all these results for the case $n > 1$ fail to be known hitherto. Concerning this topic, up to now certain results were obtained only for matrix functions with discontinuities of semi-almost periodic type. In KARLOVICH,

SPITKOVSKIĬ [1] it is shown that, under certain conditions, the Φ-factorability problem for such matrix functions can be reduced to the problem of existence of a so-called P-factorization of an almost-periodic function F given on the real line, i.e., of a representation

$$F(t) = F_+(t)(\delta_{kj}e^{i\nu_j t})_{k,j=1}^{n}F_-(t) ,$$

where $F_+^{\pm 1}$ ($F_-^{\pm 1}$) are almost-periodic matrix functions with non-negative (non-positive) Fourier exponents and $\nu_j \in \mathbb{R}$.

Some classes of matrix functions admitting a P-factorization are selected in GRUDSKIĬ, KHEVELEV [1], KARLOVICH, SPITKOVSKIĬ [1], SPITKOVSKIĬ [15], KHLGATYAN [1]. However, the existence problem of a P-factorization is very complicated even for second-order triangular matrix functions with almost-periodic polynomials as elements.

It should be mentioned that P-factorization may be viewed as a factorization with, generally speaking, infinite partial indices. The study of such a type of factorization and of the corresponding Riemann problem is very complicated even in the scalar case (see the pioneer work of GOVOROV [1,2], but also ALEKHNO [1], GRUDSKIĬ [2], GRUDSKIĬ, DYBIN [1,2], DYBIN, DODOKHOVA [1,2]). On the other hand, in the matrix case research related to this subject is only beginning to evolve.

In connection with Corollary 8.1 we want to note that, if $G \notin L_\infty^-$, then the algebra $L_\infty^-[G]$ agrees with the smallest closed subalgebra of L_∞ containing L_∞^- and G. It is not hard to derive this result from Lemmas 5.2 and 5.3 and the fact proved in HOFFMAN, SINGER [1] that every closed subalgebra of L_∞ containing H_∞ and not coinciding with H_∞ contains C, too. For the scalar case on the circle Corollary 8.1 is, in fact, established in LEE [1] [1].

[1] LEE [1] has discussed the case $G \in H_\infty + C$, where the algebra $H_\infty[G]$ was defined as generated by H_∞ and G; it was claimed that the demand $G^{-1} \in H_\infty[G]$ is equivalent to the Φ-factorability of the function G. However, in the proof of the necessity the condition $G \in H_\infty + C$ was not used, but the sufficience is a simple consequence of Theorem 5.2 which does not need any additional arguments.

Finally, we intend to note that the class of smooth contours, apparently, does not seem to be the most general class for which Theorems 8.3, 8.4 and Corollary 8.1 remain valid. The problem of extending this class is of some significance.

CHAPTER 9. THE GENERALIZED RIEMANN BOUNDARY VALUE PROBLEM

The generalized Riemann boundary value problem (also known as the Markushevich problem or the general boundary value problem of linear conjugation) on the unit circle $\mathbf{T}$ consists in the determination of functions φ^+ $(\in H_p^+)$ and φ^- $(\in H_p^-)$ by the boundary condition on $\mathbf{T}$

$$\varphi^+(t) + a(t)\varphi^-(t) + b(t)\overline{\varphi^-(t)} = c(t) , \qquad (9.1)$$

where c $(\in L_p)$ and a,b $(\in L_\infty)$ are given functions.

In view of the presence of conjugation in the boundary condition (9.1), the generalized Riemann boundary value problem is linear only then, when L_p and $H_p^{\pm}$ are considered as spaces over the field of real numbers. Therefore, speaking about the number of linearly independent solutions of the homogeneous $(c(t) \equiv 0)$ problem (9.1) or about the number of solvability conditions of the inhomogeneous problem, we shall understand the linear independence as the linear independence over the field of real numbers.

The study of the boundary value problem (9.1) is based on its reduction to the Riemann boundary value problem (9.4) for a two-dimensional vector with a certain matrix coefficient G. In doing so, the images of problems (9.1) and (9.4) are closed only simultaneously; the codimensions of the images and the dimensions of the kernels of problem (9.1) and (9.4), respectively, coincide, which allows us to employ the results of the preceding chapters for the study of problem (9.1).

On the basis of the mentioned relationship between problems (9.1) and (9.4) in Section 9.1 a criterion of Fredholmness of the generalized Riemann boundary value problem in L_p will be established and formulae for its defect numbers will be derived. It turns out that a necessary and sufficient condition of Fredholmness of problem (9.1) consists in the membership of the function a^{-1} to the class L_∞ and the Φ-factorability of the matrix coefficient G of problem (9.4), where the defect numbers are expressed by the partial indices of the matrix function G. In 9.1 it will also be established that the perturbation of the coefficient b by a function of class $H_\infty + C$ does not influence on the Fredholmness of problem (9.1). Hence, for $b \in H_\infty + C$, the Fredholmness of problem (9.1) is completely determined by the function a. In Section 9.2 the boundary value problem (9.1) is considered in the space L_2 under the assumption that $a = a_1 a_0 \overline{a_2}$, where $a_1^{\pm 1}$, $a_2^{\pm 1} \in H_\infty$, $a_0 \in C$. A condition of Fredholmness of the boundary value problem (9.1) in L_2 will be stated and estimates for its defect numbers as well as a stability condition for problem (9.1) understood as stability of the partial indices of G will be established. Corresponding results for the so-called elliptic, parabolic and hyperbolic case of problem (9.1) will be derived as consequences.

Section 9.3 is dedicated to the boundary value problem (9.1) with continuous coefficients a(t) and b(t). In this case the results of 9.2 allow an essential strengthening and, at the same time, may be formulated much more simply and elegantly. In particular, if the matrix function G can be factored with continuous factors, then the sufficient condition of stability of problem (9.1) presented in Section 9.2 becomes also necessary.

Section 9.4 contains comments on published papers.

9.1. CRITERION OF FREDHOLMNESS OF THE GENERALIZED RIEMANN BOUNDARY VALUE PROBLEM IN THE SPACE L_p

For functions φ of class H_p^+ ($\overset{\circ}{H}{}_p^-$), we shall denote by $\hat{\varphi}$ the function which is defined by the equation $\hat{\varphi}(z) = z^{-1}\overline{\varphi(z)}$, where obviously $\hat{\varphi} \in \overset{\circ}{H}{}_p^-$ (H_p^+).

Besides of problem (9.1) we shall consider the system

$$\begin{cases} \varphi_1^+(t) + tb(t)\varphi_2^+(t) + a(t)\varphi_1^-(t) = g_1(t) , \\ \overline{a(t)}t\varphi_2^+(t) + \overline{b(t)}\varphi_1^-(t) + t\varphi_2^-(t) = g_2(t) , \end{cases} \tag{9.2}$$

where $g_1, g_2 \in L_p$ are known and $\varphi_j^+ \in H_p^+$, $\varphi_j^- \in \overset{\circ}{H}{}_p^-$ (j=1,2) are the desired functions. System (9.2) may be rewritten in the form

$$(AP + BQ)\tilde{\varphi} = g ,$$

where $g = \begin{pmatrix} g_1 \\ g_2 \end{pmatrix}$, $\tilde{\varphi} = \begin{pmatrix} \varphi_1 \\ \varphi_2 \end{pmatrix} \in L_p^2$ $(P\varphi_j = \varphi_j^+ , Q\varphi_j = \varphi_j^- , \quad j = 1,2)$,

$$A = \begin{pmatrix} 1 & tb \\ 0 & t\bar{a} \end{pmatrix} , \quad B = \begin{pmatrix} a & 0 \\ \bar{b} & t \end{pmatrix} .$$

To the latter equation the theory developed in the preceding sections may be applied. At the same time, as we shall see now, system (9.2) is closely associated with the initial problem (9.1), a fact, which is crucial for the examination of this problem.

<u>LEMMA 9.1.</u> If $\{\varphi^+,\varphi^-\}$ is a solution of problem (9.1), then, setting $\varphi_1^{\pm} = \varphi^{\pm}$, $\varphi_2^{\pm} = \hat{\varphi}^{\pm}$, we obtain a solution of system (9.2) for $g_1 = c$, $g_2 = \bar{c}$. Vice versa, if $\left\{\begin{pmatrix} \varphi_1^+ \\ \varphi_2^+ \end{pmatrix} , \begin{pmatrix} \varphi_1^- \\ \varphi_2^- \end{pmatrix}\right\}$ is a solution of system (9.2), then $\left\{\begin{pmatrix} \hat{\varphi}_2^- \\ \hat{\varphi}_1^- \end{pmatrix} , \begin{pmatrix} \hat{\varphi}_2^+ \\ \hat{\varphi}_1^+ \end{pmatrix}\right\}$ is a solution of the same system but with the right-hand side $\begin{pmatrix} \bar{g}_2 \\ \bar{g}_1 \end{pmatrix}$ and $\varphi^+ = \frac{1}{2}(\varphi_1^+ + \hat{\varphi}_2^-)$, $\varphi^- = \frac{1}{2}(\varphi_1^- + \hat{\varphi}_2^+)$ is a solution of problem (9.1).

The proof of Lemma 9.1 is reduced to a simple calculation. ≡

A similar connection exists between the solutions of the problem

$$a(t)\psi^+(t) - \overline{b(t)t^2\psi^+(t)} + \psi^-(t) = d(t) , \qquad (9.1')$$

which is called the associate to problem (9.1), and the solutions of the system

$$\begin{aligned} a(t)\psi_1^+(t) + \psi_1^-(t) - \overline{tb(t)}\psi_2^-(t) &= h_1(t) , \\ -t^2b(t)\psi_1^+(t) + t\psi_2^+(t) + t\overline{a(t)}\psi_2^-(t) &= h_2(t) . \end{aligned} \qquad (9.2')$$

Lemma 9.1 shows that system (9.2) with the right side $\begin{pmatrix} g_1 \\ g_2 \end{pmatrix}$ is solvable if and only if the same system with the right-hand side $\begin{pmatrix} \bar{g}_2 \\ \bar{g}_1 \end{pmatrix}$ is solvable. The identity

$$\begin{pmatrix} g_1 \\ g_2 \end{pmatrix} = \frac{1}{2}\begin{pmatrix} g_1 + \bar{g}_2 \\ \bar{g}_1 + g_2 \end{pmatrix} + \frac{1}{2i}\begin{pmatrix} i(g_1 - \bar{g}_2) \\ i(g_2 - \bar{g}_1) \end{pmatrix}$$

permits us to represent the set $\mathfrak{L}$ of right-hand sides $\begin{pmatrix} g_1 \\ g_2 \end{pmatrix}$ for which system (9.2) is solvable in the form of a sum $\mathcal{L}_0 + i\,\mathcal{L}_0$, where $\mathcal{L}_0$ consists of elements from $\mathcal{L}$ of the type $\begin{pmatrix} g \\ \bar{g} \end{pmatrix}$.

Clearly, $\mathcal{L}$ is a linear space over the field of complex and $\mathcal{L}_0$ over the field of real numbers, both linear spaces being closed only simultaneously.

In addition, Lemma 9.1 shows that problem (9.1) with the right-hand side c is solvable if and only if $\begin{pmatrix} c \\ \bar{c} \end{pmatrix} \in \mathcal{L}_0$. Consequently, the closedness of the image of problem (9.1) is equivalent to the closedness of $\mathcal{L}_0$ and, thus, of $\mathcal{L}$. From this we conclude the validity of the next assertion.

<u>COROLLARY 9.1.</u> The image of problem (9.1) is closed only simultaneously with the image of the operator $AP + BQ$.

Lemma 9.1 enables us also to establish relationships between solutions of the homogeneous problems (9.1) and (9.2). Owing to this lemma, the pair $\left\{\begin{pmatrix} \hat{\varphi}_2^- \\ \hat{\varphi}_1^- \end{pmatrix}, \begin{pmatrix} \hat{\varphi}_2^+ \\ \hat{\varphi}_1^+ \end{pmatrix}\right\}$ is, together with a certain pair $\left\{\begin{pmatrix} \varphi_1^+ \\ \varphi_2^+ \end{pmatrix}, \begin{pmatrix} \varphi_1^- \\ \varphi_2^- \end{pmatrix}\right\}$, also a solution of the homogeneous problem (9.2), which allows for writing the lineal M of solutions of the homogeneous problem (9.2) in the form $M = M_0 + iM_0$, where $M_0 \subseteq M$ consists of pairs of vector functions of the type $\left\{\begin{pmatrix} \varphi^+ \\ \hat{\varphi}^- \end{pmatrix}, \begin{pmatrix} \varphi^- \\ \hat{\varphi}^+ \end{pmatrix}\right\}$ (compare with similar considerations concerning the image of problem (9.2)). Furthermore, $\{\varphi^+, \varphi^-\}$ is a solution of the homogeneous problem (9.1) iff $\left\{\begin{pmatrix} \varphi^+ \\ \hat{\varphi}^- \end{pmatrix}, \begin{pmatrix} \varphi^- \\ \hat{\varphi}^+ \end{pmatrix}\right\}$ is a solution of the homogeneous problem (9.2). Thus, an isomorphism between the solution space of the homogeneous problem (9.1) and M_0 is established (both of them are regarded over the field of real numbers). Since the dimension of the lineal M_0 (over the field of real numbers) coincides with the dimension of M (over the complex field), the following result is valid.

<u>COROLLARY 9.2.</u> The number l of linearly independent solutions (over the field of real numbers) of the homogeneous problem (9.1) is

equal to the number of linearly independent (over the field of complex numbers) solutions of the homogeneous problem (9.2).

The following result can be verified analogously.

COROLLARY 9.2'. The number l' of linearly independent solutions (over the field of real numbers) of the homogeneous problem (9.1') coincides with the number of linearly independent solutions (over the complex field) of the homogeneous problem (9.2'). For the space M' of solutions of the homogeneous problem (9.2) we have the representation $M' = M'_0 + iM'_0$, where M'_0 consists of all elements from M' of the form $\left\{\begin{pmatrix}\psi^+\\ \hat{\psi}^-\end{pmatrix}, \begin{pmatrix}\psi^-\\ \hat{\psi}^+\end{pmatrix}\right\}$. Moreover, $\left\{\begin{pmatrix}\psi^+\\ \hat{\psi}^-\end{pmatrix}, \begin{pmatrix}\psi^-\\ \hat{\psi}^+\end{pmatrix}\right\}$ belongs to M'_0 if and only if $\{\psi^+,\psi^-\}$ is a solution of the homogeneous problem (9.1').

Corollaries 9.1 and 9.2 illustrate that for the normal solvability and finite dimensionality of the kernel of problem (9.1), it is necessary and sufficient that $AP + BQ$ is a Φ_+-operator.
From this we deduce the following proposition.

THEOREM 9.1. For finite dimensionality of the kernel and normal solvability of problem (9.1), it is necessary that $a^{-1} \in L_\infty$, i.e., for some $\delta > 0$, the inequality

$$|a(t)| \geq \delta \tag{9.3}$$

holds a.e. on $\mathbb{T}$.

Proof. We employ Theorem 3.18, due to which for $AP + BQ$ to be a Φ_+-operator, it is necessary that the conditions $A^{-1}, B^{-1} \in L_\infty$ be fulfilled. In the case under consideration, $\det A = t\bar{a}$, $\det B = ta$, therefore, either of the requirements $A^{-1} \in L_\infty$, $B^{-1} \in L_\infty$ is equivalent to condition (9.3), which proves Theorem 9.1. ≡

Condition (9.3) is necessary, in particular, for the Fredholmness of problem (9.1). In the sequel, this condition is supposed to be fulfilled.
Assuming that $a^{-1} \in L_\infty$, we can rewrite the system (9.2) in the form

$$(P + GQ)\tilde{\varphi} = \tilde{g}, \tag{9.4}$$

where

$$G = A^{-1}B = \bar{a}^{-1}\begin{pmatrix} |a|^2-|b|^2 & -tb \\ \overline{tb} & 1 \end{pmatrix}, \quad \tilde{g} = A^{-1}g = \begin{pmatrix} g_1-b\bar{a}^{-1}g_2 \\ t^{-1}\bar{a}^{-1}g_2 \end{pmatrix}. \quad (9.5)$$

In turn, the system (9.2') may be rewritten in the form

$$(Q + G'P)\tilde{\psi} = \tilde{h} \,, \qquad (9.4')$$

where G' is the matrix function transposed to G, and

$$\tilde{h} = \begin{pmatrix} h_1 - t^{-2}\bar{b}\,\bar{a}^{-1}\,h_2 \\ t^{-1}\,\bar{a}^{-1}\,h_2 \end{pmatrix}.$$

Above (Corollary 9.1) it was established that problem (9.1) is normally solvable if and only if the image of the operator $AP + BQ$ or, which is equivalent, the image of the operator $P + GQ$ is closed, where the matrix function G is defined by equation (9.5). Now we want to complete this statement by describing the connection between the codimensions of the images of problems (9.1) and (9.4).

LEMMA 9.2 The image of problem (9.1) is a lineal (over the field of real numbers) the closure of which is the subspace defined by the condition

$$\mathrm{Re} \int_{\mathbb{T}} c(t)\psi^{+}(t)dt = 0 \,, \qquad (9.6)$$

where ψ^{+} runs over the "+"-components of the solutions of the homogeneous problem (9.1').

Proof. According to Lemma 9.1, problem (9.1) with the right-hand side c is solvable simultaneously with problem (9.2) with the right-hand side $\begin{pmatrix} c \\ \bar{c} \end{pmatrix}$ or, in other words, with problem (9.4) in which the right-hand side is equal to $\tilde{g} = \begin{pmatrix} c - b\bar{c}\,\bar{a}^{-1} \\ t^{-1}\,\bar{a}^{-1}\,\bar{c} \end{pmatrix}$.

As mentioned in Theorem 3.1, a necessary condition for solvability of problem (9.4) consists in the orthogonality of its right-hand side to the "+"-components of all solutions of the homogeneous associate problem, i.e. of problem (9.4') with zero right-hand side. Evidently, the solutions of the homogeneous problems (9.4') and (9.2') coincide.

The orthogonality of the vector function $\tilde{g}$ to the "+"-components of all vector functions from M' is equivalent to its orthogonality to the "+"-components of all vector functions from M'_0, i.e. to vector functions of the type $\begin{pmatrix} \psi^+ \\ \hat{\psi}^- \end{pmatrix}$, where $\{\psi^+, \psi^-\}$ is a solution of the homogeneous problem (9.1').

But

$$\int_{\mathbb{T}} \tilde{g}' \begin{pmatrix} \psi^+ \\ \hat{\psi}^- \end{pmatrix} dt = \int_{\mathbb{T}} ((c - b\bar{c}\bar{a}^{-1})\psi^+ + t^{-1}\bar{a}^{-1}\bar{c}\ \overline{t\psi^-})dt =$$

$$\int_{\mathbb{T}} \bar{a}^{-1}t^{-2}(c\bar{a}t^2\psi^+ - \bar{c}bt^2\psi^+ + \overline{c\psi^-})dt .$$

Using the fact that $\overline{bt^2\psi^+ - \psi^-} = a\psi^+$, we may rewrite the right-hand side in the following manner:

$$\int_{\mathbb{T}} \bar{a}^{-1}t^{-2}(c\bar{a}t^2\psi^+ - \bar{c}\overline{a\psi^+})dt = \int_{\mathbb{T}} c\psi^+dt - \int_{\mathbb{T}} \bar{c}\overline{\psi^+}t^{-2}dt .$$

Since the contour under consideration is the unit circle, we have $\overline{dt} = -t^{-2}dt$, so that we finally get:

$$\int_{\mathbb{T}} \tilde{g} \begin{pmatrix} \psi^+ \\ \hat{\psi}^- \end{pmatrix} dt = 2 \operatorname{Re} \int_{\mathbb{T}} c(t)\psi^+(t)dt .$$

Thus, condition (9.6) coincides precisely with the condition of orthogonality of $\tilde{g}$ to the "+"-components of the solutions of the homogeneous problem (9.4). Therefore, it is necessary for the solvability of problem (9.1).

Now, let the function c $(\in L_p)$ satisfy the condition (9.6). Then in every neighbourhood of the vector function $\tilde{g} = \begin{pmatrix} c - b\bar{c}\,\bar{a}^{-1} \\ t^{-1}\ \bar{a}^{-1}\ \bar{c} \end{pmatrix}$ there exists a vector function $\tilde{f}$ belonging to the image of problem (9.4). The vector function $f = A\tilde{f} = \begin{pmatrix} f_1 \\ f_2 \end{pmatrix}$ belongs to the image of problem (9.2). With regard to Lemma 9.1, the function $\frac{1}{2}(f_1+\bar{f}_2)$ lies in the image of problem (9.1). Due to the closeness of the vector functions $\tilde{g}$ and $\tilde{f}$ it is possible to make the vector functions $A\tilde{g} = \begin{pmatrix} c \\ \bar{c} \end{pmatrix}$ and $A\tilde{f} = f$ as close as desired and, consequently, also the functions c and $\frac{1}{2}(f_1+\bar{f}_2)$. Hence, the image of problem (9.1) is dense in the subspace generated by conditions (9.6), which proves Lemma 9.2. ≡

COROLLARY 9.3. The codimension l' of the closure of the image of problem (9.1) (over the field of real numbers) agrees with the codimension of the closure of the image of problem (9.4) (over the complex field).

Indeed, Lemma 9.2 demonstrates that the number l' coincides with the number of linearly independent (over the field of real numbers) solutions of the homogeneous problem (9.1'), which, according to Corollary 9.2, is equal to the dimension (over the complex field) of the kernel of problem (9.2'). It remains to use the coincidence of the kernels of problems (9.2') and (9.4').

Now we are going to formulate the main theorem of the present section.

THEOREM 9.2. The generalized Riemann boundary value problem (9.1) is Fredholm in L_p iff

1) condition (9.3) is fulfilled,
2) the matrix function G defined by equation (9.5) is Φ-factorable in L_p.

If the conditions 1) and 2) are satisfied, then the index of problem (9.1) coincides with the total index of G, the number l of linearly independent solutions of the homogeneous problem is equal to the sum of positive partial indices, whereas the number l' of solvability conditions for the inhomogeneous problem is opposite to the sum of negative partial indices.

Proof. The necessity of condition 1) for the Fredholmness of problem (9.1) has been stated by Theorem 9.1. If this condition is satisfied, then, owing to Corollaries 9.1 - 9.3, problem (9.1) is Fredholm only simultaneously with problem (9.4). Thanks to Theorem 3.16, the Fredholmness of problem (9.4) is equivalent to the Φ-factorability of the matrix function G.

According to Corollary 9.2, the number l coincides with the dimension of the kernel of problem (9.4), i.e. with the sum of positive partial indices of G (by Theorem 3.16). The formula for l' is established on the strength of Corollary 9.3. Now, from the formulae for l and l

it is not hard to obtain the formula for the index of problem (9.1). Hence, Theorem 9.2 is proved. ≡

We would like to mention that the formulae for l and l' can be written in the following manner: Let $\varkappa_1$ and $\varkappa_2$ be the partial indices of the matrix function G. Then

$$\begin{aligned} l &= \max\{0,\varkappa_1+\varkappa_2\}, \quad l' = \max\{0,-\varkappa_1-\varkappa_2\}, \text{ if } \varkappa_1\varkappa_2 \geq 0; \\ l &= \varkappa_1, \quad l' = -\varkappa_2, \quad \text{if } \varkappa_1\varkappa_2 < 0. \end{aligned} \tag{9.7}$$

<u>DEFINITION 9.1.</u> The problem (9.1) will be said to be stable, if the partial indices of the corresponding matrix function (9.5) are stable.

If $\varkappa_1\varkappa_2 \geq 0$, then, due to formulae (9.7), the numbers l and l' are defined uniquely by the total index of the matrix function (9.5). Therefore, they remain unchanged under small perturbations of the coefficients a and b. Thus, the condition $\varkappa_1\varkappa_2 \geq 0$ (all the more, the condition of stability of problem (9.1.)) is sufficient for the stability of the numbers l and l'.

The next statement, which often proves to be useful in the study of problem (9.1), results directly from Theorem 9.2.

<u>THEOREM 9.3.</u> The substitution of the coefficient b in problem (9.1) by $b-\varphi$ with $\varphi \in H_\infty + C$ does not influence neither on the Fredholmness of the problem not on the value of its index. If $t\varphi \in H_\infty$, then the numbers l and l' of problem (9.1) also remain unchanged.

<u>Proof.</u> We take advantage of the identity

$$\bar{a}^{-1}\begin{pmatrix} |a|^2-|b-\varphi|^2 & -t(b-\varphi) \\ \overline{t(b-\varphi)} & 1 \end{pmatrix} = \begin{pmatrix} 1 & t\varphi \\ 0 & 1 \end{pmatrix} \bar{a}^{-1} \begin{pmatrix} |a|^2-|b|^2 & -tb \\ \overline{tb} & 1 \end{pmatrix} \begin{pmatrix} 1 & 0 \\ -\overline{t\varphi} & 1 \end{pmatrix}, \tag{9.8}$$

which is true for any choice of the function φ. If $\varphi \in H_\infty + C$, then the matrix function $\begin{pmatrix} 1 & t\varphi \\ 0 & 1 \end{pmatrix}$, together with its inverse, belongs to the class $H_\infty^+ + C$, whereas $\begin{pmatrix} 1 & 0 \\ -\overline{t\varphi} & 1 \end{pmatrix}$ belongs to $H_\infty^- + C$ together with its inverse.

Applying Theorem 5.5, we observe that the matrix functions

$$\bar{a}^{-1}\begin{pmatrix} |a|^2-|b-\varphi|^2 & -t(b-\varphi) \\ \overline{t(b-\varphi)} & 1 \end{pmatrix} \quad \text{and} \quad \bar{a}^{-1}\begin{pmatrix} |a|^2-|b|^2 & -tb \\ \overline{tb} & 1 \end{pmatrix}$$

are Φ-factorable in L_p only simultaneously, where their total indices differ by the sum of total indices of the matrix functions $\begin{pmatrix} 1 & t\varphi \\ 0 & 1 \end{pmatrix}$ and $\begin{pmatrix} 1 & 0 \\ -\overline{t\varphi} & 1 \end{pmatrix}$. The latter two matrix functions are triangular and the indices of their diagonal elements are equal to zero. According to Theorem 4.6, their partial (and, thus, also total indices) are equal to zero. Hence, the total indices of the matrix functions

$$\bar{a}^{-1}\begin{pmatrix} |a|^2-|b-\varphi|^2 & -t(b-\varphi) \\ \overline{t(b-\varphi)} & 1 \end{pmatrix} \quad \text{and} \quad \bar{a}^{-1}\begin{pmatrix} |a|^2-|b|^2 & -tb \\ \overline{tb} & 1 \end{pmatrix}$$

coincide. This proves the first statement of the theorem.

If $t\varphi \in H_\infty$, then identity (9.8) shows that not only the total but also the partial indices of the matrix functions

$$\bar{a}^{-1}\begin{pmatrix} |a|^2-|b-\varphi|^2 & -t(b-\varphi) \\ \overline{t(b-\varphi)} & 1 \end{pmatrix} \quad \text{and} \quad \bar{a}^{-1}\begin{pmatrix} |a|^2-|b|^2 & -tb \\ \overline{tb} & 1 \end{pmatrix}$$

agree. Making use of formulae (9.6), we get the second statement of the theorem. ≡

COROLLARY 9.4. If $b \in H_\infty + C$, then problem (9.1) is Fredholm if and only if the function a is Φ-factorable in L_p. In the latter case the index of problem (9.1) coincides with the double p-index of the function a.

In fact, for $b \in H_\infty + C$, problem (9.1) is Fredholm only along with problem (9.1) with the same coefficient a and $b \equiv 0$, whereas the indices of these problems coincide. At the same time, for $b \equiv 0$, problem (9.1) changes into the ordinary Riemann problem $\varphi^+ + a\varphi^- = c$. In order to avoid possible misunderstandings, note that the index of problem (9.1), for $b \equiv 0$, is the double index of the corresponding Riemann problem (i.e., does not agree with it), because the problem (9.1) is considered over the field of real numbers, while the Riemann problem is regarded over the complex field.

Thus, the index of problem (9.1) with a coefficient b from the class $H_\infty + C$ (in particular, with a continuous coefficient b) is necessarily even.

9.2. SUFFICIENT CONDITIONS OF FREDHOLMNESS AND STABILITY. ESTIMATES FOR THE DEFECT NUMBERS OF THE GENERALIZED RIEMANN BOUNDARY VALUE PROBLEM IN THE SPACE L_2

In this section we shall be concerned with problem (9.1) in the class L_2, assuming that the coefficient a can be represented in the form $a = a_1 a_0 \overline{a}_2$, where a_1 and a_2 are invertible elements of H_∞ and a_0 is a continuous, non-vanishing function. Applying Theorem 3.10 and Lemma 6.3, we recognize that the matrix function (9.5) is Φ-factorable only simultaneously with the matrix function

$$G_0(t) = \begin{pmatrix} |a(t)|^2-|b(t)|^2 & -tb(t) \\ \overline{tb(t)} & 1 \end{pmatrix}, \tag{9.9}$$

and the partial indices of the matrix function (9.5) are obtained from the partial indices of the matrix function (9.9) by shift on the index $\varkappa$ of the function a (coinciding with the index of a_0). Furthermore, multiplying the matrix function G_0 by the constant matrix $\begin{pmatrix} -1 & 0 \\ 0 & 1 \end{pmatrix}$ from the left, it turns into the Hermitian matrix function

$$F(t) = \begin{pmatrix} |b(t)|^2-|a(t)|^2 & tb(t) \\ \overline{tb(t)} & 1 \end{pmatrix} \tag{9.10}$$

with negative determinant. In accordance with formula (7.5), the partial 2-indices of F (and, consequently, also of G_0) are equal to $k\ (\geq 0)$ and $-k$.

Therefore, the partial indices of the matrix function G are equal to $\varkappa_1 = \varkappa + k$ and $\varkappa_2 = \varkappa - k$. Formulae (9.7) for the numbers l and l' may be rewritten in the following form:

$$\begin{aligned} l &= \max\{0, 2\varkappa\}, \quad l' = \max\{0, -2\varkappa\}, \quad \text{if } k \leq |\varkappa|, \\ l &= k + \varkappa, \quad l' = k - \varkappa, \quad \text{if } k > |\varkappa|. \end{aligned} \tag{9.11}$$

Thus, on condition that $k \le |\varkappa|$, the numbers l and l' of problem (9.1) are stable, while the stability of problem (9.1) itself is equivalent to the requirement $k = 0$.

THEOREM 9.4. Suppose the functions $\varphi_1, \varphi_2 \in M_\infty^+$ to have no common poles in $\mathcal{D}^+$, to satisfy the inequalities

$$\operatorname*{ess\,sup}_{t\in\mathbb{T}} \left| \frac{tb(t)-\varphi_1(t)}{a(t)} \right| < 1 < \operatorname*{ess\,inf}_{t\in\mathbb{T}} \left| \frac{tb(t)-\varphi_2(t)}{a(t)} \right| \tag{9.12}$$

and moreover,

$$(\varphi_1 - \varphi_2)^{-1} \in M_\infty^+ . \tag{9.13}$$

Then problem (9.1) is Fredholm in L_2 and for the numbers l and l' the relations

$$\begin{aligned} &l = \max(0, 2\varkappa), \ l' = \max(0, -2\varkappa), \quad \text{if } N \le |\varkappa| , \\ &\max(0, 2\varkappa) \le l \le \varkappa + N, \quad l' = l - 2\varkappa, \ \text{if } N > |\varkappa| \end{aligned} \tag{9.11'}$$

hold. Here N is the number of zeros (taking into account their multiplicity) of the function $\varphi_1 - \varphi_2$ in $\mathcal{D}^+$.

Proof. We represent the matrix function (9.9) in the form

$$G_0 = |\varphi_1 - \varphi_2|^{-2} \begin{pmatrix} \psi_2 & -\psi_1 \\ -\chi_2 & \chi_1 \end{pmatrix} G_1 \begin{pmatrix} \overline{\psi_2} & \overline{\chi_2} \\ \overline{\psi_1} & \overline{\chi_1} \end{pmatrix} .$$

Here $\varphi_1 = \psi_1/\chi_1$, $\varphi_2 = \psi_2/\chi_2$ ($\psi_j \in H_\infty$, χ_j is a polynomial the zeros of which are located in $\mathcal{D}^+$ and do not coincide with the zeros of ψ_j, $j=1,2$) are representations existing thanks to the condition $\varphi_1, \varphi_2 \in M_\infty^+$, and

$$G_1 = \begin{pmatrix} |\chi_2|^{-2}(|a|^2 - |tb-\varphi_1|^2) & -\overline{\chi_1}^{-1}\chi_2^{-1}(|a|^2-|b|^2+\varphi_1\overline{tb}+\overline{\varphi_2}tb-\varphi_1\overline{\varphi_2}) \\ \chi_1^{-1}\overline{\chi_2}^{-1}(|a|^2-|b|^2+\overline{\varphi_1}tb+\varphi_2\overline{tb}-\overline{\varphi_1}\varphi_2) & |\chi_1|^{-2}(|tb-\varphi_2|^2-|a|^2) \end{pmatrix} .$$

Since $\operatorname{Re} G_1 = \operatorname{diag}[|\chi_2|^{-2}(|a|^2-|tb-\varphi_1|^2), |\chi_1|^{-2}(tb-\varphi_2|^2-|a|^2)]$, the conditions (9.12) guarantee the uniform positiveness of $\operatorname{Re} G_1$ and, according to Theorem 7.8, the Φ-factorability of the matrix function G_1 in L_2 with zero partial indices. The factor $A_+ = \begin{pmatrix} \psi_2 & -\psi_1 \\ -\chi_2 & \chi_1 \end{pmatrix}$ belongs to H_∞, while its inverse, in virtue of condition (9.13), belongs to M_∞^+ and, therefore, has no influence on the Φ-factorability

(Theorem 3.10). The estimates for the partial indices given in Theorem 2.11 allow us to claim that the partial indices of the matrix function A_+G_1 are included between 0 and N . Analogously, it can be verified that the product $A_+G_1A_-$ with $A_- = \begin{pmatrix} \overline{\psi_2} & \overline{\chi_2} \\ \overline{\psi_1} & \overline{\chi_1} \end{pmatrix}$ is Φ-factorable in L_2 and the partial indices of this product do not exceed N taken absolutely. In view of condition (9.13) , $|\varphi_1 - \varphi_2|^{-2}$ is a positive function belonging to the class L_∞ together with its inverse. Applying Lemma 6.5 to this function, we find that the matrix functions G_0 and $A_+G_1A_-$ are Φ-factorable only simultaneously and their partial indices coincide.

Summarizing, we have proved that the matrix function G_0 is Φ-factorable in L_2 , so that problem (9.1) is Fredholm and, in addition, $k \leq N$. In order to complete the proof of the theorem, it remains only to use formulae (9.11). ≡

COROLLARY 9.5. Let the function $\varphi \in M_\infty^+$ satisfy the requirement

$$\operatorname*{ess\,sup}_{t\in\mathbb{T}} \left| \frac{tb(t) - \varphi(t)}{a(t)} \right| < 1 . \tag{9.14}$$

Then problem (9.1) is Fredholm and for the numbers l and l' the relations

$$l = \max\{0, 2\varkappa\}, \quad l' = \max\{0, -2\varkappa\} , \quad \text{if } \mathcal{P}(\varphi) \leq |\varkappa| ,$$
$$\max\{0, 2\varkappa\} \leq l \leq \varkappa + \mathcal{P}(\varphi) , \quad \text{if } \mathcal{P}(\varphi) > |\varkappa|$$

are fulfilled.

Here $\mathcal{P}(\varphi)$ is the number of poles (counting their multiplicity) of the function φ in $\mathcal{D}^+$.

Corollary 9.5 follows immediately from Theorem 9.4 by setting $\varphi_1 = \varphi$ and taking as φ_2 a sufficiently large constant C . In fact, in this case the conditions (9.12) and (9.13) are satisfied and the number of zeros of the function $\varphi_1 - \varphi_2$ is the number of points $t \in \mathcal{D}^+$ with $\varphi(t) = C$ and, therefore (if C is large enough), coincides with the number of poles of φ . ≡

COROLLARY 9.6. Suppose that the functions $\varphi_1, \varphi_2 \in M_\infty^+$ have no common poles in $\mathcal{D}^+$, coincide (with reference to the multiplicity) at no more than $|\varkappa|$ points and satisfy the conditions (9.12) and (9.13). Then problem (9.1) is Fredholm and the numbers l and l' of this problem are stable.

Let us emphasize a sufficient condition for stability of problem (9.1).

<u>THEOREM 9.5.</u> Assume that the functions $\varphi_1, \varphi_2 \in M_\infty^+$ have no common poles in $\mathfrak{D}^+$ and satisfy the inequalities (9.5) and the requirement $(\varphi_1-\varphi_2)^{-1} \in H_\infty$. Then problem (9.1) is Fredholm and stable in L_2.

Indeed, the requirements imposed on φ_1 and φ_2 in Theorem 9.5 coincide with the requirements involved in Theorem 9.4 completed by the condition $N = 0$. In Theorem 9.4 it was proved that $k \leq N$. Consequently, under the conditions of Theorem 9.5, $k = 0$. ≡

From Corollary 9.5 we may derive a sufficient condition of stability of problem (9.1) (which is, of course, a special case of Theorem 9.5).

<u>COROLLARY 9.7.</u> Let the function $\varphi \in H_\infty$ satisfy the condition (9.14). Then problem (9.1) is Fredholm and stable in L_2.

In particular, problem (9.1) will be Fredholm and stable in L_2 in the so-called elliptic case:[1] $\operatorname{ess\,sup}_{t\in\mathbb{T}} |b(t)/a(t)| < 1$ (it suffices to choose $\varphi \equiv 0$ in Corollary 9.7).

Now we proceed to the so-called parabolic case

$$|b(t)| = |a(t)| \quad \text{a.e. on } \mathbb{T}. \tag{9.15}$$

In this case the matrix function G_0 becomes triangular, is Φ-factorable only simultaneously with the matrix function

$$\begin{pmatrix} tb & 0 \\ 1 & \overline{tb} \end{pmatrix} \tag{9.16}$$

and has the same tuple of partial indices.

<u>THEOREM 9.6.</u> Let the condition (9.15) be satisfied. Then the following assertions are equivalent:

[1] The concept of elliptic case in this sense and the concepts of parabolic and hyperbolic cases as well were introduced in MIKHAILOV [3,4].

1) problem (9.1) is Fredholm in L_2 ,

2) $T_Q(tb)$ is a Φ_--operator ,

3) there exists a function $\varphi \in M_\infty^+$ for which

$$\operatorname*{ess\,sup}_{t\in\mathbb{T}} \left| \frac{tb(t) - \varphi(t)}{b(t)} \right| < 1 , \qquad (9.14')$$

4) there exists a function $\varphi \in H_\infty + C$ satisfying the condition (9.14').

If the statements just mentioned are satisfied, then the defect numbers l and l' of problem (9.1) can be calculated by formulae (9.11) in which the quantity k may be replaced either by the defect number $\beta(T_Q(tb))$ or by the smallest possible (with regard to the multiplicity) number ν of poles in the unit disk of the function $\varphi \in M_\infty^+$ satisfying the relation (9.14).

<u>Proof.</u> Problem (9.1) is Fredholm if and only if the matrix function (9.16) is Φ-factorable. Therefore, the equivalence of assertion 1) and 2) results from Theorem 4.4. According to just this theorem, the relation $k = \beta(T_Q(tb))$ is valid. Thus, it remains to prove the equivalence of assertions 2) - 4) and to verify the equation $\nu = \beta(T_Q(tb))$. Assume $T_Q(tb)$ to be a Φ_--operator. Then, owing to Theorem 3.13, $b^{-1} \in L_\infty$. Applying Lemma 7.4 (which is trivial in the considered scalar case), we write down the representation $tb = u\psi$, where $\psi^{\pm 1} \in H_\infty$ and the absolute value of the function u is equal to 1 a.e. on $\mathbb{T}$. Then $T_Q(tb) = T_Q(\psi)T_Q(u)$ and, since the operator $T_Q(\psi)$ is invertible, the property of $T_Q(tb)$ to be a Φ_--operator implies that $T_Q(u)$ is a Φ_--operator, too. Moreover, the defect numbers of these operators coincide.

Now we are going to consider separately the cases of zero and positive values of $\beta(T_Q(tb))$.

a) Suppose $\beta(T_Q(tb)) = 0$. Then $T_Q(tb)$ is not only a Φ_--operator but also invertible from the right. Together with it the operator $T_Q(u)$ is also invertible from the right and, by Theorem 7.12, there exists a function $\chi \in H_\infty$ such that $\|u-\chi\| < 1$. Setting $\varphi = \chi\psi$ and taking into account that $u = tb\psi^{-1}$ and $|\psi| = |b|$, we deduce inequality (9.14') with a function $\varphi \in H_\infty$. Thus, condition 3) is satisfied, where $\nu = 0$.

b) Let $\beta(T_Q(tb)) > 0$. Then, due to assertion 3) of Theorem 3.1, $\alpha(T_Q(tb)) = 0$ and the Φ_--operator $T_Q(tb)$ proves to be Fredholm. Thus, the operator $T_Q(u)$ is also Fredholm. In addition, its index (i.e, the 2-index of the function u) is opposite to the number $\beta(T_Q(tb))$. By virtue of the equivalence of assertions 1) and 3) of Theorem 7.13 (the case of non-positive indices), we find that there exists a function χ of class M_∞^+ such that $\chi^{-1} \in H_\infty$ and $\|u-\chi\| < 1$. According to the same theorem, the number of poles of such a functions in the unit disk is opposite to the index of the function u, i.e., they coincides with $\beta(T_Q(tb))$. Setting $\varphi = \chi\psi$, we come, as also in case a), to inequality (9.14'), however, already with a function φ of class M_∞^+ and a number of poles in the unit disk which is equal to $\beta(T_Q(tb))$. In summary, we have proved the implication 2) ==> 3) as well as the inequality $\nu \leq \beta(T_Q(tb))$.

The implication 3) ==> 4) is obvious.

Now we want to assume that condition 4) is fulfilled. Then, of course, $b^{-1} \in L_\infty$. Therefore, there exists again a representation $tb = u\psi$. With the aid of this representation condition (9.14) may be rewritten as follows: $\|1-\bar{u}\chi\| < 1$, where $\chi = \varphi\psi^{-1} \in H_\infty + C$. In view of the contracting mapping principle, the last inequality guarantees the invertibility of the operator $T_Q(\bar{u}\chi)$. But this operator differs from the product $T_Q(\chi)T_Q(\bar{u})$ only by the compact summand $Q\chi P\bar{u}Q$. Due to Theorem 1.5, the product $T_Q(\chi)T_Q(\bar{u})$ is Fredholm. This fact as well as Theorem 1.9 imply that $T_Q(\bar{u})$ is a Φ_+-operator. Now, Corollary 1.2 allows us to conclude that $T_Q(u)$ is a Φ_--operator. Finally, together with this operator, the operator $T_Q(tb)$ obtained from $T_Q(u)$ by multiplying it by the invertible operator $T_Q(\psi)$ is also a Φ_--operator.

Finally, it must be said that if the operator $T_Q(tb)$ turned out to be Fredholm, then the operator $T_Q(\chi)$ is also Fredholm and, moreover, the indices of these operators coincide. If, in addition, $\varphi \in M_\infty^+$, then $\chi \in M_\infty^+$ and, by Theorem 2.6, the number of poles of the function χ in the unit disk is an upper bound for the value $-\operatorname{ind} T_Q(\chi)$. Taking into consideration that the number of poles of the function χ and φ is one and the same and that, for $\beta(T_Q(tb)) > 0$, $\operatorname{ind} T_Q(\chi) = -\beta(T_Q(tb))$, we deduce that the quantity $\beta(T_Q(tb))$ is a lower bound for the number of poles in the unit disk for an arbitrary function $\varphi \in M_\infty^+$ satisfying condition (9.14').

Thus, the implication 4) ==> 1) and the inequality $\nu \geq \beta(T_Q(tb))$ have been proved and, together with them, the complete Theorem 9.6. ≡

<u>COROLLARY 9.8.</u> Let condition (9.15) be satisfied (i.e., the parabolic case holds) and the function b be Φ-factorable in L_2 with the index λ. Then problem (9.1) is Fredholm and its defect numbers are calculated by formulae (9.11) in which one has to set $k = \max\{0,-\lambda-1\}$.

In fact, the Φ-factorability of the function b is equivalent to the Fredholmness of the operator $T_Q(tb)$, i.e. to a property which is stronger than the property to be a Φ_--operator. The equation $k = \max\{0,-\lambda-1\}$ follows from the relation $k = \beta(T_Q(tb))$ already mentioned and the formulae for the defect numbers proved in Chapter 3. ≡

<u>COROLLARY 9.9.</u> Under the condition (9.15) (i.e., in the parabolic case) for the stability of problem (9.1) in L_2, it is necessary and sufficient that there exists a function $\varphi \in H_\infty$ satisfying inequality (9.14'). The Φ-factorability of the function b in L_2 with index $\lambda \geq -1$ is a sufficient condition for stability.

<u>Proof.</u> It suffices to remember that the stability of problem (9.1) in L_2 is equivalent to its Fredholmness under the additional condition $k = 0$. After that we have to use the equality $k = \nu$ in the general case and the relation $k = \max\{0,-\lambda-1\}$ in case of a Φ-factorable function b.

We would like to note that in the parabolic case, i.e., under the additional condition (9.15), inequality (9.14) is equivalent to inequality (9.14'). Thus, in the parabolic case the sufficient conditions of Fredholmness and stability indicated in Corollaries 9.5 and 9.7 are necessary and sufficient. Moreover, the bounds for the defect numbers stated in Corollary 9.5 are exact in the parabolic case.

For the so-called hyperbolic case

$$\operatorname*{ess\,inf}_{t\in\mathbb{T}} |b(t)/a(t)| > 1 \tag{9.17}$$

we do not have such a complete solution of the question of Fredholness of problem (9.1) and the values of l and l' as for the elliptic and parabolic case. Here we only want to present a criterion of stability of the problem based on the representation

$$|b(t)|^2 - |a(t)|^2 = |\psi(t)|^2 \tag{9.18}$$

($\psi^{\pm 1} \in H_\infty$), which is possible by Theorem 7.10 in virtue of condition (9.17).

<u>THEOREM 9.7.</u> Assume the condition (9.17) to be satisfied, and let the convex hull of the range of the function $tb\psi^{-1}$ have a distance from zero which is greater than 1. Then the problem (9.1) is Fredholm and stable in L_2.

<u>Proof.</u> We represent the matrix function (9.9) in the form

$$G_0 = \begin{pmatrix} -\psi & 0 \\ 0 & 1 \end{pmatrix} \begin{pmatrix} tb\psi^{-1} & 1 \\ 1 & \overline{tb\psi^{-1}} \end{pmatrix} \begin{pmatrix} 0 & 1 \\ \overline{\psi} & 0 \end{pmatrix}.$$

From this representation it is clear that the matrix function G_0 is Φ-factorable only simultaneously with the matrix function

$$G_1 = \begin{pmatrix} tb\psi^{-1} & 1 \\ 1 & \overline{tb\psi^{-1}} \end{pmatrix}$$

and has the same tuple of partial indices. Furthermore, it should be noticed that the function ψ from the representation (9.18) is defined up to a constant coefficient with absolute value 1. Due to the choice of this multiplier and the condition imposed on the range of the function $tb\psi^{-1}$, we can ensure that $\operatorname{Re}(tb(t)\psi(t)^{-1}) \geq 1+\delta$ for some $\delta > 0$ a.e. on $\mathbb{T}$. Since

$$\operatorname{Re} G_1 = \begin{pmatrix} \operatorname{Re}(tb\psi^{-1}) & 1 \\ 1 & \operatorname{Re}(tb\psi^{-1}) \end{pmatrix},$$

the condition $\operatorname{Re}(tb(t)\psi(t)^{-1}) \geq 1+\delta$ guarantees the uniform positiveness of the matrix function $\operatorname{Re} G_1$ and, hence, the Φ-factorability of G_1 (and, along with it, also of G_0) in L_2 with zero partial indices, which proves Theorem 9.7. ≡

9.3. THE GENERALIZED RIEMANN BOUNDARY VALUE PROBLEM WITH CONTINUOUS COEFFICIENTS. A STABILITY CRITERION

Let us consider problem (9.1) with continuous coefficients a and b. From Theorem 9.2 and Corollary 9.4 we deduce directly the following criterion of Fredholmness of problem (9.1).

THEOREM 9.8. Problem (9.1) with continuous coefficients a and b is Fredholm (independently of the value $p \in (1,\infty)$) iff $a(t) \neq 0$ for all $t \in \mathbb{T}$. If this condition is satisfied, then the defect numbers l and l' of problem (9.4) do not depend on the parameter p either, and the index of the problem is equal to the double index $\varkappa$ of the function a.

We would like to explain that the independence of the numbers l and l' of the parameter value p can be seen from formulae (9.7) and from the fact that the partial indices of a continuous non-singular matrix function do not depend on p.

Thus, the numbers l and l' of problem (9.1) are, for all values of p, the same as for $p = 2$. Therefore, they can be calculated by formulae (9.11).

The estimates for the numbers l and l' mentioned in Theorem 9.4, 9.5 and their corollaries may be written in a somewhat simpler form.

THEOREM 9.4'. Let the functions a and b be continuous on $\mathbb{T}$, and assume the functions φ_1 and φ_2 to be meromorphic in $\mathcal{D}^+$, continuous extendable onto $\mathbb{T}$, to have no common poles and to coincide at no more than N points (counting the multiplicities) of $\mathcal{D}^+$.

If, for all $t \in \mathbb{T}$,

$$|tb(t) - \varphi_1(t)| < |a(t)| < |tb(t) - \varphi_2(t)|, \tag{9.19}$$

then for the numbers l and l' the relations (9.11') are valid.

Proof. Condition (9.19) means, especially, that $a(t) \neq 0$ and $\varphi_1(t) \neq \varphi_2(t)$ for all $t \in \mathbb{T}$. This implies that problem (9.1) is Fredholm and the function $(\varphi_1-\varphi_2)^{-1}$ is meromorphic in $\mathcal{D}^+$ and continuously extendable onto $\mathbb{T}$. Thus, condition (9.13) is satisfied. In addition, in view of the continuity of the functions a, b, φ_1, φ_2 on $\mathbb{T}$, from (9.19) we deduce (9.12).
Consequently, Theorem 9.4 can be applied and, hence, formulae (9.11) are valid for the case under study. This completes the proof of Theorem 9.4'. ≡

COROLLARY 9.6. Let the functions φ_1 and φ_2 be meromorphic in $\mathcal{D}^+$, continuously extendable onto $\mathbb{T}$, not have any common poles in $\mathcal{D}^+$, coincide at no more than $|\varkappa|$ points of $\mathcal{D}^+$ (with regard to the multiplicities) and satisfy the requirements (9.19). Then the numbers l and l' of problem (9.1) are calculated by the formulae $l = \max\{0,2\varkappa\}$, $l' = \max\{0,-2\varkappa\}$ (and, hence, are stable).

From Theorem 9.4 we derive a sufficient condition of stability of problem (9.1) with continuous coefficients a and b.

THEOREM 9.5'. Let the functions φ_1 and φ_2 be meromorphic in $\mathcal{D}^+$, continuously extendable onto $\mathbb{T}$ and satisfy inequalities (9.19) and the requirement $\varphi_1(z) \neq \varphi_2(z)$ for all $z \in \mathcal{D}^+$. Then problem (9.1) is stable.

It appears that the sufficient condition established in Theorem 9.5 becomes also necessary, if we require that the factorization factors of the matrix function G_0 defined via formula (9.9) be continuous (in other words, that G_0 be factorable in C). This condition is satisfied [1], e.g., if the functions a and b satisfy a Hölder condition.

[1] as results from Theorem 5.8. Other theorems of Section 5.2 yield other sufficient factorability conditions of G_0 in C.

THEOREM 9.9. Let the coefficients a and b of problem (9.1) be such that the matrix function (9.9) is factorable in C. For the stability of problem (9.1), it is necessary and sufficient that there exist rational functions r_1 and r_2 satisfying the requirements

$$|tb(t)-r_1(t)| < |a(t)| < |tb(t)-r_2(t)| \quad \text{for} \quad t \in \mathbb{T}, \tag{9.19'}$$

$$r_1(z) \neq r_2(z) \quad \text{for} \quad z \in \mathfrak{D}^+. \tag{9.20}$$

Proof. Sufficiency. The condition (9.19') shows, in particular, that the function r_1 does not have poles on $\mathbb{T}$. If the function r_2 has also no poles on $\mathbb{T}$, then, setting $\varphi_j = r_j$ $(j=1,2)$, from Theorem 9.5 we may conclude the stability of problem (9.1).

Now we assume that the function r_2 has poles on $\mathbb{T}$. Using the representations $r_j = p_j/q_j$, where p_j, q_j $(j=1,2)$ are polynomials, we rewrite the second of the inequalities (9.19') and the condition (9.20) in the form

$$|a(t)q_2(t)| < |tb(t)q_2(t)-p_2(t)| \quad \text{for} \quad t \in \mathbb{T}, \tag{9.21}$$

$$p_1(z)q_2(z) - p_2(z)q_1(z) \neq 0 \quad \text{for} \quad z \in \mathfrak{D}^+. \tag{9.22}$$

Moreover, from (9.19') we deduce that $r_1(t) \neq r_2(t)$ for all $t \in \mathbb{T}$, so that inequality (9.22) continues to be valid for any $z \in \mathfrak{D}^+ \cup \mathbb{T}$. Now we introduce a polynomial $\tilde{q}_2$ that has no zeros on $\mathbb{T}$ but is arbitrarily close to q_2 in such a way that conditions (9.21) and (9.22) (the latter for any $z \in \mathfrak{D}^+ \cup \mathbb{T}$) remain true, if we replace q_2 by $\tilde{q}_2$.

Then the conditions of the theorem are preserved, if we replace r_2 by $\tilde{r}_2 = p_2/\tilde{q}_2$. The function $\tilde{r}_2$ does not have poles on $\mathbb{T}$. Hence, due to what was proved above, problem (9.1) is stable.

Necessity. Let the problem (9.1) be stable, i.e., the partial indices of the matrix function (9.9) are equal to zero. Then the partial indices of the Hermitian matrix function (9.10) with negative determinant are also equal to zero. According to Theorem 7.4, the matrix function (9.10) can be factored in the form

$$F = A_0 \begin{pmatrix} 1 & 0 \\ 0 & -1 \end{pmatrix} A_0^* . \qquad (9.23)$$

Since the matrix function G_0 (and, together with it, the matrix function F) is factorable in C, then by Corollary 7.7 the matrix function

$$A_0 = \begin{pmatrix} \alpha & \beta \\ \gamma & \delta \end{pmatrix} ,$$

which is analytic in $\mathfrak{D}^+$, is continuous and non-singular in $\mathfrak{D}^+ \cup \mathbb{T}$. Consequently

$$\alpha(z)\delta(z) - \beta(z)\gamma(z) \neq 0 \quad \text{for any } z \in \mathfrak{D}^+ \cup \mathbb{T} . \qquad (9.24)$$

Writing the factorization (9.23) elementwise, we find that

$$|b|^2-|a|^2 = |\alpha|^2-|\beta|^2 , \quad tb = \alpha\overline{\gamma} - \beta\overline{\delta} , \quad 1 = |\gamma|^2-|\delta|^2 . \qquad (9.25)$$

After some transformations we obtain from this

$$|\alpha\delta-\beta\gamma| = |a| , \quad tb\gamma-\alpha = (\alpha\delta-\beta\gamma)\overline{\delta} , \quad tb\delta-\beta = (\alpha\delta-\beta\gamma)\overline{\gamma} ,$$

so that

$$|tb\gamma-\alpha| = |a\delta| , \quad |tb\delta-\beta| = |a\gamma|. \qquad (9.26)$$

By virtue of (9.25), $|\gamma| > |\delta|$, therefore, from (9.26) it follows that

$$|tb\gamma-\alpha| < |a\gamma| , \quad |tb\delta-\beta| > |a\delta| . \qquad (9.27)$$

The functions $\alpha, \beta, \gamma, \delta$, which are analytic in $\mathfrak{D}^+$ and continuous in $\mathfrak{D}^+ \cup \mathbb{T}$, can be uniformly approximated by polynomials with arbitrary degree of accuracy. We choose polynomials p_1, q_1, p_2, q_2 sufficiently close to $\alpha, \beta, \gamma, \delta$, respectively, in such a way that, replacing $\alpha, \beta, \gamma, \delta$ by p_1, q_1, p_2, q_2, inequalities (9.24) and (9.26) remain valid. Setting $r_j = p_j/q_j$ $(j=1,2)$, we can satisfy requirements (9.19) and (9.20), which proves Theorem 9.9. ≡

9.4. COMMENTS

Beginning with the paper of MARKUSHEVICH [1] the boundary value problem (9.1) (for $\mathbb{T}$ replacing by a general contour Γ) was considered by a number of authors under various assumptions for Γ and the coefficients a, b, c (BOJARSKI [1,2], LITVINCHUK [1], MIKHAĬLOV [1-4], NIKOLAĬCHUK [1], NIKOLAĬCHUK, SPITKOVSKIĬ [1,2], PRIMACHUK [1], SABITOV [1,2]).

A detailed historical explanation of this subject is contained in Ch. 5 of the monograph LITVINCHUK [3].

The method of reducing problem (9.1) on the unit circle to the Riemann problem for two pairs of functions was proposed by one of the authors (LITVINCHUK [1,2]).[1] Lemmas 9.1, 9.2 and the corollaries of them yield a detailed description of this method. We would like to note that the approaches used in the proof of Lemmas 9.1 and 9.2 go back to MANDZHAVIDZE's paper [1]. Theorems 9.1 - 9.3 in the form stated here are borrowed from the article SPITKOVSKIǏ [8]. In the special case $t\varphi \in H_\infty$ Theorem 9.3 generalizes a result of NIKOLAǏCHUK [1]. The definition of stability of problem (9.1) was introduced by LITVINCHUK [1].

Section 9.2 is essentially the paper SPITKOVSKIǏ [8], in which the case $p \neq 2$ was considered, too. Here we restricted ourselves to the theory of problem (9.1) in the space L_2, since for this case the results obtained are more complete and easy to formulate. Formulae for the numbers l and l' in the elliptic case of problem (9.1) were first obtained by BOJARSKI [2] for Hölder coefficients a and b and, for coefficients a, b from the class L_∞, by MIKHAǏLOV [3,4].

Corollary 9.9 on the solvability of problem (9.1) in the parabolic case under the factorability condition of the coefficient b generalizes results of MIKHAǏLOV [3,4] and LITVINCHUK [1,2].

The proof of the equivalence of statements 2) and 4) from Theorem 9.6, in principle, reproduces the proof of the criterion for $T_Q(G)$ to be a Φ_--operator, which was discussed in the Comments from Chapter 4. Theorem 9.7 associated with the hyperbolic case, generalizes the correspondent result from SPITKOVSKIǏ [1].
Let us remark that the assumption concerning the coefficient a stipulated in Section 9.2 can be replaced by the weaker requirement $a \in \mathfrak{M}$ [2], as it was done in SPITKOVSKIǏ [8]. In the authors' papers LITVINCHUK, SPITKOVSKIǏ [1,2] there was obtained a strengthening of Theorem 9.4

[1] By the way, notice that the results obtained with the help of this method in the Fredholm theory of problem (9.1) may be transferred to the case of a smooth contour.

[2] The definition of the class $\mathfrak{M}$ can be found in the Comments to Chapter 6.

consisting in the claim that, actually, $l = \varkappa + \max\{\varkappa, -\varkappa, N\}$ as soon as there exists a function φ_1 satisfying conditions (9.12), while the number N is chosen to be the smallest possible. Also, there is given a complete classification of problem (9.1) (which generalizes that proposed by Mikhaĭlov and makes it more precise), a stability criterion for the defect numbers and a stability criterion for the problem itself generalizing from the parabolic case Corollary 9.9. In the papers LITVINCHUK, SPITKOVSKIĬ [1,2] the question of relationships between problem (9.1) and certain problems from theory of approximation by functions with partially fixed poles is discussed either. Here we have no possibility to deal with these questions in detail which require large volumes of additional information (mainly from ADAMYAN, AROV, KREĬN [1,2] and other papers of these authors). Thus, the reader is referred to the article LITVINCHUK, SPITKOVSKIĬ [2], in which all background information needed as well as a systematic explanation of the results themselves is contained.

Theorem 9.8 is due to BOJARSKI [1] (for Hölder coefficients) and MIKHAĬLOV [3,4] (in the general case). Another way of proving this theorem was proposed by VEKUA in [4], see also VEKUA, KVESELAVA [1].

Theorems 9.4, 9.5' and 9.9 extend corresponding results from NIKOLAĬCHUK, SPITKOVSKIĬ [1,2] to the case of continuous coefficients a and b. Several special cases of Theorem 9.4' were discussed by SABITOV [1,2] and PRIMACHUK [1].

Problem (9.1) with a piecewise continuous coefficient a (and $b \in L_\infty$) was considered for the first time in MIKHAĬLOV [4]. Employing Theorem 9.2 and the criterion of Φ-factorability of piecewise continuous matrix functions (Theorem 5.6), A.I. YATSKO and S.I. YATSKO [1,2] obtained a criterion of Fredholmness for problem (9.1) with piecewise continuous coefficients a and b in the classes L_p $(1 < p < \infty)$. In doing so, they discovered that unlike the situation described by Theorem 9.8, the Fredholmness of problem (9.1) depends on the coefficient b (and, naturally, on the parameter value p). These results have also been extended to more general coefficients a and b.

Problem (9.1) on a composite contour was studied in YATSKO [1] and in spaces with weight - in YATSKO [2]. The paper of KHVOSHCHINSKAYA [2] is devoted to the explicit solution of problem (9.1) with piecewise constant coefficients which have no more than three points of discontinuity. She made use of the reduction of (9.1) to the vector-valued Riemann boundary value problem (9.4) and applied those results from KHVOSHCHINSKAYA [1] which were mentioned in the Comments to Chapter 5.

Finally, we would like to remark that problem (9.1) with matrix coefficients a and b of dimension $n \times n$ can be reduced to a Riemann boundary value problem for $2n$ pairs of functions with the aid of the method described in the present chapter (see KAPANADZE [1,2], LITVINCHUK [1,2]). The method of reduction to a vector-valued Riemann boundary value problem proved to be effective for the study of other multi-term boundary value problems from the theory of analytic functions, too. Thus, recently the authors together with Yu.D. LATUSHKIN succeeded in obtaining new results concerning Fredholmness and bounds for the defect numbers of a generalized Carleman boundary value problem with the help of this method (the statement of this problem and related results known previously may be found in Ch. 5 of the book LITVINCHUK [3]; the latest results themselves have not been published yet).

REFERENCES

Abrahamse, M. B.

[1] The spectrum of a Toeplitz operator with a multiplicatively periodic symbol.- J. Funct. Anal., 1979, 31, N.2, p. 224-233.

Adamyan, V.M., Arov D.Z., Krein,M.G(Адамян В.М., Аров Д.З., Крейн М.Г.)

[1] О бесконечных ганкелевых матрицах и обобщенных задачах Каратеодори-фейера и Ф. Рисса. - Функц. анализ и его прил., I968, 2, Но. I, с. I-I9.

[2] Бесконечные блочно-ганкелевы матрицы и связанные с ними проблемы продолжения.- Изв. АН Арм.ССР, I97I, 6, Но. 2-3, с. 87-II2.

Alekhno, A. G.(Алехно А. Г.)

[1] О краевой задаче Римана с конечным числом точек завихрения - ДАН БССР, I979, 23, Но. I2, с. I069-I072.

Ambartsumyan, G. V.(Амбарцумян Г. В.)

[1] О методе редукции для одного класса теплицевых матриц.- Матем. исслед., I973, 8, Но. 2, с. I6I-I69.

[2] Об обращении некоторых теплицевых матриц.- Матем. исслед., I973, 8, Но. 3, с. I40-I44.

Antontsev, S.N. (Антонцев С. Н.)

[1] Обобщенная система Коши-Римана с почти регулярными коэффицентами. - В кн.: Матрич. вопр. теории функций и отображений. Вып. 7: Материалы 4-го Коллоквиума, Донецк, I974 г. Киев: Наук. думка, I975, с. 3-I7.

Azoff, E., Clancey, K. F.

[1] Toeplitz operators with sectorial matrix-valued symbols. - Indiana Univ. Math. J., 1977, 26, N.5, p. 933-938.

Babaev, A.A., Salaev, V. V. (Бабаев А. А., Салаев В. В.)

[1] Краевые задачи и сингулярные уравнения на спрямляемом контуре.- Матем. заметки, I982, 3I, Но. 4, с. 57I-580.

Babeshko, V.A. (Бабешко В. А.)

[1] Факторизация одного класса матриц-функций и её приложения. - ДАН СССР, I975, 223, Но. 5, с. I094-I097.

[2] К факторизации одного класса матриц-функций, встречающихся в теории упругости. - ДАН СССР, I975, 223, Но. 6, с. I333-I356.

[3] Эффективный способ приближенной факторизации матриц-функций. - ДАН СССР, I979, 247, Но. 5, с. I089-I093.

Ball, J., Helton, J.

[1] Factorization results related to shifts in an indefinite metric.- Integr. Equat. and Oper. Theory, 1982, 5, N 5, 632-658

Bart, H., Gohberg, I., Kaashoek, M. A.

[1] Minimal factorization of matrix and operator functions. - Operator Theory: Advances and Applications, Vol. 1. Basel - Boston - Stuttgart 1979: Birkhäuser-Verlag

[2] Wiener-Hopf factorization of analytic operator functions and realization. - Vrije Universiteit Amsterdam, Rapport nr. 231 1983

[3] Exponentially dichotomous operators and inverse Fourier transforms.- Report 8511/M (1985). Erasmus University Rotterdam.

Belarmino G.D. (Белармино Г. Д.)

[1] Краевая задача Римана и сингулярные интегральные уравнений. - Науч. тр. Азерб. ун-та Сер. физ.-мат.н., Но. 6, с. 76-85.

Belokopytova L.V., Ivanenko O.A., Fil'shtinskiĭ L.A. (Белокопытова Л. В., Иваненко О.А., Фильшеинский Л.А.)

[1] Передача нагрузки от упругого ребра к полубесконечной пьезокерамической пластине. - Изв. АН Арм. ССР. Механика, 1981, Но. 5 с. 41-51.

Bonsall F.F., Gillespie T.A.

[1] Hankel operators with PC symbols and the space $H_\infty +$ PC. - Proc. Royal Soc. Edinburg, 1981, 89 A, p. 17-24.

Bojarski B. (Боярский Б. В.)

[1] Об устойчивости задачи Гильберта для голоморфного вектора. - Сообщ. АН СССР, 1958, 21, Но. 4, с. 391-398.

[2] Об обобщенной граничной задаче Гильберта. - Сообщ. АН Груз. ССР, 1960, 25, Но. 4, с. 385-390.

[3] Анализ разрешимости граничных задач теории функций. - В кн.: Исслед. по совр. пробл. теории функций комплексного переменного. М.: Физматгиз, 1961, с. 57-59.

[4] On the index problem for systems of singular integral equations.- Bull. Acad. polon. Sci., 1963, XI No. 10; 653-655; 1965, XIII, No. 9, 627-631, 633-637.

Böttcher,A., Silbermann B.

[1] Invertibility and asymptotics of Toeplitz matrices, Berlin: Akademie-Verlag, 1983.

Bourbaki N. (Бурбаки Н.)

[1] Функции действительного переменного.- М.: Наука, 1965 transl. from: Eléments de matématique.Paris: Herman.

Budyanu M.S. (Будяну М.С.)

[1] Одна теорема о факторизации оператор-функций. - Изв. АН Молд. ССР, 1965, Но. 7, с. 22-31

Budyanu M.S., Gohberg I. (Будяну М.С., Гохберг И.Ц.)

[1] О задаче факторизации в абстрактных банаховых алгебрах. I. Распадающиеся алгебры. - Матем. исслед., 1967, 2, Но. 2, с. 25-51.

[2] О задаче факторизации в абстрактных банаховых алгебрах. II. Нераспадающиеся алгебры. - Матем. исслед., 1967, 2, Но. 3, с. 3-19.

[3] Общие теоремы о факторизации матриц-функций. I. Основная теорема. - Матем. исслед., 1968, 3, Но. 2, с. 87-103. English transl.: Amer. Math. Soc. Transl. 1973, 102, 1-14

[4] Общие теоремы о факторизации матриц-функций. II. Некоторые признаки и их следствия. - Матем. исслед., 1968, 3, Но. 3, с. 3-18. English transl.: Amer. Math. Soc. Transl. 1973, 102, 15-26.

Calderon A.P.

[1] Cauchy integrals on Lipschitz curves and related operators.- Proc. Nat. Acad. Sci. USA, 1977, 74, No. 4, p. 1324-1327.

Chakrabarti A.

[1] Derivation of the solution of certain singular integral equations. - J: Indian Inst. Sci., 1980, 62, No. 10, p. 147-157.

[2] On a system of Wiener-Hopf equations associated with a mixed diffraction problem. - Z. angew. Math. und Mech., 1981, 61, No. 7, p. 339-341.

Chebotarev G.N. (Чеботарев Г. Н.)

[1] К решению в замкнутой форме краевой задачи Римана для системы n пар функций. - Уч. зап. Казанск. Ун-та, I956, II6, Но. 4, с. 3I-58.

[2] Частные индексы краевой задачи Римана с треугольной матрицей второго порядка. - Успехи матем. наук, I956, II, вып. 3, с. I92-202.

[3] Некоторые матричные уравнения и их применение к решению в замкнутой форме краевой задачи Римана. - Тр. Всесоюзного матем. Съезда, I956, I, с. III.

Chebotarev G.N., Gakhov F.D. (Чеботарев Н. Г., Гахов Ф. Д.)

[1] О краевой задаче Римана для системы n пар функций. - Уч. зап. Казанск. ун-та, I950, II0, Но.7,с.45-50.

Chebotaru I.S. (Чеботару И.С.)

[1] Сведение уравнений Винера-Хопфа к системам с нулевыми индексами. Изв. АН Молд. ССР, I967, Но. 8, с. 54-66.

Chelidze V.G., Dzhvarsheĭshvili G.G. (Челидзе В.Г., Джваршейшвили А.Г.)

[1] Интеграл типа Коши. - Изв. вузов. Математика, I978, Но. 6, с. II7-I28.

Cherepanov G.P. (Черепанов Г. П.)

[1] Решение одной линейной краевой задачи для двух функций и её приложение к некоторым смешанным задачам теории упругости. - ПММ, I962, 26, Но. 5, с. 907-9I2.

[2] Задача Римана-Гильберта для внешности разрезов вдоль прямой или вдоль окружности. - ДАН СССР, I964, I56, Но. 2, с. 275-277.

[3] Об одном интегрируемом случае краевой задачи Римана для нескольких функций. - ДАН СССР, I965, I6I, Но. 6, с. I285-I288.

Choriev Kh. (Чориев Х.)

[1] Устойчивость частичных индексов и численный метод решения задач Римана и Римана-Гильберта для матриц. - Научн. труны Ташк. ун-та, I974, 460,с. I63-I68.

[2] Регулярная устойчивость краевой задачи Римана и вычисление частичных индексов. - Сб. научн. трудов. Ташк. ун-та, I977, 548, с. I20-I23.

[3] Регулярная устойчивость краевой задачи Римана и вычисление частичных индексов. - В кн.: Вопр. метрич. теории отображений и её применение: Материалы 5-го коллоквиума, Донецк, I976, г. Киев: Наук. думка, I978, с. I46-I48.

Clancey K.F.

[1] A local result for systems of Riemann-Hilbert barrier problems.- Trans. Amer. Math. Soc., 1974, 200, p. 315-325.

Clancey K.F., Gohberg I.

[1] Local and global factorizations of matrix valued functions. - Trans. Amer. Math. Soc., 1977, 232, p. 155-167.

[2] Localization of singular integral operators. - Math. Zeitschr., 1979, 169, p. 105-117.

[3] Factorization of matrix functions and singular integral operators. Basel-Boston-Stuttgart: Birkhäuser-Verlag, 1981.

Clancey K.F., Gosselin J.A.

[1] On the local theory of Toeplitz operators. - Ill. J. Math., 1978, 22, N 3, p. 449-458.

Coburn L.A.

[1] Weyl's theorem for non-normal operators. - Michigan Math. J., 1966, 13, p. 285-286.

Cotlar M.A.

[1] A unified theory of Hilbert transforms and ergodic theorems. - Rev. Mat. Guyana, 1955, 1, N 2, p. 105-167.

Danilov E.A. (Данилов Е.А.)

[1] Зависимость числа решений однородной задачи Римана от контура и модуля коэффициента. - ДАН СССР, I982 г., 264, Но. 6, с. I305-I308.

Danilyuk I.I. (Данилюк И.И.)

[1] О задаче Гильберта с измеримыми коэффициентами. - Сибирск. матем. ж., I960, I, Но. 2, с. I7I-I97.

[2] К теории одномерных сингулярных уравнений. - В кн.: Проблемы механики сплошной среды. М.: Изд-во АН СССР, I96I, с. I35-I44.

[3] Лекции по краевым задачам для аналитических функций и сингулярным интегральным уравнениям. - Новосибирск: Новосиб. ун-т I964.

[4] Про загальну линийну граничну задачу з вимирними коэффициентами для багатьок аналитичных функций в класси E_p. - ДАН УРСР, сер. А, I97I, 9, с. 774-777.

[5] Об общей линейной граничной задаче с измеримыми коэффициентами для многих аналитических функций в классе E_p. - В кн.: Мех. сплошн. среды и родств. пробл. анализа. М.: Наука, I972, с. I75-I79.

[6] Нерегулярные граничные задачи на плоскости. - М.: Наука, I975.

Danilyuk I.I., Shelepov V.Yu. (Данилюк И.И., Шелепов В.Ю.)

[1] Об ограниченности в L_p сингулярного оператора с ядром Коши вдоль кривой ограниченного вращения. - ДАН СССР, I974, Но. 3, с. 5I4-5I7.

[2] При обмежнисть у зважених просторах L_p сингулярных интегральних операторов вздовж линий з обмежним обертанням. - ДАН УРСР, сер. А., I969, 3, с. I99.

Devinatz A.

[1] Toeplitz operators on H^2 spaces. - Trans. Amer. Math. Soc., 1964, 112, p. 304-307.

Devinatz A., Shinbrot M.

[1] General Wiener-Hopf operators. - Trans. Amer. Math. Soc., 1969, 145, p. 467-494.

Didenko V.D. (Диденко В.Д.)

[1] Сходимость коллокационного метода и метода механических квадратур для систем сингулярных интегральных уравнений. - Изв. вузов Математика, I979, Но. I, с. 76-79.

[2] Метод механических квадратур для систем сингулярных интегральных уравнений. - Изв. вузов. Математика, I979, Но. 3, с. 73-76.

[3] Приближенное решение систем сингулярных интегральных уравнений в полупространствах непрерывных функций. - ДАН СССР, I980, 254, Но. 6, с. I3I4 - I3I7.

Didenko V.D., Chernetskiĭ V.A. (Диденко В.Д., Чернецкий В.А.)

[1] Краевая задача Римана с комплексной ортогональной матрицей. - Матем. заметки, I978, 23, вып. 3, с. 405-4I6.

Didenko V.D., Tikhonenko N.Ya. (Диденко В.Д., Тихоненко Н.Я.)

[1] О приближенном решении матричной задачи Римана. - Изв. вузов. Математика, I980, Но. I, с. I6-I9.

[2] О приближенном решении матричной задачи Римана. - Дифференц. уравнения, I98I, I7, Но. II, с. 2087-2089.

Douglas R.G.

[1] Toeplitz and Wiener-Hopf operators in H^{∞} + C. - Bull. Amer. Math. Soc., 1968, 74, N 5, p. 895-899.

[2] Banach algebra techniques in operator theory. - New York & London: Academic Press, 1972.

Douglas R.G., Sarason D.E.

[1] Fredholm Toeplitz Operators. - Proc. Amer. Math. Soc., 1970, 26, p. 117-120.

Douglas R.G., Widom H.

[1] Toeplitz operators with locally sectorial symbols. - Indiana Univ. Math. J., 1970, 20, Nr. 4. p. 385-388.

Dukhovnyĭ A.M. (Духовный А.М.)

[1] Одномерное случайное блуждание, зависящее от параметра. - ДАН УССР. Сер. А., I962, Но. 3, с. 9-I2.

Duracz T., Zelazny R.

[1] Solutions to some Riemann-Hilbert problems in two-group neutron transport theory. - Trans. Theory and Statist. Phys., 1982, 11, Nr. 1, p. 5-28.

Duren P.L.

[1] Theory of H^p spaces. - New York & London: Academic Press, 1970.

Duren P.L., Shapiro H.S., Shields A.L.

[1] Singular measures and domains not of Smirnov type. - Duke Math. J., 1966, 13, Nr. 2, p. 247-254.

Dybin V.B., Dodokhova G.V. (Дыбин В. Б., Додохова Г.В.)

[1] Корректная постановка краевой задачи Римана на прямой с почти-периодическим разрывом коэффициента. - ВИНИТИ, Но. I497-8I Деп., I98I.

[2] Корректная постановка краевой задачи Римана на замкнутом контуре в случае почти-периодических разрывов у её коэффициента. - ВИНИТИ, Но. 3579-82 Деп., 1982.

Dzhvarsheĭshvili A.G. (Джваршейшвили А.Г.)

[1] Замечание об одной граничной задаче. - Сообщ. АН Груз ССР, 1981, 102, Но. 2, с. 297-299.

Emets Yu.P. (Емец Ю.П.)

[1] О некоторых приложениях краевой задачи Римана в магнитной гидродинамике. - В кн.: Материали Всес. конференции по краев. задачам. Казань: Казанск. ун-та, 1970, с. 101-106.

Engibaryan N.B. (Енгибарян Н.Б.)

[1] Факторизация матриц-функций и нелинейные интегральные уравнения. - Изв. АН Арм. ССР. Математика, 1980, 15, Но. 3, с. 233-244.

Faour Nazih S.

[1] The Fredholm index of a class of vector-valued singular integral operators. - Indian J. Pure and Appl. Math., 1980, 11, Nr. 2, p. 135-146.

Feinstein J., Shamash Y.

[1] Spectral factorization of a rational matrix. - IEEE Trans. Inform. Theory, 1977, 23, Nr. 4, p. 534-538.

Frolov V.D. (Фролов В.Д.)

[1] К теории сингулярных интегральных уравнений с измеримыми коэффициентами. - ДАН СССР, 1968, 189, Но. 6, с. 1185-1188.

[2] О сингулярных интегральных уравнениях с измеримыми коэффициентами в пространствах L_p с весом. - Матем. исслед., 1970, 5, Но. 1, с. 141-151.

Gabdulkhaev B.G. (Габдулхаев Б.Г.)

[1] Некоторые вопросы теории приближенных методов. - Изв. вузов, Математика, 1971, Но. 6, с. 15-23.

Gakhov F.D. (Гахов Ф.Д.)

[1] О краевой задаче Римана для систем n пар. функций. - ДАН СССР, 1949, 67, Но. 4, с. 601-606.

[2] Один случай краевой задачи Римана для системы n пар функций. - Изв. АН СССР, сер. матем., 1950, 14, Но. 6, с. 549-568.

[3] Краевая задача Римана для систем n пар функций. - Успехи матем.наук, 1952, 7, Но. 4, с. 3-54.

[4] Краевые задачи. - М.: Наука, 1977.

Gakhov F.D., Cherskiĭ Yu.I. (Гахов Ф.Д., Черский Ю.И.)

[1] Уравнения типа свертки. - М.: Наука, 1978.

Ganin M.Ts. (Ганин М.Ц.)

[1] Об интегральном уравнении Фредгольма с ядром, зависящим от разности аргументов. - Изв. вузов. Математика, 1963, Но. 2, с. 31-43.

Gantmakher F.R. (Gantmacher F.R., Гантмахер Ф. Р.)

[1] Теория матриц. - М.: Наука, I967.
English transl. of 1st ed.: New York, Chelsea, 1959;
German transl: Berlin, Deutscher Verlag der Wiss., 1970

Gavdzinskiĭ V.N., Spitkovskiĭ I.M. (Гавдзинский В.Н., Спитковский И.М.)

[1] Об одном способе эффективного построения факторизации. - Украинск. матем. ж., I982, 34, Но. I, с. I5-I9.

Gerus O.F. (Герус О.Ф.)

[1] Об одном особом интегральном уравнении и краевой задаче Римана. - Украинск. матем. ж., I98I, 33, Но. 3, с. 282-285.

Glazman I.M., Lyubich Yu.I. (Глазман И.М., Любич Ю.И.)

[1] Конечномерный линейный анализ в задачах. - М.: Наука, I969.

Glebov A.V., Deundyak V.M. (Глебов А.В., Деундяк В.М.)

[1] Гомотопическая классификация нетеровых систем сингулярных интегральных операторов с разрывными коэффициентами. - Функцион. анализ и его прил., I978, I2, Но, 3, с. 76-77.

Gohberg I. (Гохберг И.Ц.)

[1] О числе решений однородного сингулярного интегрального уравнения с непрерывными коэффициентами. - ДАН СССР, I958, I22, Но. 3, с. 327-330.

[2] О границах частных индексов матриц-функций. - Успехи матем. наук, I959, I4, вып. 4, с. I59-I64.

[3] Общая задача о факторизации матриц-функций в нормированных кольцах и её приложения. - ДАН СССР, I962, I46, Но. 2, с. 284-287.

[4] Задача факторизации в нормированных кольцах, функции от изометрических и симметрических операторов и сингулярные интегральные уравнения. - Успехи матем. наук, I964, I9, вып. I, с, 7I-I24.

Gohberg I., Feldman I.A. (Гохберг И.Ц., Фельдман И.А.)

[1] Уравнения в свертках и проекционные методы их решения. - М.: Наука, I97I.
English transl.; Convolution equations and projection methods for their solution. Transl. Math. Mon., vol. 41, 1974, Providence, R.I.
German transl.: Faltungsgleichungen und Projektionsverfahren zu ihrer Lösung. Akademie-Verlag Berlin 1974.

[2] Одно замечание о стандартной факторизации матриц-функций. - Уч. зап. Бельцкого педин-та, I960, 5, с. 65-69.

[3] Об индексах кратных расширений матриц-функций. - Изв. АН Молд. ССР, I967, 8, с. 76-80.

Gohberg, I., Kaashoek M.A., van Shagen F.

[1] Similarity of operator blocks and canonical forms. II. Infinite dimensional case and Wiener-Hopf factorization. - Top. Mod. Oper. Theory. 5 Int. Conf. Oper. Theory. Timisoara and Herculane. June 2-12 1980, - Basel e.a., 1981, p. 121-170.

Gohberg I., Krein M.G. (Гохберг И.Ц., Крейн М.Г.)

[1] Основные положения о дефектных числах, корневых векторах и индексах, линейных операторов. - Успехи матем. наук, 1957, 12, вып. 2, с. 43-118. (English transl. Amer. Math. Soc. transl., 1960, 13, p. 185-264).

[2] Об устойчивой системе частных индексов Гильберта для нескольких неизвестных функций. - ДАН СССР, 1958, 119, Но. 5, с. 854-857.

[3] Системы интегральных уравнений на полупрямой с ядрами, зависящими от разности аргументов. - Успехи матем. наук, 1958, 13, вып. 2, с. 3-72. (English trans.: Amer. Math. Soc. transl. 1960, 2, p. 217-287).

[4] Введение в теорию линейных несамосопряженных операторов. - М.: Наука, 1965. (English transl.: Transl. Math. Monographs, vol. 18, 1969, Providence, R.I.)

[5] Теория вольтерровых операторов в гильбертовом пространстве и её приложения. - М.: Наука, 1967. (English transl.: Transl. Math. Monographs, 1970, Providence, R.I.)

Gohberg I., Krupnik N.Ya. (Гохберг И.Ц., Крупник Н.Я.)

[1] Система сингулярных интегральных уравнений в пространстве L_p с весом. - ДАН СССР, 1969, Но. 5, с. 998-1001.

[2] Сингулярные интегральные операторы с кусочно-непрерывными коэффициентами и их символы. - Изв. АН СССР, сер. матем. 1971, 35, Но. 4, с. 940-964.

[3] Об одном локальном принципе и алгебрах, порожденных теплицевыми матрицами. - Annakle ştiintifice ale univ. "Al. I. Cuza", Iaşi, Sect. I.a) Matematica, 19, 1, 1973.

[4] Введение в теорию одномерных сингулярных интегральных операторов. - Кишинев: Штиинца, 1973. German transl.: Einführung in die Theorie der eindimensionalen singulären Integraloperatoren - Basel - Boston - Stuttgart, Birkhäuser Verlag, 1979.

Gohberg I., Leiterer J. (Гохберг И.Ц., Лайтерер Ю.)

[1] Общие теоремы о канонической факторизации оператор-функций относительно контура. - Матем. исслед., 1972, 25, Но. 3, с. 87-134.

[2] Факторизация оператор-функций относительно контура. - I. Mathem. Nachr. 1972, N. 52, 259-282. - II. Mathem. Nachr. 1972, N 54, 41-74. - III. Mathem. Nachr. 1973, N 55, 33-61.

Gohberg I., Sementsul A.A. (Гохберг И.Ц., Семенцул А.А.)

[1] Теплицевы матрицы, составленные из коэффициентов фурье функций с разрывами почти-периодического типа. - Матем. исслед., 1970, 5, Но. 4, с. 63-83.

Goluzin G.M. (Голузин Г.М.)

[1] Геометрическая теория функций комплексного переменного. - М.: Наука, 1966.

Gordadze E.G. (Гордадзе Э.Г.)

[1] О задаче Римана-Привалова в случае негладкой граничной линии. - Тр. Тбилисс. мат. ин-та ГрузССР, 1967, 33, с. 25-31.

[2] О сингулярном интегральном операторе с кусочно-непрерывным коэффициентом. - Сообщ. АН ГрузССР, 1971, 63, Но. 2, с. 277-280.

[3] О сингулярных интегралах с ядром Коши. - Тр. Тбилисс. мат. ин-та АН ГрузССР, 1972, 42, с. 5-17.

[4] О сингулярном интегральном операторе на негладких линиях. - Тр. Симпоз. по мех. сплошн. среды и родствен. проб. анализа, 1971. Тбилиси: Мецниереба, 1974, 2. с. 74-85.

[5] О граничной задаче линейного сопряжения. - Сообщ. АН ГрузССР, 1976, 81, Но. 3, с. 545-548.

[6] О граничной задаче линейного сопряжения для линий Радона. - Сообщ. АН ГрузССР, 1976, 84, Но. 1, с. 29-32.

Gordadze E.G., Khvedelidze B.V. (Гордадзе Э.Г., Хведелидзе Б. В.)

[1] О сингулярных интегральных уравнениях и задаче регуляризации. - Тр. Тбилис. мат. ин-та АН ГрузССР, 1976, 53, с. 15-17.

Gordienko V.N. (Гордиенко В.Н.)

[1] Факторизация матриц-функций частного вида. - Украинск. матем. ж., 1971, 23, Но. 1, с. 81-88.

Gorodzha L.V., Emets Yu.P., Zhukova N.I., Zverovich E.I.
(Городжа Л.В., Емец Ю.П., Жукова Н.И., Зверович Э.И.)

[1] О применении обобщенной краевой задачи Римана к расчету электрических полей. - ДАН БССР, 1979, 23, Но. 2, с. 118-120.

Govorov N.V. (Говоров Н.В.)

[1] О краевой задаче Римана с бесконечным индексом. - ДАН СССР, 1964, 154, Но. 6, с. 1247-1249.

[2] Об ограниченных решениях однородной краевой задачи Римана с бесконечным индексом степенного порядка. - Теория функций, функц. анадиз и их прил., 1969, 11, с. 3-34.

Grudskiĭ S.M. (Грудский С.М.)

[1] О компактности одного интегрального оператора. - ВИНИТИ, Но. 4856-80 Деп., 1980, 10 с.

[2] Краевая задача Римана с разрывами почти-периодического типа в классе $L_\infty(\Gamma)$. - Дифференц. и интеграль. уравнений и их прил. Элиста, 1982, с. 30-41.

Grudskii S.M., Dybin V.B. (Грудский С.М., Дыбин В.Д.)

[1] Краевая задача Римана с разрывами почти-периодического типа у её коэффициента. - ДАН СССР, 1977, 237, Но. 1, с. 21-24.

[2] Краевая задача Римана в пространстве $L_p(\Gamma,\rho)$ с почти-периодическими разрывами у её коэффициента. - Матем. исслед., 1980, 54, с. 36-49.

Grudskii S.M., Khevelev A.V. (Грудский С.М., Хевелев А.В.)

[1] Об обратимости в $L_2(R)$ сингулярных интегральных операторов с периодическими коэффициентами и сдвигом. - ДАН СССР, 1983, 269, Но. 6, с. 1303-1306.

Gusak D.V. (Гусак Д.В.)

[1] Метод факторизации в граничных задачах для одного класса процессов на цепи Маркова. I - Киев: Препринт ИМ АН УССР 78.6, 1978.

[2] Метод факторизации в граничных задачах для одного класса процессов на цепи Маркова. II - Киев: Препринт ИМ АН УССР 78.II, 1978.

Halmos P.

[1] A Hilbert space problem book. - Princeton, New Jersey: Van Nostrand Company, Inc., 1967.

Halmos P., Sunder V.S.

[1] Bounded operators on L^2 spaces. - Berlin: Springer, 1978.

Hartman P.

[1] On completely continuous Hankel matrices. - Proc. Amer. Math. Soc., 1958, 9, N 6, p. 862-866..

Hartman P., Wintner A.

[1] The spectra of Toeplitz's matrices. - Amer. Math. J., 1954, 76, p. 867-882.

Heinig G. (Хайниг Г.)

[1] Об обратимости сингулярных интегральных операторов. - Сообщ. АН ГрузССР, 1979, 96, No. I, с. 29-32.

[2] Struktur des Kernes und partielle Indizes bei allgemeinen linearen Operatoren.- Preprint P - 36/80. IMath. Berlin, 1980.

Heinig G., Rost K.

[1] Algebraic methods for Toeplitz-like matrices and operators. - Mathematical Research vol. 19, Berlin: Akademie Verlag, 1984. - Operator Theory: Advances and Appl., vol. 13, Basel-Boston-Stuttgart: Birkhäuser Verlag 1984.

Heinig G., Silbermann B.

[1] Factorization of matrix functions in algebras of bounded functions. - In: Spectral theory of linear operators and related topics. Operator Theory: Advances and Appl., vol. 14, p. 157-177; Basel: Birkhäuser Verlag, 1984.

Heins A.E.

[1] The Sommerfeld half-plane problem revisited, I. The solution of a pair of coupled Wiener-Hopf integral equations. - Math. Meth. Appl. Sci., 1982, 4, N.1, p. 74-90.

[2] The Sommerfeld half-plane problem revisited, II. The factoring of a matrix of analytic functions. - Math. Meth. Appl. Sci., 1983, 5, N.1, p. 14-21.

Hoffman K.

[1] Banach spaces of analytic functions. - Prentice-Hall, Englewood Cliffs, New Jersey, 1962.

Hoffman K., Singer I.M.

[1] Maximal algebras of continuous functions. - Acta Math., 1960, 103, p. 217-241.

Hunt R.A., Muckenhoupt B., Wheeden R.L.

[1] Weighted norm inequalities for the conjugate function and Hilbert transform. - Trans. Amer. Math. Soc., 1973, 176, p. 227-252.

Hurd R.A.

[1] A note on the solvability of simultaneous Wiener-Hopf equations. - Canad. J. Phys., 1979, 57, N 3, p. 402-403.

Idemen M.

[1] A new method to obtain exact solutions of vector Wiener-Hopf equations. - Z. angew. Math. Mech., 1979, 59, p. 656-658.

[2] On an integral transform with kernel and its applications to second order canonical problems of GTD. - SIAM J. Appl. Math., 1982, 42, N.3, p. 636-652.

Isakhanov R.S. (Исаханов Р.С.)

[1] Об одной задаче линейного сопряжения для кусочно-голоморфных векторов. - Сообщ. АН ГрузССР, 1969, 53, Но. 3, с. 537-540.

Ishchenko E.V. (Ищенко Е.В.)

[1] Скалярная разрывная граничная задача линейного сопряжения в случае общей кусочно-гладкой кривой. - Сообщ. АН Груз. ССР, 1982, 108, Но. 3, с. 489-492.

[2] О факторизации кусочно-непрерывных матриц-функций. - Сообщ. АН ГрузССР, 1982, 106, Но. 2, с. 245-248.

[3] Векторная разрывная задача линейного сопряжения. - Сообщ. АН ГрузССР, 1983, 109, Но. 1, с. 37-40.

Iskenderov B.B. (Искендеров Б.Б.)

[1] Об одной нелинейной краевой задаче для кусочно-аналитических функций. - Науч. тр. Азерб. ун-та. Сер. физ.-мат. наук, 1979, 5, с. 30-38.

Its A.R. (Итс А.Р.)

[1] Асимптотика решений нелинейного уравнения Шредингера и изомонодромные деформации систем линейных дифференциальных уравнений. - ДАН СССР, 1981, 261, Но. 1, с. 14-18.

Ivanov V.V. (Иванов В.В.)

[1] Некоторые свойства особых интегралов типа Коши и их приложения. - ДАН СССР, 1958, 121, Но. 5, с. 793-794.

Jonckheere E., Delsarte P.

[1] Inversion of Toeplitz operators, Levinson equations, and Gohberg-Krein factorization - A simple and unified approach for the rational case. - J. Math. Anal. and Appl., 1982, 87, N.1, p. 295-310.

Jonckheere E., Silverman L.

[1] Spectral theory of the linear-quadratic optimal control problem: analytic factorization of rational matrix-valued functions. - SIAM J. Control and optimization, 1981, 19, N 2, p. 262-281.

Kadushin V.P. (Кадушин В.П.)

[1] К прямым методам решения одного класса сингулярных интегральных уравнений. - Изв. вузов. Математика, 1976, Но. II, с. 109-III.

Kapanadse G.A. (Капанадзе Г.А.)

[1] Обобщенная граничная задача Гильберта для нескольких неизвестных функций. - Сообщ. АН ГрузССР, 1973, 72, Но. 3, с. 537-540.

[2] Об одной обобщенной граничной задаче Гильберта для нескольких неизвестных функций в случае круга единичного радиуса. - Сообщ. АН ГрузССР, 1974, Но. I, с. 25-28.

Karapetyants N.K., Samko S.G. (Карапетянц Н.К., Самко С.Г.)

[1] Сингулярные интегральные операторы со сдвигом Карлемана в случае кусочно-непрерывных коэффициентов. I - Изв. вузов Математика, 1975, Но. 2, с. 43-54.

Karlovich Ju.I., Spitkovskiĭ I.M. (Карлович Ю.И., Спитковский И.М.)

[1] О нетеровости некоторых сингулярных интегральных операторов с матричными коэффициентами класса SAP и связанных с ними систем уравнений свертки на конечном промежутке. - ДАН СССР, 1983, 269, Но. 3,

Kato T.

[1] Perturbation theory for linear operators.-Berlin: Springer-Verlag, 1966.

Kats B.A. (Кац Б.А.)

[1] Краевая задача Римана с осциллирующим коэффициентом. - Труды семинара по краевым задачам. Казань, 1977, вып. I4, с. II0-I20.

[2] О краевой задаче Римана с коэффициентом, допускающим разрывы колебательного типа. - ДАН СССР, 1979, 244, Но. 3, с. 521-525.

[3] Задача Римана на замкнутой жордановой кривой. - Изв. вузов. Математика, 1983, Но. 4, с. 68-80.

Khavin V.P. (Хавин В.П.)

[1] Граничные свойства интегралов типа Коши и гармонически сопряженных функций в областях со спрямляемой границей. - Матем. сб., 1965, 68, Но. 4, с. 499-517.

Khavinson S.Ya. (Хавинсон С.Я.)

[1] Экстремальные задачи для некоторых классов аналитических функций в конечно-связных областях. - Матем. сб., 1955, 36, Но. 3, с. 445-478.

Khlgatyan R.G. (Хлгатян Р.Г.)

[1] Некоторые вопросы теории систем разностных уравнений Винера-Хопфа. - ВИНИТИ, Но. 227-80 Деп., 1980.

Khrapkov A.A. (Храпков А.А.)

[1] Некоторые случаи упругого равновесия бесконечного клина с несимметричным надрезом в вершине под действием сосредоточенных сил. - Прикл. матем. и мех., 1971. 35, Но. 4, с. 677-689.

Khvedelidze B.V. (Хведелидзе Б.В.)

[1] Линейные разрывные граничные задачи функций, сингулярные интегральные уравнения и некоторые их приложения. - Тр. Тбилисс. мат. ин-та ГрузССР, 1956, 23, с. 3-158.

[2] Замечания к моей работе "Линейные разрывные граничные задачи теории функций, сингулярные интегральные уравнения и некоторые их приложения", - Сообщ. АН ГрузССР, 1958, 21 Но. 2, с. 129-130.

[3] Граничная задача Римана-Привалова с кусочно-непрерывным коэффициентом. - Тр. Груз. политех. ин-та, 1962, 81, Но. 1, с. 11-29.

[4] Метод интегралов типа Коши в разрывных граничных задачах теории голоморфных функций одной комплексной переменной. - В кн. Итоги науки и техники, Сер. "Современные проблемы математики". М., 1975, 7, с. 5-162.

[5] О задаче линейного сопряжения и характеристических сингулярных интегральных уравнениях. - В кн. Комплексный анализ и его приложения. М.: Наука, 1978, с. 577-585.

Khvedelidze B.V., Ishchenko E.V. (Хведелидзе Б.В., Ищенко Е.В.)

[1] О разрывной задаче линейного сопряжения с кусочно-непрерывным коэффициентом. - Сообщ. АН ГрузССР, 1980. 97, Но. 3, с. 529-532.

[2] Разрывная граничная задача линейного сопряжения с кусочно-непрерывным коэффициентом. - Тр. Тбилисс. мат. ин-та АН ГрузССР, 1982, 69, с. 108.

Khvoshchinskaya L.A. (Хвощинская Л.А.)

[1] Однородная краевая задача Римана для двух функций с кусочно-постоянной матрицей в случае двух или трех особых точек. - ВИНИТИ, Но. 5157-81 Деп., 1981, 48, с.

[2] Решение некоторых задач, сводящихся к краевой задаче Римана для двух пар функций с кусочно постоянной матрицей. - ВИНИТИ, Но. 6531-82, Деп., 1982.

Kirillov A.A., Gvishiani A.D. (Кириллов А.А., Гвишиани А.Д.)

[1] Теоремы и задачи функционального анализа. - М.: Наука, 1979.

Kokilashvili V.M. (Кокилашвили В.М.)

[1] О весовых неравенствах для сингулярных интегралов с ядром Коши на гладких контурах. - Сообщ. АН ГрузССР, 1978, 90, Но. 3, с. 537-540.

Kokilashvili V.M., Paatashvili V.A. (Кокилашвили В.М., Пааташвили В.А.)

[1] О краевой задаче линейного сопряжения с измеримыми коэффициентами. - ДАН СССР, 1975, 224, Но. 5, с. 1008-1011.

[2] Краевая задача линейного сопряжения с измеримыми коэффициентами. - Тр. Тбилисс. матем. ин-та АН ГрузССР, 1977, 55, с. 59-92.

[3] Задача линейного сопряжения с измеримыми коэффициентами для одного класса граничных кривых. - Сообщ. АН ГрузССР, 1978, 91, Но. 1, с. 25-27.

[4] О разрывной задаче линейного сопряжения и сингулярных интегральных уравнениях. - Дифференц. уравнения, 1980, 16, Но. 9, с. 1650-1659.

Kolmogorov A.N., Fomin S.V. (Колмогоров А.Н., Фомин С.В.)

[1] Элементы теории функций и функционального анализа. - М.: Наука, 1972.

Komyak I. I. (Комяк И.И.)

[1] Об интегральном уравнении типа свертки с N ядрами. - ДАН СССР, 1968, 179, Но. 2, с. 279-282.

[2] О решении в квадратурах интегрального уравнения Фредгольма с ядром, зависящим от разности аргументов. - Изв. АН БССР. Сер. физ-мат. н., 1973, Но. 2, с. 37-42.

[3] Об одном случае разрешимости в квадратурах интегрального уравнения с ядром, зависящим от разности. - Вестник БГУ. Сер. I, 1974, Но. I, с. 70-72.

[4] Формула обращения для уравнения Вольтерра с разностным ядром. ДАН БССР, 1976, 20, Но. 2, с. 106-109.

Koppelman W.

[1] Singular integral equations, boundary value problem and the Riemann-Roch theorem. - J. Math. Mech., 1961, 10, N2, p.247-277.

Krasnosel'skiĭ M.I., Zabreĭko P.P., Pustyl'nik E.I., Sobolevskiĭ P.E. (Красносельский М.И., Забрейко П.П., Пустыльник Е.И., Соболевский П.Е.)

[1] Интегральные операторы в пространствах суммируемых функций. - М.: Наука, 1966. English version: Integral operators in spaces of summable functions. Leyden: Noordhoff Intern. Publish. 1976.

Kravchenko V.G., Nikolaĭchuk A.M. (Кравченко В.Г., Николайчук А.М.)

[1] Об эквивалентности одного типа краевой Римана для системы n пар функций полному сингулярному уравнению с ядром Коши. - Дифференц. уравнения, 1973, 9, Но. 2, с. 343-376.

[2] О частных индексах задачи Римана для двух пар функций. - ДАН СССР, 1974, 216, Но. I, с. 53-56.

Kreĭn M.G. (Крейн М.Г.)

[1] Интегральные уравнения на полупрямой с ядром, зависящим от разности аргументов. - Успехи матем. наук, 1958, 13, вып. 5, с. 3-120.

Kreĭn M.G., Krasnosel'skiĭ M.I., Mil'man D.P. (Крейн М.Г., Красносельский М.А., Мильман Д.П.)

[1] О дефектных числах линейных операторов в банаховом пространстве и о некоторых геометрических вопросах. - Сб. трудов ин-та матем. АН УССР, 1948, II, с. 97-112.

Kreĭn M.G., Melik-Adamyan F.E. (Крейн М.Г., Мелик-Адамян Ф.Э.)

[1] К теории S-матриц канонических дифференциальных уравнений с суммируемым потенциалом. - ДАН СССР, 1968, 46, Но. 4, с. 150-155.

[2] Некоторые приложения теоремы о факторизации унитарной матрицы. - Функцион. анализ и его прил., 1970 4, с. 73-75.

Krein M.G., Spitkovskiĭ I.M. (Крейн М.Г., Спитковский И.М.)

[1] О факторизации матриц-функций на единичной окружности. ДАН СССР, 1977, 234, Но. 2, с. 287-290.

[2] О факторизации α -секториальных матриц-функций на единичной окружности. - Матем. исслед., 1978, Но. 47, с. 41-63.

[3] О некоторых обобщениях первой предельной теоремы Сегё. - Anal. Math. 1983, 9, N. I, 23-41.

Kreĭn S. G. (Крейн С.Г.)

[1] Линейные уравнения в Банаховом пространстве. - М.: Наука, 1971.

Krichever I.M. (Кричевер И.М.)

[1] Аналог формулы Даламбера для уравнений главного кирального поля и уравнения - ДАН СССР, 1980, 253, Но. 2, с. 288-292.

[2] Авнеомодельные решения уравнений типа Кортевега-де Фриза. - Функц. анализ и его прил., 1980, 14, Но 3, с. 83-84.

Kruglov V.E. (Круглов В.Е.)

[1] Решение задачи Римана на одной n-листной римановской поверхности. - Матем. исслед., 1974, 9, Но. 2, с. 230-236.

[2] Абелевы дифференциалы и уравнения поверхности, заданные циклической группой подстановок. - Сообщ. АН ГрузССР, 1978, 92, Но. 3, с. 537-540.

[3] Частные индексы, абелевы дифференциалы I рода и уравнение поверхности, заданные конечной абелевой группой подстановок.- Сибирск. матем. ж., 1981, 22, Но. 6, с. 87-101.

[4] Частные индексы и одно приложение факторизации некоторых матриц подстановочного типа не выше четвертого порядка, I. - ВИНИТИ, Но. 3278-82 Деп., 1982,

[5] Частные индексы и одно приложение факторизации некоторых матриц подстановочного типа не выше четвертого, II. - ВИНИТИ, Но 3279-82, Деп., 1982.

[6] Частные индексы и одно приложение факторизации некоторых матриц подстановочного типа не выше четвертого порядка, III.- ВИНИТИ, Но. 3280-82, Деп., 1982,

Krupnik N.Ya. (Крупник Н.Я.)

[1] О фактор-норме сингулярного оператора. - Матем. исслед., 1975, Но.2, 10, с. 255-263.

[2] Критерий нетеровости сингулярных интегральных операторов с измеримыми коэффициентами. - Сообщ. АН ГрузССР, 1975, 80, Но. 3, с. 533-536.

[3] Некоторые общие вопросы теории одномерных сингулярных интегральных уравнений с матричными коэффициентами. - Матем. иссл. 1976, Но. 42, с. 91-112.

[4] О сингулярных интегральных операторах с матричными коэффициентами. - Матем. иссл., 1977, Но. 45, с. 93-99.

[5] Некоторые следствия из теоремы Ханта, Макенхаупта и Видена. - Матем. иссл., 1978, Но. 47, с. 64-70.

[6] Банаховы алгебры с символом и сингулярные интегральные операторы. – Кишиниев: Штиинца, 1984.

Krupnik N.Ya., Nyaga V.I. (Крупник Н.Я., Няга В.И.)

[1] О сингулярных интегральных операторах в случае некладкого контура. – Матем. исслед., 1975, 10, Но. 1, с. 144-164.

Krupnik N.Ya., Polonskii E.P. (Крупник Н.Я., Полонский Е.П.)

[1] О норме оператора сингулярного интегрирования. – Функцион. анализ и его прил., 1975, 9, Но. 4, с. 73-74.

Krupnik N.Ya., Rosenberg S.M. (Крупник Н.Я., Розенберг С.М.)

[1] Один контрпример в теории сингулярных интегральных операторов. В кн.: Исслед. по алгебре, мат. анализу и их прил. Мат. науки. Кишинев: Штиинца, 1977, с. 89-90.

Kuchment P.A. (Кучмент П.А.)

[1] О нетеровых операторах в паре Банаховых пространств. – Труды НИММ ВГУ, вып. 3, Воронеш, 1971, с. 61-77.

Kuliev V.D., Sadykhov A.E. (Кулиев В.Д., Садыхов А.Э.)

[1] Проблема Римана для двух пар функций и её применение в теории упругости. – Изв. АН АрмССР. Механика, 1979, 32, Но. 2, с. 26-37.

Kvitko A.N. (Квитко А.Н.)

[1] Векторно-матричная задача Римана для эллиптической системы дифференциальных уравнений первого порядка в случае сложного контура на римановой поверхности. – Мат. методы и физ.-мех. поля, 1980, Но. 11, с. 12-15.

Lagvilava E.T. (Лагвилава Э.Т.)

[1] О факторизации матриц-функций. – Сообщ. АН ГрузССР, 1978, 89, Но. 2, с. 317-319.

Lappo-Danilevskiĭ I.A. (Лаппо-Данилевский И.А.)

[1] Теория функции от матриц и системы линейных дифференциальных уравнений. – М.-Л.: ГИТЛ, 1934.

Lax P.D.

[1] On the factorization of matrix-valued functions. – Comm. pure and appl. math., 1976, 29, N.6, p. 683-688.

Lee M.

[1] On Toeplitz-operators. – J. Math. Soc. Jap., 1971, 23, N.2, p. 320-322.

Le-Din Zon (Ле-Динь Зон)

[1] Об одном случае решения в замкнутом виде задачи Гильберта для вектор-функции. – Математический анализ и его приложения. Ростов-на-Дону, 1974, 5, с. 189-193.

Lee M., Sarason D.

[1] The spectra of some Toeplitz operators. – J. Math. Anal. and Appl., 1971, 33, N.3, p. 529-543.

Leiterer J. (Лайтерер Ю.)

[1] О факторизации матриц- и оператор функций. – Сообщ. АН Груз ССР, 1977, 88, Но. 3, с. 541-544.

Lindenstrauss J., Tzafriri L.

[1] On the complemented subspace problem. - Israel J. Math., 1971, 9, p. 263-269.

Litvinchuk G.S. (Литвинчук Г.С.)

[1] Об устойчивости одной краевой задачи теории аналитических функций. - ДАН СССР, I967, I74, Но. 6, с. I268-I270.

[2] Две теоремы об устойчивости частных индексов краевой задачи Римана и их приложения. - Изв. вузов Математика, I967, Но. I, с. 47-54.

[3] Краевые задачи и сингулярные интегральные уравнения со сдвигом. - М.: Наука, I977,

Litvinchuk G.S., Nikolaǐchuk A.M., Spitkovskiǐ I.M. (Литвинчук Г.С., Николайчук А.М., Спитковский И.М.)

[1] Некоторые теоремы об устойчивости и оценках частных индексов краевой задачи Римана и их приложения. - В кн.: Проблемы развития прикладных математических исследованиях: IУ Республиканская конференция математиков Белоруссии: Тезисы докладов. Минск, I975, ч. 2, с. I05.

Litvinchuk G.S., Spitkovskiǐ I.M. (Литвинчук Г.С., Спитковский И.М.)

[1] Точные оценки дефектных чисел краевой задачи Римана - ДАН СССР, I980, 255, Но. 5, с. I042-I046.

[2] Точные оценки дефектных чисел обобщенной краевой задачи Римана, факторизация эрмитовых матриц-функций и некоторые проблемы приближения мероморфными функциями. - Матем. об., I982, II7, Но. 2, с. I96-2I5.

Lowengrub M., Walton J.

[1] Systems of generalized Abel equations. - SIAM J. Math. Anal., 1979, 10, N.4, p. 794-807.

Magnaradze L.G. (Магнарадзе Л.Г.)

[1] Об одной линейной граничной задаче теории функций комплексного переменного. - ДАН СССР, I949, 88, Но. 4, с. 657-660.

Maister A.V. (Майстер А.В.)

[1] О разрешимости систем интегральных уравнений с обобщенными степенными ядрами. - ВИНИТИ, Но. 578I-82 Деп., I982.

Mal'tsev A.A. (Мальцев А.А.)

[1] Основы линейной алгебры. - М.: Наука, I970.

Malyshev V.A. (Малышев В.А.)

[1] Уравнения Винера-Хопфа и их применения в теории вероятностей. В кн.: Итоги наука и техники, Сер. "Теория вероятностей, математическая статистика, теоретическая кибернетика". М., I976, I3, с. 5-35.

Mandzhavidze G.F. (Манджавидзе Г.Ф.)

[1] Об одном сингулярном интегральном уравнении с разрывными коэффициентами и его применении в теории упругости. - Прикл. матем. и мех., 1951, 15, вып. 3, с. 279-296.

[2] О приближенном решении граничных задач теории функций комплексного переменного. - Сообщ. АН ГрузССР, 1953, 14, Но. 10, с. 577-582.

[3] О приближенном решении граничных задач теории аналитических функций. - Тр. Третьего Всесоюзного матем. съезда, 1956, I, с. 88.

[4] Приближенное решение граничных задач теории аналитических функций. - В кн.: Исслед. по совр. пробл. теории функций комплексного переменного. М.: Физматгиз, 1960, с. 365-370.

[5] О поведении решений граничной задачи линейного пояржения. - Тр. Тбилис. мат. ин-та ГрузССР, 1969, 35, с. 173-182.

[6] Граничная задача линейного сопряжения с кусонно-непрерывным матричным коэффициентом. - В кн.: Механика сплошной среды и родств. пробл. анализа, М.: Наука, 1972, с. 297-304.

[7] О применении теории обобщенных аналитических функций к граничной задаче сопряжения со смещением. - ДАН СССР, 237, 1977 Но. 6, с. 1285-1288.

[8] Об одном семействе граничных задач линейного сопряжения. - Сообщ. АН ГрузССР, 1979, 95, Но. 2, с. 289-292.

[9] Применение теории обобщенных аналитических функций к изучению граничных задач сопряжения со смещением. - Диффер. и интегральн. уравнения и краев. задачи - Тбилиси 1979, с. 165-186.

Mandzhavidze G.F., Khvedelidze B.V. (Манджавидзе Г.Ф., Хведелидзе Б. В.)

[1] О задаче Римана-Привалова с непрерывными коэффициентами. - ДАН СССР, 1958, 123, Но. 5, с. 791-794.

[2] О задаче линейного сопряжения сингулярных интегральных уравнениях с ядром Коши с непрерывными коэффициентами. - Тр. Тбилис. мат. ин-та ГрузССР, 1962, 28, с. 85-105.

Markus A.S., Matsaev V.I. (Маркус А.С., Мацаев В.И.)

[1] О спектральных свойствах голоморфных оператор-функций в гильбертовом пространстве. - Матем. исслед., 1974, 9, Но. 4, с. 79-91.

[2] Два замечания о факторизации матриц-функций. - Матем. исслед. 1976, Но. 42, с. 216-223.

Marcus M., Minc H.

[1] A survey of matrix theory and matrix inequalities. - Boston: Allyn and Bacon, Inc., 1964.

Markushevich A.I. (Маркушевич А.И.)

[1] Об одной граничной задаче теории аналитических функций. - Уч. зап. Моск. ун-та, 146, 100, с. 20-30.

Maskudov F.G., Veliev S.G. (Маскудов Ф. Г., Велиев С.Г.)

[1] Факторизация матрицы рассеяния для системы уравнений Дирака на всей оси.- Изв. АН АзССР. Серия физ.-техн. и мат. наук, 1975, Но. 2, с. 70-75.

van der Mee C.V.M.

[1] Spectral analysis of the transport equation. I. Nondegenerate and multigroup case. - Integr. Equat. and Oper. Theory, 1980, 3, p. 529-573.

[2] Semigroup and factorization methods in transport theory. - Amsterdam: Math. Centre Tract., 1981, p. 146-169.

Meister E.

[1] Das Riemannsche Randwertproblem. - Neuere Ergebnisse und Anwendungsgebiete, Sitzungsber. - Berlin Math. Ges., 1969 - 1971, s.l.s.a., p. 36-37.

Meunargiya O.V. (Меунаргия О.В.)

[1] Об одной системе граничных задач линейного сопряжения. - Сообщ. АН ГрузССР, 1976, 82, Но. I, с. 17-20.

[2] О построении эффективного решения задачи линейного сопряжения для нескольких неизвестных функций. - Теорет. и мат. физика, 1978, 37, Но. 3, с. 423-426.

[3] Многокомпонентная граничная задача линейного сопряжения. - Теорет. и мат. физика, 1981, 47, Но. 3, с. 419-424.

Miamee A.G., Salehi H.

[1] On the factorization of a nonnegative operator valued function. - Lect. Notes Math., 1978, 656, p. 129-137.

Mikhailov L.G. (Михайлов Л.Г.)

[1] Общая краевая задача о бесконечно малых изгибаниях склеенных поверхностей. - Изв. вузов. Математика, 1960, Но. 5, с.99-109.

[2] Об одной граничной задаче линейного сопряжения. - ДАН СССР, 1961, 139, Но. 2, с. 294-297.

[3] Общая задача сопряжения аналитических функций и её применения.- Изв. АН СССР, Сер. Матем. 1963, 27, Но. 5, с. 969-992.

[4] Новый класс особых интегральных уравнений и его применение к дифференциальным уравнениям с сингулярными коэффициентами. - Душанбе: Изд-во АН ТаджССР, 1963.

Mikhlin S.G. (Михлин С.Г.)

[1] Сингулярные интегральные уравнения с непрерывными коэффициентами. ДАН СССР, 1948, 59, Но. 3, с. 435-438.

Michlin S., Prößdorf S.

[1] Singuläre Integraloperatoren. - Berlin: Akademie Verlag, 1980

Mishchishin I.I. (Мищишин И.И.)

[1] Об одной задаче линейного сопряжения с кусочно постоянными коэффициентами. - Исслед. по соврем. пробл. суммир. и приближ. функций и их прил., 6. Днепропетровск, 1975, с. 188-194.

[2] Метод нахождения частных решений однородной задачи линейного сопряжения. - В кн.: Актуальн. пробл. ЭВМ и программир. Днепропетровск, 1981, с. 96-100.

Morozov V.V. (Морозов В.В.)

[1] О коммутативных матрицах.- Уч. зап. Казанск. ун-та, 1952, 112, Но. 9, с. 17-20.

Mossakovskiǐ V.I., Mishchishin I.I. (Моссаковский В.И., Мищишин И.И.)

[1] Об одной задаче линейного сопряжения. - Гидроаэромеханика и теория упругости. 8, Харьков, 1968.

Mückenhoupt S.

[1] Weighted norm inequalities for the Hardy maximal function. - Trans. Amer. Math. Soc., 1972, 165, p. 207-226.

Muskhelishvili N.I. (Мусхелишвили Н.И.)

[1] Сингулярные интегральные уравнения. - М.: Наука, 1968. English transl. (of 1st ed.), Groningen:Noordhoff, 1953. German transl. Berlin: Akademie-Verlag, 1965.

Muskhelishvili N.I., Vekua N.P. (Мусхелишвили Н.И., Векуа Н.П.)

[1] Краевая задача Римана для нескольких неизвестных функций и её приложение к системам сингулярных интегральных уравнений.- Тр. Тбилис. мат. ин-та АН ГрузССР, 1943, 12, с. 1-46.

Sz.-Nagy B., Foiaş C.

[1] Analyse harmonique des opérateurs de l'espace de Hilbert. - Masson et Cie, Académia Kiadó, 1967.

Nakazi Takahiko

[1] A note on the factorization of operator valued functions.- Proc. Amer. Math. Soc., 1981, 81, N.4, p. 591-594.

Nehari Z.

[1] On bounded bilinear forms. - Ann. Math., 1957, 65, N.1, p. 153-162.

Nikolaǐchuk A.M. (Николайчук А.М.)

[1] Некоторые оценки для частных индексов краевой задачи Римана.- Украинск. матем. ж., 1971, 23, Но. 6, с. 793-798.

[2] Об устойчивости краевой задачи Маркушевича. - Украинск. матем. ж., 1974, 26, Но. 4, с. 558-559.

Nikolaǐchuk A.M., Spitkovskiǐ I.M. (Николайчук А.М., Спитковский И.М.)

[1] О краевой задаче Римана с эрмитовой матрицей. - ДАН СССР, 1975, 221, Но. 6, с. 1280-1283. English Transl. Sov. Math. Dokl. 1975, N 6, p. 533-536

[2] Факторизация эрмитовых матриц-функций и её приложения к граничным задачам. - Украинск. Матем. ж., 1975, 27, Но. 6, с. 767-779.

Nikolenko P.V. (Николенко П.В.)

[1] Об устойчивости системы частных индексов семейства матриц-функций. - ВИНИТИ, Но. 5525-81 Деп., 1981, 17 с.

[2] Об отсутствии стандартной факторизации матриц-функций. - ВИНИТИ, Но. 3666-82, Деп., 1982, 6 с.

Novikov S.P. (ed.)

[1] Теория солитонов: Метод обратной задачи/Под ред. С.П. Новикова. М.: Наука, 1980.

Novokshenov V.Yu. (Новокшенов В.Ю.)

[1] Уравнения в свертках на конечном отрезке и факторизация эллиптических матриц. - Матем. заметки, 1980, 27, Но. 6, с. 935-946.

[2] Асимптотика при $t \to \infty$ решения задачи Коши для нелинейного уравнения Шредингера. - ДАН СССР, 1980, 251, Но. 4, с. 799-802.

[3] Асимптотика решений при $t \to \infty$ двумерного обобщения цепочки Тоды: ДАН СССР, 1982, 265, Но. 6, с. 1320-1324.

Nyaga V.I. (Няга В.И.)

[1] О символе сингулярных интегральных операторов в случае кусочно-ляпуновского контура. - Матем. исслед., 1974, 9, Но. 2, с. 109-125.

[2] О сингулярных интегральных операторах в случае неограниченного контура. - В кн.: Исслед. по алгебре, матем. анализу и их прил. Кишинев: Штиинца, 1977, с. 91-95.

[3] О сингулярных интегральных уравнениях в пространствах с весом.- Изв. вузов Математика. - 1979, Но. 2, с. 73-75.

Paatashvili V.A. (Пааташвили В.А.)

[1] О сингулярных интегралах Коши. - Сообщ. АН ГрузССР, 1969, 53, Но.3, с. 529-532.

Paatashvili V.A., Khuskivadze G.A. (Пааташвили В.А., Хускивадзе Г.А.)

[1] Об ограниченности сингулярного оператора Коши в пространствах Лебега в случае негладких контуров. - Тр. Тбилис. мат. ин-та, АН ГрузССР, 1982, 69, с. 93-107.

Page B.L.

[1] Bounded and compact vectorial Hankel operators. - Trans. Amer. Math. Soc., 1970, 150, N 2, p. 529-539.

Palais R.

[1] Seminar on the Atiyah-Singer index theorem. - Princeton: Princeton Univ. Press, 1965.

Pal'tsev B.V. (Пальцев Б.В.)

[1] Об одном признаке непрерывности канонической матрицы решения задачи Гильберта. - ДАН СССР, 1976, 226, Но. 6, с. 1271-1274.

[2] О задаче Дирихле для одного дифференциального оператора, встречающегося в теории случайных процессов. - Изв. АН СССР. Сер. матем., 1977, 41, Но. 6, с. 1348-1387.

[3] Об одном классе уравнений свертки на конечном интервале. - ДАН СССР, 1979, 247, Но. 1, с. 41-44.

[4] Уравнения свертки на конечном интервале для одного класса символов, имеющих степенную асимптотику на бесконечности. - Изв. АН СССР, Сер. матем., 1980, 44, Но. 2, с. 322-394.

[5] Обобщение метода Винера-Хопфа для уравнений свертки на конечном интервале с символами, имеющими степенную асимптотику на бесконечности. - Матем. сб., 1980, 113, Но. 3, с. 355-399.

[6] Об одном методе построения канонической матрицы решений задачи Гильберта, возникающей при решении уравнений свертки на конечном интервале. - Изв. АН СССР. Сер. матем., 1981, 45, No. 6, с. 1332-1390.

Panchenko I.F. (Панченко И.Ф.)

[1] Задача Римана для n пар функций на римановой поверхности в случае сложного контура в пространстве $L_p^n(\Gamma)$. - Исслед. по функц. анализу и дифференц. уравнениям, Кишинев, 1978, с.68-75.

Peller V.V.

[1] Vectorial Hankel operators, commutators and related operators of the Schatten- von Neumann class.-Integr. Equat. and Oper. Theory, 1982, 5, N.2, p. 244-272.

Petrov N.N. (Петров Н.Н.)

[1] О разрешимости линейных уравнений в произведении банаховых пространств. - В кн.: Топологические пространства и их отображения. Рига, 1979, с. 102-106.

Pichorides S.K.

[1] On the best values of the constants in the theorems of M. Riesz, Zygmund and Kolmogorov. - Studia Math., 1972, 19, N.2, p. 165-179.

Pilidi Z.S. (Пилиди В.С.)

[1] Априорные оценки для некоторого класса одномерных сингулярных интегральных операторов с разрывными коэффициентами. - Матем. заметки, 1979, 26, No. 2, с. 227-234.

Plemelj J.

[1] Riemannsche Funktionenscharen mit gegebener Monodromiegruppe. - Monatsh. für Math. und Phys., 1908, 19, p. 211-245.

Popov Yu.A. (Попов Ю.А.)

[1] Задача дифракции плоской электромагнитной волны на трех полубесконечных плоскостях. - В кн.: Материалы 5-й конф. молод. ученных, 1980,: МФТИ, ВИНИТИ, No. 2849 - 80 Деп., 1980, с.88-91.

Pousson H.R.

[1] Systems of Toeplitz operators on H^2. - Proc. Amer. Math. Soc., 1968, 19, N.3, p. 603-608.

[2] Systems of Toeplitz operators on H^2. II.-Trans. Amer. Math. Soc., 1968, 133, N.2, p. 527-536.

Power S.C.

[1] Fredholm Toeplitz operators and slow oscillation. - Can. J. Math., 1980, 32, N 5, p. 1058-1071.

Primachuk L.P. (Примачук Л.П.)

[1] О краевой задаче с сопряжением. - Изв. АН БССР, сер. физ.-мат. наук, 1967, No. 4, с. 59-62.

[2] О частных индексах задачи Римана с треугольной матрицей. - ДАН БССР, 1970, 14, No. 1, с. 5-7.

[3] Задача Римана для внешности разрезов вдоль прямой с подстановочными матрицами. - ДАН БССР, 1978, 22, No. 2, с. 115-118.

[4] Об одном интегрируемом случае задачи Римана с подстановочной матрицей. - ДАН БССР, 1978, 22, No. 4, с. 310-313.

Privalov I.I. (Привалов И.И.)

[1] Об одной граничной задаче в теории аналитических функций. - Матем. сб., 1934, 41, No. 4, с. 519-526.

[2] Граничные свойства аналитических функций. - М.-Л.: ГИТЛ, 1950.

Prößdorf S.

[1] Einige Klassen singulärer Gleichungen. - Berlin: Akademie-Verlag, 1974.

Prößdorf S., Schmidt G.

[1] Necessary and sufficient conditions for the collocation method for singular integral equations.- Math.Nachr. 1979, 89, p.203-215.

Prößdorf S., Silbermann B.

[1] Projektionsverfahren und die näherungsweise Lösung singulärer Gleichungen.-Leipzig: Teubner-Text, 1977.

Rabindranathan M.

[1] On the inversion of Toeplitz operators. - J. Math. and Mech., 1969, 19, N 3, p. 195-206.

Radon I. (Радон И.)

[1] О краевых задачах для логарифмического потенциала. - Успехи матем. наук, 1946, 1, вып. 3-4, с. 96-124.

Rajamaki M.

[1] The fundamental matrix of the Hilbert problem in transport theory. Transp. Theory and Statist. Phys., 1973, 3, N.4, p. 175-197.

Rawlins A.D., **Williams** W.E.

[1] Matrix Wiener-Hopf factorization. - Quart. J. Mech. and Appl. Math., 1981, 34, N.1, p. 1-8.

Reyman A.G., Semenov-Tian-Shansky M.A.

[1] Reduction of Hamiltonian systems, Affine Lie algebras and Lax equations. - Inventiones Math., 1979, 54, p. 81-100.

[2] Reduction of Hamiltonian systems, Affine Lie algebras and Lax equations. II.- Inventiones Math., 1981, 63, p. 423-432.

Rochberg R.

[1] Toeplitz operators on weighted H^p spaces. - Indiana Univ. Math. J., 1977, 26, N.2, p. 291-298.

Rodin Yu. L. (Родин Ю.Л.)

[1] Структура общего решения краевой задачи Римана для голоморфного вектора на компактной римановой поверхности. - ДАН СССР, 1977, 232, No. 5, с. 1019-1022.

[2] Структура общего решения краевой задачи Римана для голоморфного вектор-функции на компактной римановой поверхности. - В кн.: Вопр. матрич. теории отображения и её применение: Материалы 5-го Коллоквиума, Донецк, 1976, г. Киев: Наук. Думка, 1978, с. 103-120.

Rodin Yu.L., Turakulov A. (Родин Ю.Л., Туракулов А.)

[1] Краевая задача Римана для обобщенных аналитических функций с сингулярными коэффициентами на компактной римановой поверхности. - Сообщ. АН ГрузССР, 1979, 96, No. 1, с. 21-24.

Röhrl H.

[1] On holomorphic families of fiber bundles over the Riemannian sphere. - Mem. Coll. Sci. Univ. Kyoto, 1961,33,N.3, p. 435-477.

[2] Über das Riemann-Privalovsche Randwertproblem. - Math. Ann., 1963, 151, p. 365-423.

Rozanov Yu.A. (Розанов Ю.А.)

[1] Стационарные случайные процессы. - М.: Наука, 1963.

Sabitov I.Kh. (Сабитов И.Х.)

[1] Об общей краевой задаче линейного сопряжения на окружности. - Сибирск. матем. ж., 1964, 5, No. 1, с. 124-129.

[2] Об одной граничной задачей линейного сопряжения. - Матем. сб., 1964, No. 2, с. 262-274.

Saginashvili A.I. (Сагинашвили А.И.)

[1] Сингулярные интегральные операторы с коэффициентами, имеющими разрывы полу-почти-периодического типа. - Сообщ. АН ГрузССР, 1979, 94, No. 2, с. 289-291.

[2] Сингулярные интегральные операторы с полу-почти-периодическими разрывами у коэффициентов. - Сообщ. АН ГрузССР, 1979, 95, No.3, 541-543.

[3] Сингулярные интегральные уравнения с коэффициентами, имеющими разрывы полу-почти-периодического типа. - Тр. Тбилисс. мат. ин-та АН ГрузССР, 1980, 66, с. 84-95.

Sarason D.E.

[1] Generalized interpolation on H^{∞}. - Trans. Amer. Math. Soc., 1967, 127, p. 179-203.

[2] Algebras of functions on the unit circle. - Bull. Amer. Math. Soc., 1972, 79, N.2, p. 286-299.

[3] Toeplitz operators with semi-almost-periodic symbols. - Duke Math. J., 1977, 44, N.2, p. 357-364.

[4] Toeplitz operators with piecewise quasicontinuous symbols. - Indiana Univ.Math. J., 1977, 26, N.5, p. 817-838.

Seifullaev R.K. (Сейфуллаев Р.К.)

[1] Краевая задача Римана на негладкой разомкнутой кривой. - Матем. сб., 1980,112, No. 2, с. 147-161.

Seleznev V.A. (Селезнев В.А.)

[1] Системы сингулярных уравнений на ГQ -контурах. - Динамика сплошной среды, 1974, вып. 18, с. 243-248.

[2] Линейные сингулярные уравнения на квазиконформых контурах и задача линейного сопряжения для голоморфного вектора на замкнутых римановых поверхностях. - В кн.: Вопр. матрич. теории отображения и её применение. Материалы 5-го Коллоквиума, Донецк, 1976, г. Киев: Наук. думка, 1978, с. 125-135.

[3] Краевая задача Римана в классах жордановых границ. - В кн.: Метрич. вопросы теории функций. Киев: Наук. думка, 1980, с. 125-152

Sementsul A.A. (Семенцул А.А.)

[1] О сингулярных интегральных уравнениях с коэффициентами, имеющими разрывы почти-периодического типа. - Матем. исслед. 1971, 6, Но. 3, с. 92-114.

Sergeev A.G. (Сергеев А.Г.)

[1] Факторизация оператор-функций, непрерывных по Гельдеру. - УМН, 1972, 27, Но. 6, с. 253.

[2] Локальный L_p-принцип в задаче факторизации. - УМН, 1974, 29, Но. 6, с. 175-176.

Shamir E. (Шамир Е.)

[1] Решение систем Римана-Гильберта с кусочно-непрерывными коэффициентами в L_p.- ДАН СССР, 1966, 167, Но. 5, с. 1000-1003.

Shelepov V.Yu. (Шелепов В.Ю.)

[1] О задаче Римана в областях, граница которых имеет ограниченное вращение. - ДАН СССР, 1968, 181, Но. 3, с. 565-568.

Sherman D.I. (Шерман Д.И.)

[1] О приемах решения некоторых сингулярных интегральных уравнений. - ПММ, 1948, 12, Но. 4, с. 423-452.

Shestopalov V.P. (Шестопалов В.П.)

[1] Метод задачи Римана-Гильберта в теории дифракций. - Харьков: ХГУ, 1971.

Shmul'yan Yu.L. (Шмульян Ю.Л.)

[1] Задачи Римана с положительно определенной матрицей. - Успехи матем. наук, 1953, 8, вып. 2, с. 143-145.

[2] Задача Римана с эрмитовой матрицей. - Успехи матем. наук, 1954, 9, вып. 4, стр. 243-248.

Shneiberg I.Ya. (Шнейберг И.Я.)

[1] О разрешимости линейных уравнений в интерполяционных семействах банаховых пространств. - ДАН СССР, 1973, 212, Но. 1, с. 57-59.

[2] Спектральные свойства линейных операторов в интерполяционных семействах банаховых пространств. - Матем. исслед., 1974, 9, Но. 2, 214-229.

Shubin M.A. (Шубин М.А.)

[1] Факторизация зависящих от параметра матриц-функций в нормированных кольцах и связанные с ней вопросы теории нетеровых операторов. - Матем. сб., 1967, 73, Но. 4, с. 610-629.

[2] О локальном принципе в задаче факторизации. - Матем. исслед., 1971, 6, Но. 1, с. 174-180.

[3] Псевдодифференциальные операторы и спектральная теория. - М., Наука, 1978.

F.-O.

General Wiener-Hopf factorization methods. - Boston - London - Melbourne: Pitman; Research Notes in Mathematics, vol. 19, 1985.

vskiĭ, I.M. (Спитковский И.М.)

Устойчивость частных индексов краевой задачи Римана со строго невырожденной матрицей. - ДАН СССР, 1974, 218, Но. I, с. 46-49.

Задача факторизации измеримых матриц-функций. - ДАН СССР, 1976, 227, Но. 3, с. 576-579.

О частных индексах непрерывных матриц-функций. - ДАН СССР, 1976, 229, Но. 5, с. 1059-1062.

О множителях, не влияющих на факторизуемость. - ДАН СССР, 1976, 231, Но. 6, с. 1300-1303.

О факторизации матриц-функций, хаусдорфово множество которых расположено внутри угла. - Сообщ. АН ГрузССР, 1977, 86, Но.3 с. 561-564.

К вопросу о факторизуемости измеримых матриц-функций. - ДАН СССР, 1978, 240, Но. 3, с. 541-544.

Факторизация измеримых матриц-функций и её приложения к краевым задачам для аналитических функций. - Канд. дисс., Одесса, 1978.

К теории обобщенной краевой задачи Римана в классах L_p. - Украинск. матем. ж., 1979, 31, Но. I, с. 63-73.

О множителях, не влияющих на факторизуемость. - Матем. заметки, 1980, 27, Но. 2, с. 291-299.

] Некоторые оценки для частных индексов измеримых матриц-функций. - Матем. сб., 1980, III, Но. 2, с. 227-248.

] О блочных операторах и связанных с ними вопросах теории факторизации матриц-функций. - ДАН СССР, 1980, 254, Но. 4, с. 816-820.

2] Критерий нетеровости блочно-треугольных операторов и связанные с ними вопросы теории факторизации матриц-функций. - ВИНИТИ, Но. 2543-81, Деп., 1981.

3] Факторизация измеримых матриц-функций, связанные с ней теории систем сингулярных интегральных уравнений и векторной краевой задачи Римана, I.- Дифференц. уравнения, 1981, 17, Но. 4, с. 697-709.

4] Факторизация измеримых матриц-функций,связанные с ней вопросы вопросы теории систем сингулярных интегральных уравнений и векторной краевой задачи Римана, II. - Дифференц. уравнения, 1982, 18, Но. 3.

5] Факторизация некоторых классов полу-почти-периодических матриц-функций и её приложения к системам уравнений типа свертки. - Изв. вузов. Математика, 1983, Но. 4, с. 88-94.

16] О факторизации матриц-функций из классов $\tilde{A}_n(p)$ и TL. - Украинск. матем. ж., 1983, 33, Но. 4.

Siewert C.E., Kelley S.T.

[1] An analytical solution to a matrix Riemann-
Z. angew. Math. und Phys., 1980, 31, N 3, p.

Siewert C.E., Kelley C.T., Garcia R.D.M.

[1] An analytical expression for the H matrix re
Rayleigh scattering. - J. Math. Anal. and Ap
N 2, p. 509-518.

Silbermann B.

[1] Lokale Theorie des Reduktionsverfahrens für
Math. Nachr., 1981, 104, p. 137-146.

Simonenko I.B. (Симоненко И.Б.)

[1] Краевая задача Римана с непрерывным коэффиц
1959, 124, Но. 2, с. 278-281.

[2] Краевая задача Римана для n пар функций с н
коэффициентами. - Изв. вузов. Математика, 1
с. 140-145.

[3] Краевая задача Римана для n пар функций с из
фициентами и её применение к исследованию си
гралов в пространствах L_p с весами. - ДАН СС
Но. 1, с. 36-39.

[4] Краевая задача Римана для n пар функций с из
фициентами и её применение к исследованию си
гралов в пространствах L_p с весами. - Изв. АН
1964, 28, Но. 2, с. 277-306.

[5] Новый общий метод исследования линейных опер
ных уравнений. - I.- Изв. АН СССР, сер. мат.,
с. 567-586.

[6] Некоторые общие вопросы теории краевой задач
АН СССР, сер. мат., 1968, 32, Но. 5, с. 1138-

[7] Эквивалентность локальной факторизуемости изм
и локальный нетеровости порожденного ею сингу
тора. - ВИНИТИ, Но.2194-80, Деп., 1980, 21 с.

[8] Эквивалентность факторизуемости и локальной ф
измеримых функций, определенных на контурах т
Но. 2193-80, Деп., 1980, 51 с.

[9] О факторизуемости и локальной факторизуемости
функций. - ДАН СССР, 1980, 250, Но. 5, с. 106

[10] О связи локальной факторизуемости и локальной
Сообщ. АН ГрузССР, 1980, 98, Но. 2, с. 281-28

[11] О глобальной и локальной факторизуемости изме
функции и нетеровости порожденного ею сингуля
ра. - Изв. вузов. Математика, 1983, Но. 4, с.

Soldatov A.P. (Солдатов А.П.)

[1] К теории сингулярных интегральных операторов к
типа. - Дифференц. уравнения, 1979, 15, Но. 3,

[2] Краевая задача линейного сопряжения теории фун
АН СССР, сер. мат., 1979, 43, Но. 1, с. 184-20

Storozh O. G. (Сторож О.Г.)
[1] К вопросу об обратимости треугольных операторных матриц второго порядка. - Мат. методы и физ.-мех. поля. Киев, 1980, с. 85-87.

Stray A., Фума К.О.
[1] On interpolating functions with minimal norm. - Proc. Amer. Math. Soc., 1978, 68, N.1, p. 75-78

Suciu I.
[1] Factorization theorems for operator valued functions on multiply connected domains. - Rev. Roum. Math. pures et appl., 1979, 24, N 8, p. 1251-1269.

Tolokonnikov L.A., Pen'kov V.B. (Толоконников Л.А., Пеньков В.Б.)
[1] Приложение краевой задачи Римана с разрывным матричным коэффициентом к механике. - Изв. вузов Математика, 1980, Но. 12, с. 55-59.

Troitskiĭ V.E. (Троицкий В.Е.)
[1] Об одном способе нахождения матрицы Иоста. - Теорет. и мат. физика, 1978, 37, Но. 2, с. 243-245.

Tsitskishvili A.R. (Цицкишвили А.Р.)
[1] Эффективное решение задач сопряжения для нескольких неизвестных функций в случае кусочно-постоянной матрицы. - Тр. Тбилис. мат. ин-та АН ГрузССР, 1969, 35, с. 67-103.

[2] Об эффективном решении задачи сопряжения. - Сообщ. АН ГрузССР, 1970, 58, с. 17-23.

[3] О фильтрации в земляных плотинах. - Тр. Тбилисск. ун-та, Математика, механика, астрономия, 1980, 210, с. 12-40.

[4] О фильтрации через треугольное ядро земляной плотины. - Тр. Тбилисск. ун-та. Математика, механика, астрономия, 1980, 210, с. 42-51.

Tumarkin G.Ts., Khavinson S.Ya. (Тумаркин Г.Ц., Хавинсон С.Я.)
[1] К определению аналитических функций класса E_p в многосвязных областях. - Успехи матем. наук, 1958, 13, вып. 1, с. 201-206

[2] О теореме разложения для аналитических функций класса E^p в многосвязных областях. - Успехи матем. наук, 1958, 13, вып. 2, с. 223-228.

[3] Классы аналитических функций в многосвязных областях. - В кн.: Исслед. по совр. пробл. теории функций комплексного переменного. М.: Физматгиз, 1960, с. 45-77.

Vasil'ev I.L. (Васильев И.Л.)
[1] О единственности решения системы уравнений Абеля с постоянными коэффициентами. - ДАН БССР, 1981, 25, Но. 2, с.105-107.

[2] Системы интегральных уравнений с ядром Абеля на отрезке вещественной оси. - Изв. АН БССР. Сер. физ.-мат. наук, 1982, Но. 2, с. 47-53.

Vekua I.N. (Векуа И.Н.)
[1] Обобщенные аналитические функции. - М.: Физматгиз, 1959.

Vekua N. P. (Векуа Н. П.)

[1] Краевая задача Гильберта с рациональными коэффициентами для нескольких неизвестных функций. - Сообщ. АН ГрузССР, 1946, 7, Но. 9-10, с. 595-600.

[2] Об одной задаче теории функций комплексного переменного. - ДАН СССР, 1952, 86, Но. 3, с. 457-460.

[3] Об одной задаче Гильберта для нескольких неизвестных функций. Тр. Тбилис. мат. ин-та АН ГрузССР, 1968, 34, с. 5-12.

[4] Системы сингулярных интегральных уравнений. - М.: Наука, 1970.

Vekua N.P., Kveselava D.A. (Векуа Н. П., Квеселава Д. А.)

[1] Об одной краевой задаче теории функций комплексного переменного. - Сообщ. АН ГрузССР, 1941, II, Но. 3, 223-240.

[2] Об одной краевой задаче теории функций комплексного переменного и её применении к решению системы сингулярных интегральных уравнений. - Тр. Тбилис. мат. ин-та АН ГрузССР, 1941, 9, с. 38-48.

Verbitskiĭ I.E. (Вербицкий И. Э.)

[1] О мультипликаторах в пространствах l_p с весом. - Матем. исслед., 1977, 45, с. 3-16.

Verbitskiĭ I.E., Krupnik N.Ya. (Вербицкий И.Э., Крупник Н.Я.)

[1] Точные константы в теоремах Бабенко К.И., Хведелидзе Б.В. об ограниченности сингулярного оператора. - Сообщ. АН ГрузССР, 1977, 85, Но. I, с. 21-24.

[2] Точные константы в теоремах об ограниченности сингулярных операторов в пространствах L_p с весом и их приложения. - Матем. исслед., 1980, 54, с. 21-35.

Virozub A.I., Matsaev V.I. (Вирозуб А.И., Мацаев В.И.)

[1] Одна теорема о канонической факторизации оператор-функций. - Матем. исслед., 1973, 8, Но. 3, с. 145-150.

Vladimirov V.S., Volovich I.V. (Владимиров В.С., Волович И.В.)

[1] Уравнение Винера-Хопфа, задача Римана-Гильберта и ортогональные многочлены. - ДАН СССР, 1982, 266, Но. 4, с. 788-791.

[2] Об одной модели статистической физики. - Теорет. и мат. физика., 1983, 54, Но. I, с. 8-22.

Vorob'ev V.L. (Воробьев В.Л.)

[1] Изгиб полубесконечной пластинки, лежащей на линейно-деформируемом основании с учетом сил сцепления: Дис. Канд. физ-мат. наук. - Одесса, 1973.

Vorovich I.I., Aleksandrov V.M., Babeshko V.A.
(Ворович И.И., Александров В.М., Бабешко В.А.)

[1] Неклассические смешанные задачи теории упругости. - М.: Наука, 1974.

Walsh J.L. (Уалш Дж.Л.)

[1] Интерполяция и аппроксимация рациональными функциами в комплексной области.-М.: Изд-во иностр. лит., 1961.

Walton J.R.

[1] Systems of generalized Abel integral equations with applications to simultaneous dual relations. - SIAM J. Math. Anal., 1979, 10, N.4, p. 808-822.

Widom H.

[1] Inversion of Toeplitz matrices. II. - Illinois J. Math., 1960, 4, p. 88-99.

[2] On the spectrum of Toeplitz matrices. - Pacif. J. Math., 1964, 14, p. 365-375.

[3] Perturbing Fredholm operators to obtain invertible operators.- J. Funct. Anal., 1975, 20, N.1, p. 26-31.

[4] Asymptotic behavior of block Toeplitz matrices and determinants. II. - Advanc. Math., 1976, 21, N.1, p. 1-29.

Wiener N., Masani P.

[1] The prediction theory of multivariate stochastic processes. II. - Acta Math., 1958, 99, p. 93-137.

Yandarov O.V. (Яндаров О.В.)

[1] Об одном классе B - пространств и его применении к теории сингулярных интегральных операторов. - Дифференц. уравнения, 1982, 18, Но. 2, с. 356-359.

Yatsko A.I. (Яцко А.И.)

[1] Критерий нетеровости обобщенной краевой задачи Римана для сложного контура. - Изв. вузов. Математика, 1981, Но. 11, 86-87.

[2] Обобщенная краевая задача Римана в пространствах L_p с весом.- Украинск. матем. ж., 1981, 33, с. 283-286.

Yatsko A.I., Yatsko S.I. (Яцко А.И., Яцко С.И.)

[1] К теории обобщенной краевой задачи Римана с кусочно-непрерывными коэффициентами. - В кн.: Теоретические и прикладные вопросы дифференциальных уравнений и алгебры. Киев: Наук. думка, 1978, с. 250-253.

[2] Обобщенная краевая задача Римана с кусочно-непрерывными коэффициентами. - Украинск. матем. ж., 1978, 30, Но. 5, с. 646-653.

Zakharov V.E., Shabat A.B. (Захаров В.Е., Шабат А.Б.)

[1] Интегрирование нелинейных уравнений математической физики методом обратной задачи рассеяния. II-Функц. анализ и его прил. 1979, 13, вып. 3, с. 13-22.

Zalcman L.

[1] Bounded analytic functions on domains of infinite connectivity, Trans. Amer. Math. Soc., 1969, 144, p. 241-269.

Zapuskalova T.A., Kats B.A. (Запускалова Т.А., Кац Б.А.)

[1] Краевая задача Римана на спиралеобразном контуре. - В кн.: Тр. семинара по краев. задачам. Казань: Казанск. ун-т, 1978, с. 53-61.

Zolotarevskiĭ V.A.   (Золотаревский В.А.)

[1] Сходимость коллокационного метода и метода механических квадратур для систем сингулярных уравнений. - Изв. вузов Математика, 1976, Но. 6, с. 105-108.

Zolotarevskiĭ V.A., Chebotaru I.S.   (Золотаревский В.А., Чеботару И.С.)

[1] О приближенном решении систем уравнений Винера-Хопфа с ненулевыми частными индексами. - Изв. вузов. Математика, 1978, Но. 7, с. 102-105.

Zverovich È.I.   (Зверович Э.И.)

[1] Краевые задачи теории аналитических функций в гельдеровских классах на римановых поверхностях. - Успехи матем. наук, 1971, 26, вып. 1, с. 113-179.

Zverovich L.F.   (Зверович Л.Ф.)

[1] Задача Римана для нескольких неизвестных функций в случае сложного контура. - В кн.: Теория функций комплексного переменного и краевые задачи. - Чебоксары.: Чувашский ун-т 1974, вып. 2, с. 21-38.

[2] Задача Римана для n пар функций на римановой поверхности в случае сложного контура. - Сибирск. матем. ж., 1975, 16, Но. 3, - с. 510-519.

Zverovich È.I., Panchenko V.F.   (Зверович Э.И., Панченко В.Ф.)

[1] О краевой задаче Римана для n пар функций на римановой поверхности в пространстве $L_n(\rho(t))$. - Вестн. Белорус. ун-та. сер. 1. Математика, физика, механика, 1978, Но. 3, с.71-72.

Zverovich È.J., Pomerantseva L.I.   (Зверович Э.И., Померанцева Л.И.)

[1] Задача Римана для n пар функций с матрицами подстановочного типа. - ДАН СССР, 1974, 217, Но. 1, с. 20-23.

SUBJECT INDEX

NOTATION INDEX